Biochemical
Evolution

Biochemical Evolution

EDITED BY

H. GUTFREUND

PROFESSOR OF PHYSICAL BIOCHEMISTRY
UNIVERSITY OF BRISTOL

CAMBRIDGE UNIVERSITY PRESS

CAMBRIDGE

LONDON   NEW YORK   NEW ROCHELLE

MELBOURNE   SYDNEY

Published by the Press Syndicate of the University of Cambridge
The Pitt Building, Trumpington Street, Cambridge CB2 1RP
32 East 57th Street, New York, NY 10022, USA
296 Beaconsfield Parade, Middle Park, Melbourne 3206, Australia

First published 1981

Printed in Great Britain by
J. W. Arrowsmith Ltd., Bristol BS3 2NT

British Library Cataloguing in Publication Data
Biochemical evolution.
1. Chemical evolution
I. Gutfreund, Herbert
575   QH325   80-49949
ISBN 0 521 23549 9 hard covers
ISBN 0 521 28025 7 paperback

Contents

1

Some problems in molecular evolution

H. GUTFREUND

DEPARTMENT OF BIOCHEMISTRY, UNIVERSITY OF BRISTOL,
BRISTOL BS8 1TD, UK

Nothing in biology makes sense except in the light of evolution
(Dobzhansky, 1973).

In recent years there has been a considerable increase in the literature on
many aspects of the study of biological evolution. This has ranged from
philosophical (Popper, 1972) to mathematical, functional and chemical
aspects (J. M. Smith *et al.*, 1979). Even restriction of the discussion to
biochemical evolution requires considerable selection of topics from a
great range of possible approaches to present some of the salient ideas in
a small volume. Since the present work is intended to interest and instruct
biological scientists including honours and graduate students, it is
important to emphasise that for each conclusion drawn here, from the
facts presented, there are likely to be very different views expressed by
other authors. Even among the authors of the present volume alternative
or conflicting proposals were put forward, sometimes by the same author
and retained by the editor. Any attempt to discuss more than a small
selection of the possible interpretations of the evidence would confuse
rather than educate and stimulate the reader.

Biochemical evolution can be treated in three inter-related ways.
Firstly, one can think about the evolution of chemical reactions occurring
in biological systems from the prebiotic processes discussed by Schuster
in chapter 2, to the catalysis and control of enzyme mediated metabolic
sequences discussed by Clarke in chapter 4. Secondly, one can use
biochemistry as a tool to study the sequence of events in evolution.
Chapter 3 on computer methods for the construction of phylogenetic
trees is dependent on the enormous amount of information which
is becoming available from ever-improving biochemical techniques
for the analysis of amino-acid and nucleotide sequences. To varying
extents biochemical techniques and approaches have provided key

information about the evolution of different physiological functions, of which only a small and necessarily arbitrary selection can be included in this volume. In this respect modern biochemistry is an equal partner with palaeontology and other classical biological subjects in its contribution to the sum total of our present knowledge about the sequence of evolutionary events.

A third aspect of biochemical evolution, neglected here because it could easily fill a separate volume, is the use of arguments from the theory of evolution as a tool to elucidate biochemical pathways and mechanisms. Krebs (1979) in a splendid article entitled 'On asking the right kind of question in biological research' applies the rule that non-functional characters do not survive in the course of evolution. He states this as 'a rule and not a law. A law has no exceptions but a rule may have exceptions'. The applicability of this 'rule' is likely to depend on whether one considers evolution at the level of function or near the gene (single bases or amino acids).

The fundamental problems of biochemical evolution 'evolve' into those of morphological evolution. In prebiotic and prokaryotic evolution the competition is first between single molecules in their efficiency to replicate, utilising the available natural products. In bacteria one can still define the evolutionary advantages of single mutations and primary gene products (enzymes) which enable the organism to live in a new environment. We are, however, already confronted with the problem of, often quite long, sequences of metabolic pathways. The possible backwards evolution of these pathways is discussed by Clarke in chapter 4. As the complexities increase during the evolution of higher organisms the competition is more and more between functions determined by many genes. However, even in vertebrates a single mutation can be lethal if, for instance, it results in a non-functional haemoglobin. Alternatively a mutation can enable animals to spread into a new niche if they become fitter in an environment of lower oxygen tension.

Discussions about the theory of evolution have often been confused by the introduction of beliefs instead of reason and facts and by lack of care in the language used to express the ideas. The essentially stochastic nature of evolution is a gamble where the winner takes all. It is usually argued that in a functional complex organism a random change in a gene (mutation) will most likely be deleterious. On rare occasions a mutation will result in off-spring fitter for their environment which will produce more progeny than other individuals. This increased fecundity can be brought about in a variety of ways, such as longer reproductive life, greater success in attracting mates or greater fertility. It is important to emphasise that the environment has no influence on the probability of a mutation which would adapt the organism to this environment. Of course, radiation and chemicals can cause mutations, but these will cause mutations at random and not for the benefit of the organism. There are mechanisms for adaptation to the environment, but they rely on the genetic history of the organism which has provided it with the necessary control function to switch from the utilisation of one metabolite to another by the turning off and on of genes for the synthesis of different enzymes (see chapter 4). Darwinian evolution functions through muta-tion, reproduction and selection. A detailed discussion of the principles involved will be found in a number of recent elementary treatments (Smith, 1975) or popular reviews (for example, *Scientific American*, September 1978). The interested reader not familiar with the principles of genetics is also well advised to study an introductory text or relevant chapter in a general biology volume. In modern treatments (for instance, Suzuki & Griffiths, 1976) one finds the molecular as well as the biological basis of mutation, reproduction and inheritance. Another more general survey of Darwinian evolution and critical discussion of its many fallacious interpretations is found in a collection of essays by Gould (1979). It is important to emphasise that Darwin (1809–82) did not know of Mendel (1822–84) and his laws of inheritance.

In the twentieth century Darwinism, relying on the principles stated above, was fused first with population genetics and then with molecular genetics to form what is often termed Neo-Darwinism. Darwin, from the information available in his time, considered a more or less common gene pool throughout a given population. The advent of the 'one gene, one enzyme (one protein)' hypothesis and the sensitive new methods for the detection of single amino-acid displacements (mutations) in proteins have made it possible to add to Darwin's evolution by natural selection. Lewontin (1974) and Harris (1975) described in detail how the elec-trophoretic analysis of enzymes established the large degree of allelic variation in *Drosophila* and human populations respectively. In this chapter I shall discuss briefly two aspects of this extended study; first the principles and consequences of polymorphism and secondly the evolu-tionary clock.

With the information now available one has to consider four causes of evolution.
(1) Mutation pressure, which is a slow process resulting in the change of only about 10^{-5} gametes per generation.

(2) Genetic drift, which only happens in small populations, where large chance-fluctuations of gene frequencies may occur.
(3) Gene flow, where immigrants introduce alleles as in mutation pressure.
(4) Natural selection by fecundity through length of reproductive life, survival of the young, mating preference, fertility etc. Many of these factors are related to adaptation to the environment, including food supply.

A large population without mutations or immigration will not evolve. It will rapidly reach an equilibrium distribution of genes according to the Hardy–Weinberg law (see for instance, Suzuki & Griffiths (1976) *p.* 407). The reader should compare these comments with those made in the next chapter (p. 75) on 'four prerequisites for Darwinian evolution'.

The essays presented in this volume are only a small selection of topics which might interest the biochemist or are based on biochemical studies contributing to our knowledge of evolution. It was felt by the editor that a few topics treated in some depth would be more stimulating than many short treatments fitting into a small volume. A very important aspect of evolution, which does not yet receive the attention of the biochemist, but should just be mentioned, is speciation: the multiplication of species. The reader is referred to Mayr (1963) for a detailed discussion of this topic.

Function rather than structure has been emphasised and the reader should be briefly referred to some subjects of interest which have been neglected. In recent years, for instance, there has been a very active debate on the evolution of structural domains which are found as common features in different enzyme systems. The principles of divergent, convergent and parallel evolution are discussed by Peacock in chapter 3. In arguments about the evolution of three-dimensional molecular structures, one has to include the question whether common structural domains in a group of enzymes are primarily of functional significance or whether they are due to a limited number of stable configurations in proteins (Matthews, 1977).

Another interesting subject which is only briefly discussed (chapter 5), concerns the evolution of eukaryotes from prokaryotes and especially the biochemical studies of the origin of mitochondria (see, for instance, Schatz, 1979; D. C. Smith *et al.*, 1979). Some recent findings of profound differences in transcription in the present forms of the two types of cells have been used to argue against the hypothesis of the symbiotic evolution (Darnell, 1978). Another aspect of biochemical studies in this area is the comparison of the subunit structures of enzymes of gram-negative and gram-positive bacteria with those of mitochondria (Henderson, Perham & Finch, 1979).

The problem of explaining the evolution of a process involving a sequence of steps, each determined by a single gene, has already been referred to and is discussed further in chapter 4 in connection with metabolic pathways. The explanation may be different in the case of the physiological functions with which some of the other chapters are concerned. For the student of evolution 'survival of the fittest' has worked too well. The less efficient mechanisms have not survived for our perusal. The essay by Crescitelli could only cover recent aspects of the perfection of the mechanism of photoreceptor response (chapter 9). All visual systems studied so far depend on the photoisomerisation of retinal for light detection and on small differences in the protein opsin, to which the chromophore is bound, for different spectral sensitivities. Three main branches of the animal kingdom – molluscs, arthropods and vertebrates – have developed eyes which are anatomically profoundly different. It seems that the various anatomical, i.e. optical, arrangements will serve their different special purposes, but a particular principle of photochemistry was universally accepted. The ionic and chemical mechanisms involved in the transmission of the signal from rhodopsin to the optic nerve remain to be elucidated. Much of the specialised evolution may be found there, when these processes are compared between species along the phylogenetic tree.

The problem is a different one when the proteins involved in motility are considered. Actin and myosin seem to be used for many different purposes in different animals. This topic is treated in detail by Weeds & Wagner in chapter 8 and it gives some insight into the evolution of systems which are subject to functional control by calcium ions.

It seems quite possible that some steps in complex sequences may have been useful on their own for another purpose. Enzymes may have been less specific and thus able to catalyse more than one reaction or to take part in cascade mechanisms; for example, in the formation of fibrin clots, one of the many serine proteases involved may have been able to fulfil the function of all of them, albeit less efficiently.

POLYMORPHISM

Studies of the rate of mutation, to be discussed below, have shown that a population consisting of several million individuals has a reasonable chance of some mutations in most genes represented in the population.

H. Gutfreund

Although most of these mutations are maladaptive, when causing major changes, many of them will have minor consequences and may be neutral. Electrophoretic studies mentioned above suggest that normal populations have a large pool of genetic variation.

Clearly a rapid analytical method has to be used for a large population survey of different polymorphic forms of a particular enzyme from many individuals. Complete sequence analysis is used to construct phylogenetic trees and the time scale of evolution (the evolutionary clock discussed in a later section), but it is too time-consuming for the large number of samples to be analysed in a search for the number of mutations in a particular protein in a defined population or species. For such screening of enzyme samples the most widely used technique is electrophoresis. This method will, of course, only detect amino-acid substitutions involving changes in charge. About one third of all amino-acid substitutions would be expected to be of this type. This is only one of a number of factors which would cause one to under-estimate the number of mutations from comparison of proteins by electrophoresis. The main factors are given below.

(1) The degeneracy of the genetic code (Table 1.1) resulting in single base changes without amino-acid substitution.
(2) The production of proteins without activity, resulting in undetected translation or silent alleles.

Table 1.1. *The genetic code*

1st↓	2nd →	U	C	A	G	↓3rd
U		PHE	SER	TYR	CYS	U
		PHE	SER	TYR	CYS	C
		LEU	SER	STOP	STOP	A
		LEU	SER	STOP	TRP	G
C		LEU	PRO	HIS	ARG	U
		LEU	PRO	HIS	ARG	C
		LEU	PRO	GLUN	ARG	A
		LEU	PRO	GLUN	ARG	G
A		ILEU	THR	ASPN	SEB	U
		ILEU	THR	ASPN	SER	C
		ILEU	THR	LYS	ARG	A
		MET	THR	LYS	ARG	G
G		VAL	ALA	ASP	GLY	U
		VAL	ALA	ASP	GLY	C
		VAL	ALA	GLU	GLY	A
		VAL	ALA	GLU	GLY	G

(3) Mutations resulting in unstable polypeptide chains.
(4) The substitution of uncharged by uncharged, positive by positive and negative by negative charged amino acids.
(5) Lack of sensitivity of the electrophoretic method if one charged group is removed from a highly charged protein. This problem can be overcome by varying the pH of the measurements.

Harris (1975) provides a detailed survey of the enormous number of enzyme variants found in population surveys. He defines the term polymorphism or genetic polymorphism as follows: 'One may anticipate that many (mutants, alleles) will be very rare, but others will occur with an appreciable frequency, and indeed may be common enough to give rise to the well-known phenomenon usually referred to as genetic polymorphism'. In such cases the members of a population can be classified into two or more relatively common phenotypes due to their distinct alleles at certain loci.

The following different origins of polymorphism can be distinguished by genetic analysis.
(1) Multiple gene loci resulting from gene duplication and subsequent mutations can give rise to different forms of a protein in the same organism. This can also result in the formation of oligomeric proteins with different polypeptide chains in the same molecule, which has contributed to the evolution of functions as discussed below.
(2) Multiple allelism at a single locus is probably responsible for the majority of polymorphic forms of proteins.
(3) Post-translational modification of proteins causes the appearance of additional multiple forms of some proteins. This would not normally be regarded as genetic polymorphism, although the tendency could be genetically determined.

Approaching 100 allelic variants of the enzyme glucose 6-phosphate dehydrogenase are found in human populations. There appears to be no norm, no ideal enzyme with a unique amino-acid sequence. Of course, we only see those variants in which the correct active structure of the enzyme is preserved. With enzymes, just as with larger biological organisations, one could apply the criticism of Gould & Lewontin (1979) of what they call the Dr Pangloss syndrome. Voltaire makes Dr Pangloss in *Candide* expound the philosophy that this is the best of all possible worlds. The reader is referred to Gould (1979) for a wise and witty analysis of many misconceptions about evolution.

THE EVOLUTION OF OLIGOMERS (E.G. OXYGEN CARRIERS)

As mentioned above, gene duplication must have played an important role in the evolution of functions of oligomeric proteins. Many enzymes exist as aggregates of two, four and even six identical polypeptide chains, each with an independent active site. Many functions have been postulated for homogeneous oligomeric systems which are found, on careful kinetic analysis, to be artefactual (Gutfreund, 1975). Probably in some cases the early advantage of aggregation was increased stability. Subsequently interactions occurred between active sites across protomer interfaces and were preserved in those cases where cooperativity was advantageous for a particular key biological control mechanism.

Heterogeneous oligomers evolved either by enzyme units combining with unrelated regulatory units, which is not of interest in the present context, or by gene duplication and separate mutations. Of the latter phenomenon, the evolution of haemoglobin from single-chain oxygen carriers is the best-documented, if not the only, example. These proteins deserve some wider discussion than is possible here. Unfortunately we could not include a chapter devoted to them.

The large group of proteins which have a ferroporphyrin (protohaem) as a prosthetic group and act as oxygen carriers in a wide range of species have been of great interest to students of many different aspects of molecular biology. Their ubiquity, easy recognition and abundance have also contributed to the study of biochemical evolution and in turn biochemical genetics has contributed to our knowledge of their function. In vertebrates one distinguishes the tetrameric haemoglobins of the blood-stream from the monomeric myoglobins present in all forms of muscle. In other forms of animal life haemproteins occur in various oligomeric states in different tissues.

There are three distinct problems in the evolution of oxygen carriers. The evolution of tetrapyrroles (Hendry & Jones, 1980), the evolution of protoporphyrin-Fe^{2+}-globin (haemoglobin) (Ingram, 1963) and the evolution of iron or copper protein complexes. In principle it is easier to start a novel process with a protein than with another complex organic compound. Only one gene has to evolve to provide for the synthesis of a new protein while the synthesis of protohaem requires several enzymes and hence several genes. However, precursors of protohaem synthesis could have been useful for several purposes even before the formation of protoporphyrin with its ubiquitous use for plant and animal pigments in photosynthesis and respiration.

With the information available at the moment, one might hazard the guess that nature has several times solved the problem of designing oxygen carriers. There was no need for oxygen transport untii photosynthesis produced enough oxygen for aerobic life. At that time there were likely to be plenty of tetrapyrroles around for the formation of chlorophyll, and cytochrome-like compounds. At the same time copper proteins (haemocyanins) and, to a lesser extent, non-haem iron proteins are used as oxygen carriers in many early classes of the animal kingdom. Although these metalloproteins have adapted to the varied needs of their surviving 'users' they have not acquired the enormous versatility of the haemproteins (Antonini & Brunori, 1971).

Haemoglobins are a very nice example of the modulation of the function of a prosthetic group by a great variety of minor changes of the protein. The wide range of oxygen binding affinities of the protohaem are controlled by the globin variants and their response to specific ions liganding at other sites. The large number of haemoglobin variants which occur in any population (Harris, 1975) have thus 'often?' proved useful in the adaptation of a subgroup to a new niche. The same phenomenon can, as mentioned above, help to elucidate the effect of changing certain amino-acid residues on the mechanism of oxygen binding and its control (see, for instance, Morimoto, Lehmann & Perutz, 1971).

RANDOM (GENETIC) DRIFT: NEUTRALISM

It has already been mentioned that other mechanisms, in addition to Darwinian natural selection, can contribute to evolution. There is no doubt that natural selection by competition is an important, perhaps the most important, cause of evolution. However, Darwin did not exclude other mechanisms or contributory factors. In recent years there have been very hot debates about the relative importance of different factors in the overall picture of the present genetic composition of populations. In particular, the establishment of neutral genes as a major factor in evolution has been the subject of an interesting controversy. It is not my purpose, and I do not feel competent, to discuss this subject in detail and to assess the relative importance of neutralism. However, since many of the experimental data used for this debate come from biochemical investigations, the principles involved should be mentioned here. Somewhat more detail will be found in the treatment of polymorphism by Harris (1975) and Smith (1975). A proper understanding of the subject requires appreciation of its mathematical treatment by stochastic theory to which an introduction is found in Kimura & Ohta (1971).

In an infinite population the chances become negligible for a mutation to be passed on sufficiently to become established. The likelihood of a mutation becoming fixed is inversely proportional to the size of the population and directly proportional to the fecundity of its carrier. For detail and further references to the algebra involved in this reasoning the interested reader should consult Kimura & Ohta (1971). As a consequence one can postulate that in a large population even a mutation which can confer a considerable advantage on the carrier of its allele may not establish itself. It can also be argued that in a very small population a neutral or even a slightly disadvantageous mutation can establish itself due to relatively large fluctuations. The algebraic reasoning cannot be disputed and if neutral mutations arise in small populations, they will establish themselves. There remains, however, some doubt about the significant occurrence of neutral mutations and their fixation. Of course, the only test for this argument is whether some established variants are due to neutral mutations. On perusing the large amount of evidence offered, for instance by Harris (1975), one can well come to the conclusion that many different polymorphic forms of enzymes show no functional difference. Another argument for the neutrality of these polymorphs is the absence of variation in their distribution in different environments. It must be emphasised that many variants of enzymes have different functions within the same organism or in populations subjected to different environments.

Many investigations are in progress to analyse existing data and to design new experiments to test variants for their functional neutrality. It is probably fair to say that the case for or against is not a strong one and that the problem should stimulate more experimental work to provide data on mutation rates, genetic load etc. I shall conclude the discussion of this topic by mentioning two investigations which illustrate the present direction of some relevant biochemical thoughts. Place & Powers (1979) studied the catalytic efficiencies of variants of lactate dehydrogenase in the killifish in different environments. This study is typical of many designed to relate structural differences in enzymes which function at different temperatures and pressures. The authors conclude their paper with the statement: 'These catalytic differences between allozymes (another term for isozymes) ... are consistent with the selectionist hypothesis'. So far, so good, but then nobody really denies that natural selection exists as an important factor in evolution. The question is whether neutralism is also important.

It was mentioned above that the oligomeric structure of enzymes will contribute to their stability in solution. However, this stabilising effect will depend on specific contact areas and any mutation affecting the surface structure would be detrimental. Harris, Hopkinson & Edwards (1977) argue that the significantly lower incidence of polymorphism among multimeric compared to monomeric enzymes favours the neutralist hypothesis. It would be expected that more neutral mutations would occur with increasing oligomer size. This was not found to be the case in the investigations of Harris and his colleagues. In a recent paper by McConkey, Taylor & Phan (1979) there is a discussion of the bias obtained if only enzymes or only structural proteins are studied.

THE EVOLUTIONARY CLOCK

A topic which is likely to be of considerable interest, and which will require much further theoretical and experimental work for some time to come, is the study of the rate of evolution. The methods described by Peacock in chapter 3 can be extended and combined with the dating of particular branch points of the phylogenetic trees to construct what has been called an evolutionary clock. Several questions arise which are stimulating valuable investigations. In general, biologists are divided between those who believe that there have been periods of considerable evolutionary activity, followed by comparatively quiescent times, and others who consider it likely that there were only small fluctuations on a steady rate. An important contribution to these studies is the study of the rates of mutations. For this purpose, as for the construction of phylogenetic trees, complete polypeptide or nucleotide sequence information is most desirable. If the number of mutations is to be determined in a particular protein, say at different stages of primate evolution, it is of course much less work to determine the changes in sequence after the first complete analysis. Methods such as 'finger printing', first used by Ingram (1963) to define the single mutation to sickle-cell haemoglobin, select the small peptides which contain the changes. Inspection of the factors quoted earlier as contributions to the under-estimation of the number of mutations will show that only some of these are eliminated by complete sequence analysis of the gene products – polypeptide chains. Recent dramatic improvement in methods for DNA sequencing makes it likely that in future much more information about rates of evolution will be forthcoming from this method. One should point out, however, that other methods have also contributed to the available fund of knowledge about the number of nucleotide changes per codon per year or, not to put too fine a point on it, per million years. Among these methods are, in addition to electrophoretic analysis, microcomplement fixation and DNA annealing. The last of these methods relies on a comparison between the melting point of the homoduplex, double helical DNA from a single species, with that of the heteroduplex formed from specific DNA fractions from two species. The change in melting point is related to the number of substitutions (ΔT in degree Celsius to per cent substitution) as long as that number is less than about 20%.

Wilson, Carlson & White (1977) provided an excellent review of the data and ideas involved in the study of rates of biochemical evolution. They state that 'the variation in rates within a given protein class is about twice that expected for a simple Poisson process, such as radioactive decay. To this extent, the sequences in genes and proteins evolve in a clocklike manner'. The main argument about the evolutionary clock is whether it does exist, steadily ticking away with only minor hiccups, or whether it is artificially constructed from evidence accidentally selected by the methods employed. Although Wilson and his colleagues appear to champion the clock without reservation, they do quote its opponents and leave the reader to judge. The evidence depends on the reliability of the divergence times and on the availability of enough fossils. The latter appears to restrict the construction of the clock to the evolution of mammals over the last 100×10^6 years.

A layman in the field, like the present author, may be forgiven for getting the idea that the conflict is largely between biochemists and palaeontologists. Gould (1979) leaves no doubt in our minds that he does not believe in gradualism. In a talk given on the BBC on 23 June 1979, he expressed the opinion that species live for five to ten million years and do not change much, and that they appear and disappear suddenly. This, according to Gould, is not due to Darwin's argument that the records are imperfect. As a palaeontologist Gould argues that evolution is a rare but abrupt event. This brings in ideas from the catastrophe theory.

The biochemist can argue that a steady, clockwise, nucleotide replacement rate might only express itself when major environmental changes result in abrupt selection. This could be the main evidence for the argument that most resulting changes in proteins are neutral. Another scholarly and balanced essay on this topic by an author who has, at different stages of his career, taught anatomy and biochemistry (Williams,

1974) is recommended reading in this field. Williams also concerns himself with the different rates of evolution of different proteins. The arguments involved in the interpretation of the different rates of change in proteins of different functions and different size, are quite elaborate. Their proper presentation requires more space than can be given here. This introductory chapter is only concerned with pin-pointing some of the major problems in biochemical evolution which are likely to receive considerable attention in the future and do not fall within the scope of the other chapters of this volume. To conclude the discussion of this topic, reference to a recent investigation may illustrate one type of result which could add to the information needed to speculate on the rate of molecular evolution. Pies, Zwilling, Woodbury & Neurath (1980) discussed the divergence of trypsin and chymotrypsin from a common serine protease. This, like the evolution of haemoglobin from a myoglobin-like oxygen carrier, must have occurred through gene duplication and subsequent separate evolution of two enzymes with identical catalytic mechanisms, but different substrate specificities. It appears that chymotrypsin evolved more slowly than trypsin between amphibia and mammals. It seems likely that many studies of this type will have to be carried out and extended before really convincing arguments can be presented on either of the two topics discussed in this section and the previous one.

REFERENCES

Antonini, E. & Brunori, M. (1971). *Haemoglobin and myoglobin in their Reactions with Ligands*. Amsterdam-London: North-Holland Publishing Company.

Darnell, J. E. (1978). Implications of RNA·RNA splicing in evolution of eukaryotic cells. *Science*, **202**, 1257–60.

Dobzhansky, T. (1973). Nothing in biology makes sense except in the light of evolution. *American Biology Teacher*, **35**, 125–9.

Gould, S. J. (1979). *Ever since Darwin*. New York and London: W. W. Norton & Co.

Gould, S. J. & Lewontin, R. C. (1979). The spandrels of San Marco and the Panglossian paradigm: A critique of the adaptationist programme. *Proceedings of the Royal Society of London, series B*, **205**, 581–98.

Gutfreund, H. (1975). Kinetic analysis of the properties and reactions of enzymes. *Progress in biophysics and molecular biology*, **29**, 161–95.

Harris, H. (1975). *Principles of Human Biochemical Genetics*. Amsterdam-Oxford-New York: North-Holland/Elsevier Publishing Company.

Harris, H., Hopkinson, D. A. & Edwards, Y. H. (1977). Polymorphism and the subunit structure of enzymes: a contribution to the neutralist-selectionist controversy. *Proceedings of the national academy of sciences of the USA*, **74**, 698–701.

Henderson, C. E., Perham, R. N. & Finch, J. T. (1979). Structure and symmetry of *B. stearothermophilus* pyruvate dehydrogenase multienzyme complex and implication for eukaryote evolution. *Cell*, **17**, 85–93.

Hendry, G. A. F. & Jones, O. T. G. (1980). Haems and Chlorophylls: Comparison of Function and Formation. *Journal of medical genetics*, **17**, 1–14.

Ingram, V. (1963). *Haemoglobin in genetics and evolution*. New York: Columbia University Press.

Kimura, M. & Ohta, T. (1971). *Theoretical aspects of population genetics*. Princeton: University Press.

Krebs, H. A. (1979). On asking the right kind of question in biological research. In *Molecular mechanism of biological recognition*, ed. M. Balabam, pp. 27–39. Amsterdam-New York: North-Holland/Elsevier Publishing Company.

Lewontin, R. C. (1974). *The genetic basis of evolutionary change*. New York: Columbia University Press.

McConkey, E. H., Taylor, B. J. & Phan, D. (1979). Human Heterozygosity: A New Estimate. *Proceedings of the national academy of sciences of the USA*, **76**, 6500–4.

Matthews, B. W. (1977). X-Ray Structure of Proteins. In *The Proteins*, ed. H. Neurath & R. Hill, 3rd edn, vol. 3, pp. 404–590. New York: Academic Press.

Mayr, E. (1963). *Animal species and evolution*. Cambridge Mass; Harvard University Press.

Morimoto, H., Lehmann, H. & Perutz, M. F. (1971). Molecular pathology of human haemoglobin: stereochemical interpretation of abnormal oxygen affinities. *Nature*, **232**, 408–13.

Pies, W., Zwilling, R., Woodbury, R. G. & Neurath, H. (1980). Amino-terminal amino acid sequences and the evolution of frog (*Rana esculenta*) trypsin and chymotrypsin. *FEBS Letters*, **109**, 45–9.

Place, A. R. & Powers, D. A. (1979). Genetic variations and relative catalytic efficiences: lactate dehydrogenase & allozymes of *Fundulus heteroclitus*. *Proceedings of the national academy of sciences of the USA*, **76**, 2354–8.

Popper, K. (1972). *Objective knowledge*, Chapter 7. Oxford: University Press.

Schatz, G. (1979). How mitochondria import proteins from the cytoplasm. *FEBS Letters*, **103**, 203–11.

Smith, D. C. and others (1979). Royal Society Discussion on *The Cell as a Habitat*. *Proceedings of the Royal Society of London*, series B, **204**, 115–286.

Smith, J. M. (1975). *The theory of evolution*. 3rd edition. Harmondsworth, UK: Penguin Books.

Smith, J. M. and others (1979). Royal Society Discussions on *The Evolution of adaptation by natural selection*. *Proceedings of the Royal Society of London*, series B, **205**, 433–608.

Suzuki, D. T. & Griffiths, A. J. F. (1976). *An introduction to the theory of genetics*. San Francisco: W. H. Freeman & Co.

Williams, J. (1974). The primary structure of proteins in relation to evolution. *MTP International Review of Science, Biochemistry Series*, **1**, 1–56.

Wilson, A. C., Carlson, S. S. & White, T. J. (1977). Biochemical Evolution. *Annual reviews of biochemistry*, **46**, 533–639.

2

Prebiotic evolution*

PETER SCHUSTER

INSTITUT FÜR THEORETISCHE CHEMIE UND
STRAHLENCHEMIE DER UNIVERSITÄT WIEN,
WÄHRINGERSTRASSE 17, A-1090 WIEN, AUSTRIA

Prebiotic evolution is a wide and open field of research. An impressive number of scientists, with very different backgrounds, are engaged in work on various and diverse problems under this heading. Not surprisingly, there is a vast literature and a whole collection of monographs on this topic. Different people have different and often contradictory answers to questions and problems when 'hard' experimental facts and rigorous proofs are lacking. Unfortunately, this is the rule rather than the exception in the present views and theories on the origin of life and makes it a risky and difficult but, nevertheless, challenging task to write an essay about prebiotic evolution. This contribution does not, and cannot, aim at a complete presentation of the literature available and hence should be considered as a personal view which obviously attributes different weights to the accessible pieces of information. Some emphasis has been laid on the attempt to relate the results from various fields and to give hints on the continuity of development without, however, entering the realm of pure speculation.

THE HISTORICAL ROUTE AND THE PROBLEM OF ITS RECONSTRUCTION

Looking back from present-day biology to the appearance of life on the earth in the Pre-Cambrian era one is guided by a rich collection of fossils which becomes more and more fragmentary as one proceeds into the distant past (Fig. 2.1). The problem is twofold: the fossilised organisms become smaller and the older sediments inevitably are less well conserved.

The oldest rocks on the earth discovered so far are metamorphic sediments in West Greenland (Moorbath, O'Nions & Pankhurst, 1973).

* Dedicated to Professor E. Broda.

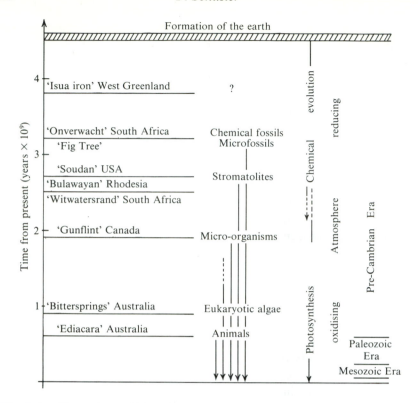

Fig. 2.1.   Chronology of biological evolution according to the fossil record. The earliest evidence for organic life on the earth is dated back to 3.2 or 3.8×10^9 years ago. The older chemical fossils from Greenland are very uncertain and presumably not of biological origin. Thus, there remain about $0.8–1.4 \times 10^9$ years for prebiotic evolution, the period to be considered here.

They are approximately 3.8×10^9 years old and seem to be free of remnants from living organisms: at least, the hints of organic life are highly uncertain (Schwartz, 1979). Much better evidence, consisting of 'molecular fossils', was obtained from some younger rocks found in Transvaal, South Africa (Onverwacht and Fig Tree series, about 3.2×10^9 years old) and Minnesota, USA (Soudan Iron Formation, about 2.7×10^9 years old). In the case of the African sediments some doubts have been raised about the biological origin of the carbon compounds (Rutten, 1971). These 'molecular fossils' are hydrocarbons of the isoprenoid series and some related compounds.

   The oldest 'real witnesses' of organic life which consist of morphologically recognisable remnants of former organisms are dated somewhat later at about 2.5×10^9 years ago (see e.g. Schopf, 1978). The best guesses on the moment when the first bacteria-like protocells started to spread over the oceans will fall between 3.0 and 3.5×10^9 years ago. Thus, there remains a period of 1.5×10^9 years from the formation of the earth which is essentially free of organic fossils. It is during this dusky epoch that prebiotic evolution took place.

How do we proceed when direct 'witnesses' like fossils are not available? Certainly we have very little chance of making a relevant guess on the details of locality and timing of the origin of life on the earth. Global information on the nature and the likelihood of processes that may have taken place and finally led to the origin of the present biosphere, however, appears to be at least as valuable as a reconstruction of the historical aspects. There are practical ways for this global approach. We mention three of them here.

(1) Simulation experiments which aim to mimic the conditions in the prebiotic atmosphere and oceans provide information on the organic material that was available at some stationary concentration and therefore could enter the long-term development.

(2) The search for 'intellectual fossils' which represent something like a 'memory' of life reveals otherwise inaccessible information on the early course of biological evolution. This information is hidden in the structures and primary sequences of those first catalysts which appear to be universal in the biosphere, such as the tRNA molecules and the ribosome. Another source of this kind may be the formal structure of the genetic code, in particular the assignment of individual amino acids to certain codons.

(3) The study of selfreplication at the molecular level is very likely to reveal some facts of relevance to questions concerning the origin of life. The dynamics of selfreplication in biology are not completely understood yet either theoretically or from the experimentalist's point of view, because of the enormous complexity of the phenomenon, although it seems that we know the essential features. Very sophisticated and elegant experiments were necessary to provide our present knowledge on RNA and DNA replication in viruses and bacteria and even more involved investigations will be needed before an ultimate answer can be given to the remaining open questions. From our point of view three problems appear to be of major relevance in prebiotic evolution: (a) the role of errors in selfreplication and their propagation from generation to generation, (b) the suppression of competition between selfreplicating elements

and (c) the requirement of spatial isolation in compartments in order to create individual organisms.

In this review we shall consider the steps of prebiotic evolution in a logical sequence, every step requiring some products of the previous ones. Such logic appeals to the intellect of a scientist or an engineer who plans and constructs in a highly organised manner. Nature, however, has no reason to search for conceptual simplicity, elegance or logic. She is not engineering, as Jacob (1977) pointed out, but she is tinkering and uses whatever is accessible whenever it is accessible. The only thing that counts is availability and efficiency. Therefore there is no need for the actual development to have followed such a sequence. Most processes may have occurred at the same time and with strong mutual interference.

PREBIOTIC CHEMISTRY

An impressive amount of excellent research work has been done on the simulation of prebiotic conditions and the reactions which presumably took place were studied extensively. There are many review articles and monographs available (see e.g. Fox & Dose, 1972; Miller & Orgel, 1974; Schwartz, 1979) and we can restrict ourselves therefore to a brief summary of the most important facts and some more recent findings. Orgel & Lohrmann (1974) formulated four conditions which they considered to be essential for relevant simulation experiments:

'... (a) All primary organic reagents must be derivable as significant products from a reducing atmosphere composed of a selection of the following simple gases: CH_4, CO, CO_2, NH_3, N_2, H_2O. Ultraviolet light, heat, or electric discharges may be used as sources of the energy involved in the synthesis of primary reagents from these elementary gases.

(b) No solvent other than water may be used. Reactions in aqueous solution must be carried out at moderate pH's, preferably between 7 and 9.

(c) Solid-state reactions must occur without excessive drying of the reactants. Solid-state reaction mixtures are preferably obtained by evaporating aqueous solutions that are initially at pH's between 7 and 9.

(d) All reactions must occur under conditions of temperature and pressure that occur on the surface of the earth today. We doubt that volcanos or thermal springs contributed much to the origin of life, so we prefer to carry out solid-state reactions at temperatures below

80°C (surface temperatures up to 90°C have been recorded in California deserts, and we have found that large areas of the desert surface are raised above 65°C for several hours on a typical summer day). . . .'

For the present purpose it seems appropriate to split the discussion into three parts: (1) reactions in the prebiotic atmosphere, dry or in contact with liquid water, (2) reactions in aqueous solutions which led to the constituents of biopolymers and the formation of polymers from monomers by condensation reactions in the presence of water and (3) reactions under prebiotic conditions which involve solid-state catalysts.

It is generally accepted today that several small molecules were abundant in the prebiotic atmosphere on the earth, namely (hydrogen), methane, carbon monoxide, carbon dioxide, ammonia, nitrogen, water and, as the source of sulphur, a small amount of hydrogen sulphide. The most important property of the primordial atmosphere was its reducing power. Due to continuous diffusion into interstellar space the actual concentration of molecular hydrogen had to be rather low. The reservoir of reducing agents thus consisted of hydrogen-containing molecules and metals, present at their low oxidation states, especially $Fe(II)$. More details on the most probable composition of the prebiotic atmosphere have been given by Holland (1962, 1978) and Schwartz (1979).

The gas phase reactions which turn out to be relevant for prebiotic chemistry fall into two classes: Miller-type and Fischer–Tropsch-type (FTT) experiments (Miller, Urey & Oró, 1976). In the former case spontaneous reactions in a gas mixture of methane, water, nitrogen and/or ammonia (methane may be replaced by carbon monoxide and hydrogen) are initiated by electric discharge, ultraviolet radiation, thermal or some other source of energy. Electric discharge experiments have been studied most extensively and in this case the essential steps of the sequence of reactions are known. The first identified products are hydrogen cyanide, formaldehyde and hydrocarbons:

$$n\,CH_4 \rightarrow C_nH_m + \left(2n - \frac{m}{2}\right)H_2 \qquad (2.1a)$$

$$CH_4 + \tfrac{1}{2}N_2 \rightarrow HCN + \tfrac{3}{2}H_2 \qquad (2.1b)$$

$$CH_4 + H_2O \rightarrow CH_2O + 2H_2 \qquad (2.1c)$$

where $m = 2n + 2, 2n, 2n - 2, \ldots$

Reactions $(2.1a)$ and $(2.1b)$ have been studied systematically in pure methane and methane–ammonia mixtures as well as in methane–nitrogen mixtures (Toupance, Raulin & Buvet, 1975). Branched hydrocarbons seem to form in higher yield than *n*-alkanes. A great variety of alkenes and alkynes appear in the reaction mixture. The hydrocarbons formed react analogously to (2.1*b*) and (2.1*c*). A complicated mixture of higher nitriles and aldehydes is the result of a typical discharge experiment in the dry vapour phase. It is worth noticing that highly unsaturated nitriles like cyanoacetylene are obtained in fairly high yields.

In contact with liquid water and in presence of ammonia these primary products react further. Aldehydes form amino- and hydroxy-nitriles which in turn are hydrolysed and finally yield a great variety of hydroxy- and amino acids typical for Miller-type experiments (Miller & Orgel, 1974). The pathway of amino-acid formation in aqueous solution follows the route of the old and well-known Strecker synthesis (Fig. 2.2).

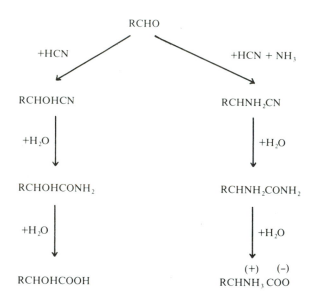

Fig. 2.2.   Prebiotic synthesis of hydroxy- and amino acids in aqueous solution. The mechanism for amino-acid formation follows essentially the pathway of the Strecker synthesis.

By means of FTT experiments Anders, Hayatsu & Studier (1974) tried to present some evidence for the formation of organic compounds in solar nebulae. There are two sources of our knowledge of the organic chemistry that is going on in 'dark' interstellar clouds: (1) chemical analysis of the organic material found in carbon-containing meteorites (Kven-volden, Lawless & Ponnamperuma, 1971; Kvenvolden, 1974) and (2) radioastronomic detection of molecules by means of their rotational spectra (Winnewisser, Churchwell & Walmsley, 1979). In FTT reactions a mixture of carbon monoxide, hydrogen and ammonia is brought into contact with naturally occurring solid catalysts like nickel-iron, magnetite, various hydrated silicates or clays. At somewhat higher temperatures (*c*. 150–250°C, additional pulse-like heating up to *c*. 500–700°C) a great variety of organic compounds is formed in the reaction mixture. In particular, Anders *et al.* (1974) identified various alkanes and arenes, including isoprenoid type, fatty acids, amino acids, purines, pyrimidines, oligomers of pyrrole, porphyrins and other organic material. The enormous diversity of organic compounds formed in FTT reactions is amazing. Some doubts have been raised, however, whether the conditions applied in these experiments are realistic in solar nebulae or on the primitive earth (Miller *et al.*, 1976). In the latter case the requirement of rather high temperatures seems to restrict FTT reactions to very limited areas around volcanoes.

The amino acids identified in reaction mixtures from electric discharge and FTT experiments as well as in the Murchison meteorite are sum-marised in Table 2.1. It seems remarkable that 14 of the 20 naturally occurring amino acids are formed in these simulation experiments at fairly high to at least detectable amounts. Addition of hydrogen sulphide to the gas mixture leads to the formation of sulphur compounds (Raulin & Toupance, 1975) and, to a certain extent, the two sulphur-containing natural amino acids are formed in aqueous solution. Possible pathways for prebiotic syntheses of phe (the abbreviations used for natural amino acids are given in Table 2.1) and trp have been proposed (Miller & Orgel, 1974). To make the list complete, the two amino acids containing amide groups, asn and gln, are available, in principle, through partial hydrolysis of the corresponding nitriles. Further, there are some vague hints that they may have come late into the genetic code like trp (Jukes, 1973; Eigen & Schuster, 1978*b*).

Although purines and pyrimidines have been identified as products of FTT reactions the major routes towards these building blocks of poly-nucleotides definitely went through aqueous solution. Oligomerisation of hydrogen cyanide at moderately high concentrations ($c_{HCN} \gtrsim 0.1$ M) in aqueous solutions containing a certain amount of ammonia leads to adenine with remarkable ease (Oró, 1961). The presence of ammonia is not essential, a photochemical reaction (Ferris, Kuder & Catalano, 1969)

Table 2.1. *Abundance of natural amino acids under prebiotic conditions*

		Simulation experiments			
		Miller type electric discharge[b]		FTT[c]	Murchison[d] meteorite
Compound[a]		Yield (μM)	Presence[e]	Presence	Presence
alanine	ala	790	+	+	+
glycine	gly	440	+	+	+
aspartic acid	asp	34	+	+	+
valine	val	19.5	+	+	+
leucine	leu	11.3	+	(+)	−
glutamic acid	glu	7.7	+	+	+
serine	ser	5.0	+	NS	NS
isoleucine	ile	4.8	+	(+)	−
proline	pro	1.5	+	(+)	+
threonine	thr	0.8	+	NS	NS
tyrosine	tyr	−	−	+	−
histidine	his	−	−	+	−
lysine	lys	−	−	+	−
arginine	arg	−	−	+	−

[a] The abbreviations for the other amino acids found in proteins are: aspargine, asn; glutamine, gln; phenylalanine, phe; cysteine, cys; methionine, met; tryptophan, trp.
[b] Miller & Orgel (1974); the yields are based on an initial amount of 336 mM methane which was applied in the experiment reported.
[c] Anders, Hayatsu & Studier (1974).
[d] Kvenvolden (1974); Miller, Urey & Oró (1976).
[e] +, identified; (+), tentative identification; −, not found; NS, not searched for.

leading to the intermediate 4-aminoimidazole-5-carbonitrile (Fig. 2.3) or the hydrolysis of hydrogen cyanide tetramer under mild conditions (Ferris, Joshi, Edelson & Lawless, 1978) giving adenine in appreciable yields. In the latter case a substantial amount of natural amino acids has been identified in the reaction mixture. Other purines, particularly guanine, hypoxanthine and xanthine seem to be present in the hydrolysate as well (Schwartz, 1979). The other, previously accepted route of prebiotic guanine synthesis starts out from 4-aminoimidazole-5-carbonitrile and cyanogen (Fig. 2.3).

The biologically important pyrimidines have not been isolated yet from reaction mixtures of hydrogen cyanide oligomerisation. The most plausible prebiotic pathway to cytosine, uracil and thymine starts from cyanoacetylene and involves a number of hydrolytic steps as well as condensation with guanidine (Fig. 2.4). Cyanoacetylene is a common

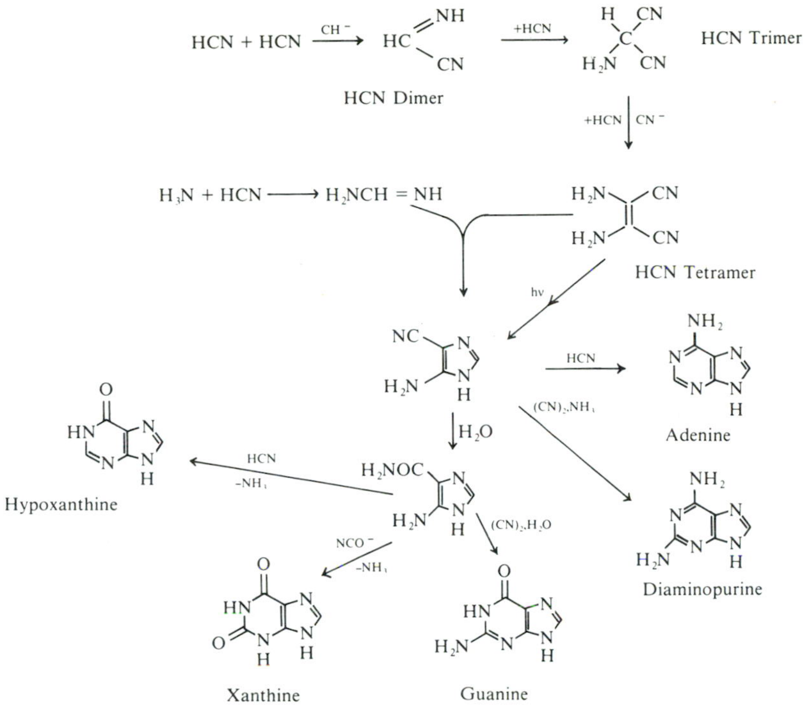

Fig. 2.3. Prebiotic synthesis of purines.

prebiotic reagent: it has been identified in interstellar space (Winnewisser *et al.*, 1979) and is a primary product of electric discharge experiments obtained in high yield from mixtures of methane and nitrogen (Toupance *et al.*, 1975).

Although there is a long-known, possibly prebiotic route to carbo-hydrates, namely the autocatalytic formose condensation of formalde-hyde in alkaline aqueous solution, the problem is not yet completely solved. Rather unrealistically high concentrations of formaldehyde are necessary to get the reaction started. Ribose, the most important sugar in prebiotic chemistry, is formed in low yield only. The reaction does not stop at a certain concentration of sugars but proceeds further and ultimately yields some unidentified high-molecular-weight material (Reid & Orgel, 1967). Solid-state catalysts may have played an important role in making the prebiotic synthesis of sugars more specific: Gabel & Ponnamperuma (1967) reported the formation of ribose together with

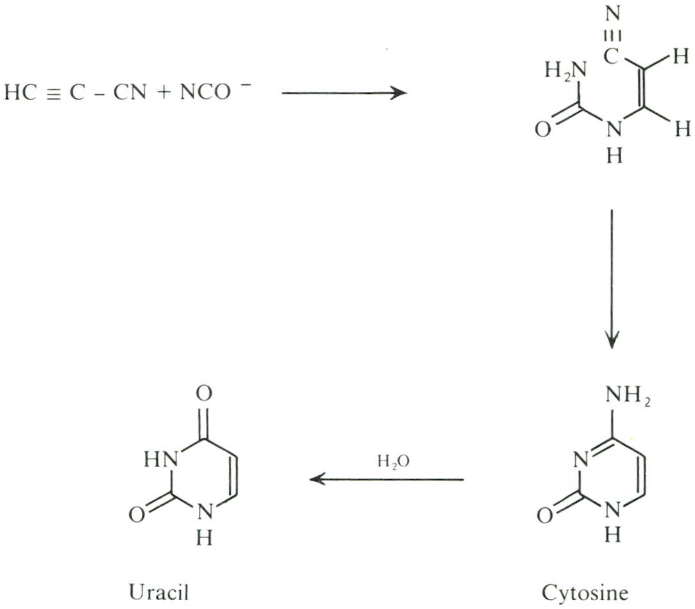

Fig. 2.4. A possible pathway of prebiotic synthesis of pyrimidines.

other sugars when they refluxed a 0.01 M aqueous solution of formaldehyde in the presence of alumina or kaolinite.

Almost all the basic building blocks of present-day biopolymers could
be formed under prebiotic conditions and presumably were present on
the primitive earth. In the logical sequence the next step consists in
combining the subunits to polymers. Many papers report various
attempts, with limited success, to synthesise oligomers and polymers
under potentially prebiotic conditions. The difficulties encountered in
these reactions are far greater than with the processes discussed before:
multiple functional groups and stereochemistry come into play.
Specificity of linkage formation is a *conditio sine qua non* for the synthesis
of uniform polymers. This problem is particularly hard to solve in the case
of oligo- or polynucleotides where we have to link specifically three
elements, base, sugar and phosphate (Fig. 2.5). Moreover, condensation
reactions are not likely to occur easily in aqueous solution and require an
appropriate source of energy.

More or less all attempts to form nucleosides from purines and
pyrimidines with ribose in aqueous solution have so far failed. Solid-state
reactions, as we shall see, are more promising in this context. Phos-

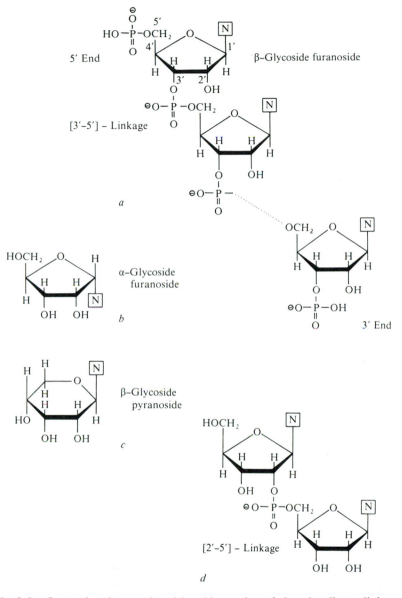

Fig. 2.5. Stereochemistry and positional isomerism of phospho-diester linkages in oligo- and polyribonucleotides. N represents one of the four pyrimidine or purine bases. *a*. The structure of present-day RNA molecules; all phospho-diester linkages are 3'–5', the ribose ring is in the furanose form and all glycosidic linkages are of the β type. *b–d*. Examples of other conformations and linkages; *b*. conformation of the α-glycoside, *c*. ribose in the pyranose form and *d*. 2'–5' phospho-diester linkage.

phorylation of nucleosides in solution was not much more successful with the exception of one type of reaction: trimetaphosphate, admittedly an agent not easily accessible under prebiotic conditions, reacts specifically with nucleosides to form cyclic-2', 3'-phosphates (Schwartz, 1969; Saffhill, 1970).

A remarkable result on prebiotic polymer formation has been reported recently by Sleeper & Orgel (1979): oligonucleotides with chain-lengths up to five and more bases are obtained in good yields from adenosine- and uridine-5'-phosphorimidazolide under specific Pb^{2+} ion catalysis. Other divalent metal ions like Mg^{2+} or Zn^{2+} do not show comparable effects. The prebiotic availability of nucleoside-5'-phos-phorimidazolides will be discussed in connection with prebiotic solid-state reactions. One major problem still remains: the oligomer is not uniform with respect to positional isomerism. Both [3'–5']- and [2'–5']-phosphodiester linkages occur (Fig. 2.5) and the latter connection, which is not found in natural polynucleotides, is even preferred. Helix formation and selective hydrolytic cleavage of the [2'–5'] linkage may help substantially to create uniform polymers with the 'correct' [3'–5']-phosphodiester linkages.

Oligopeptide formation in aqueous solution occurs to a certain degree when activated amino acids are used as starting materials. Examples are the condensation of aminoacyladenylates (Lewinsohn, Paecht-Horowitz & Katchalsky, 1967) or of amino-acid thioesters (Weber & Orgel, 1979). The latter experiment gave far better yields of oligopeptides, but even there the chains hardly exceeded five or six amino-acid residues.

With the exception of polynucleotide formation from activated nucleotides the results of prebiotic condensation reactions in homogeneous aqueous solution were not at all convincing. Now, we shall mention two kinds of reactions involving solid state catalysis: reactions in aqueous solution catalysed by clays and solid-state reactions in which reactants are evaporated to dryness and eventually heated together.

The most spectacular reaction of the first class was discovered about ten years ago: oligopeptides with a degree of polymerisation up to 50 were obtained in high yields from aqueous solutions of aminoacyladenylates in the presence of the clay montmorillonite (Paecht-Horowitz, Berger & Katchalsky, 1970; Katchalsky, 1973). Polypeptides of chain lengths up to about 100 were found too, in low yields. The clay is an alumino-silicate which specifically catalyses the polymerisation of aminoacyladenylates. Recently, Paecht-Horowitz (1978) published a systematic study on the influence of the nature of exchangeable cations in this silicate and some closely related clays with similar catalytic effects (nontronite and hectorite).

Solid-state reactions of the second type turned out to present a possible alternative to those processes which do not work properly in homogeneous aqueous solutions under prebiotic conditions: nucleoside and nucleotide formation as well as activation of amino acids. Nucleosides are formed in appreciable yields from purines and ribose or 2-deoxyribose if diluted solutions of bases and sugars are evaporated to dryness and then heated to 100°C for some time (Fuller, Sanchez & Orgel, 1972). Magnesium chloride turned out to be a very efficient catalyst for nucleoside formation. Interestingly, the salt obtained by evaporation of sea water produces even higher yields. The nucleosides formed are not uniform with respect to their stereochemistry and both α- and β-glycosidic linkages as well as furanose and pyranose rings are found. A similarly efficient condensation of pyrimidines and ribose has not yet been reported.

Inorganic phosphates, in particular ammonium phosphate, are converted quantitatively into pyrophosphate and linear polyphosphates when heated together with urea to a temperature of 85–100°C (Osterberg & Orgel, 1972). Mg^{2+} ions catalyse the reaction efficiently. Mineral phosphates like fluoro- or hydroxyapatite react in the same way but more slowly. Efficient solubilisation and activation of these minerals may be achieved by the oligomerisation products of hydrogen cyanide (Schwartz, 1974). When nucleosides are heated with inorganic phosphates and urea they are converted into a mixture of monophosphates with ester linkages at the 5'-, 2'- or 3'-hydroxygroups. Alternatively, 5'-diphosphates and to a small extent 5'-triphosphates are the major products in the presence of magnesium salts (Orgel & Lohrmann, 1974).

Solid-state chemistry also revealed an interesting scheme of possible prebiotic activation processes for oligonucleotide and oligopeptide synthesis (Lohrmann & Orgel, 1973). These reactions have parallels in contemporary biochemical mechanisms. The key step in this scheme is the solid-state conversion of imidazole and nucleoside-triphosphates or other nucleoside-oligophosphates into nucleoside-5'-phosphorimidazolides. These compounds readily undergo a whole variety of synthetic reactions either in the solid state or in aqueous solution which finally lead to both classes of biopolymers (Fig. 2.6).

We have not yet mentioned a class of biomolecules which make up major constituents of present-day cell membranes: the phospholipids. The reason for this omission is very simple: lipids have not attracted as

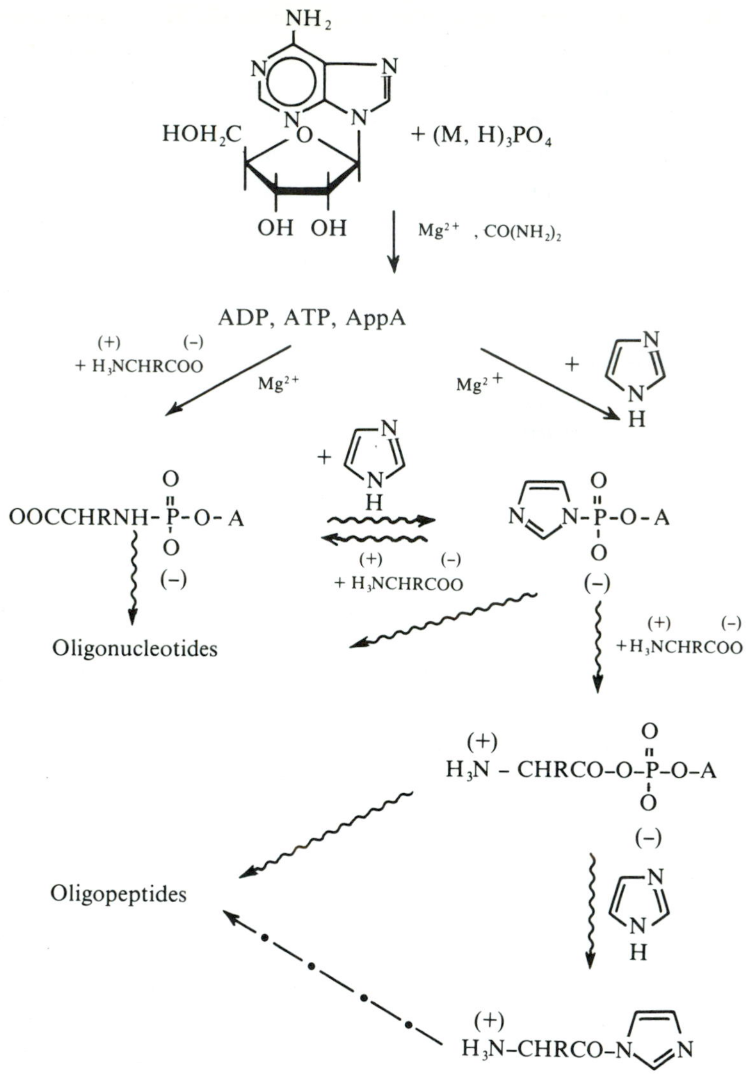

Fig. 2.6. A scheme for prebiotic nucleoside phosphorylation and activation of nucleosides and amino acids according to Lohrmann & Orgel (1973). The reactions have been selected such that they parallel present-day biochemical processes. M, a monovalent metal ion, K^+ or Na^+; A, adenosine; ADP, adenosine-5'-diphosphate; ATP, adenosine-5'-triphosphate; AppA, diadenosyl-pyrophosphate; straight arrows, solid-state reactions; wavy arrows, reactions occurring either in the solid state or in solution; broken arrow, reaction in solution which has no analogue in contemporary biochemistry.

much attention in prebiotic chemistry so far. There is little doubt, however, that they could form in principle, as a recent paper by Epps, Sherwood, Eichberg & Oró (1978) shows. These authors heated a dry mixture of ammonium palmitate, glycero-1(3)-phosphate, cyanamide and imidazole to temperatures between 60 and 90°C and found significant quantities of phospholipids.

Some efforts have been made to look at prebiotic chemistry from more general and formal points of view, in particular starting from thermo-dynamic considerations. Two papers may be mentioned as examples of these efforts: Toupance, Raulin & Buvet (1971) and Morowitz (1977). Important and interesting in themselves, they have not yet reached the point where they compete favourably with the piece-by-piece approach of preparative chemistry.

Although our picture of prebiotic chemistry is far from complete, it seems that the formation of oligonucleotides is of real prebiotic potentiality. The discussion of template-induced replication and its consequences that makes up the rest of this contribution hence has a firm basis and is not to be regarded as a purely academic exercise. We have seen further that the final result required reactions in all three stages of aggregation, in the prebiotic atmosphere, in the primordial aqueous solution and in the solid state. The progress which has been made in this field during the last two decades is very impressive but yet more efforts will be necessary to enlarge our knowledge of the details of prebiotic chemistry and to convert the 'possible' or 'plausible' pathways into 'probable' ones.

THE SELFREPLICATING MOLECULES

What are the principles that can be used to conceive a selfreplicating molecular system? Out of the various contributions to intermolecular forces, electrostatic interactions are the most promising candidates to be used in pattern recognition since they have a built-in complementarity $(+, -)$. Indeed, indirect replication by means of an intermediate negative copy is far more easily verified on the molecular level than any direct copying process. Ionic charge patterns caused by structural defects in silicate crystals were found to replicate in a complementary way in subsequently laid-down neighbouring layers. This molecular 'memory' was found to extend over as many as 20 layers (Weiss, 1975). Actually, an ancient genetic system based on replicating clays had been postulated by Cairns-Smith (1966). It is extremely difficult to visualise, however, in which way a transfer of information from clays, the alien ancestors, to nucleic acids might have taken place. Therefore, we shall start with the contemporary carriers of genetic information, the polynucleotides. The complementarity between purine and pyrimidine bases again is ultimately caused by electrostatics, although the molecular charge patterns which give rise to the specific hydrogen bond interactions are much more elaborate than those in clays. The stereochemistry of the double helix imposes a further restriction on possible candidates for complementarity (Fig. 2.7), so that the four-letter code realised in present-day biochemistry was presumably the only molecular replication system which was accessible under prebiotic conditions.

In the preceding section we saw that oligonucleotides had a certain chance to form spontaneously under prebiotic conditions. No doubt, these molecules were of rather short chain length. Usually, a mixture of di-, tri-, tetra- and pentanucleotides, with very little higher molecular weight material, was obtained in simulation experiments (Sleeper & Orgel, 1979). The largest of these oligomers may serve already as templates although the positions of the phosphate linkages are not uniform. At this stage selection comes into play already and favours the longer and sterically more homogeneous oligomers: double helix formation protects against hydrolytic cleavage and the longer and the more uniform a sequence is, the larger is its association constant with the negative strand. Oligomers with higher association constants stick together for a longer period than do their short-chain analogues. Another important feature increasing the resistance of oligonucleotides with higher molecular weights to hydrolytic degradation, is represented by preferential adsorption of longer sequences on surfaces. Hence, the larger molecules are less likely to be hydrolysed and their concentrations increase. Eventually, longer sequences may be formed by overlapping chains or by an internal selfcopying process (Fig. 2.8).

The arguments given above are firmly based on two kinds of experimental studies. Firstly, the elementary steps of base recognition and helix–coil transition have been investigated extensively by Pörschke (1977). Accurate, thermodynamic and kinetic data on double helix formation from oligonucleotides with a variety of different sequences are available. Secondly, many studies on enzyme-free, template-induced polymerisation of nucleotides under potentially prebiotic conditions have been reported. The results of these studies are briefly summarised in the following paragraph.

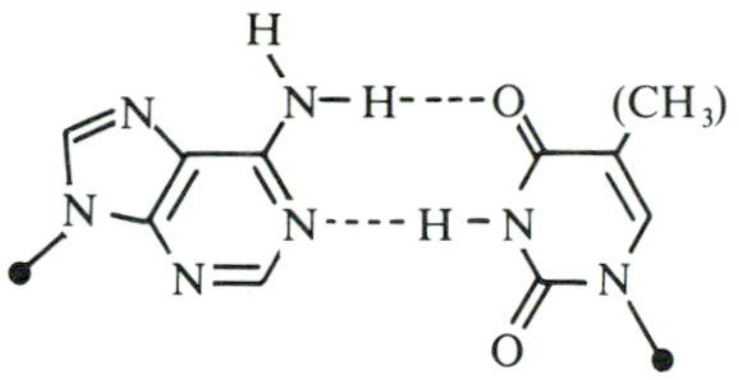

A = U(T)

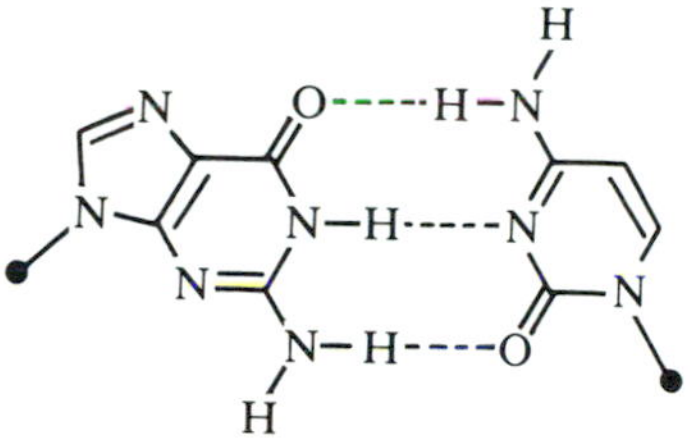

G ≡ C

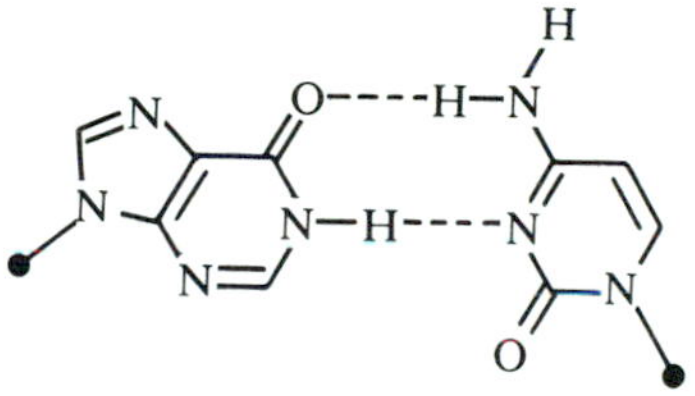

I = C

Fig. 2.7. The complementarity of purine and pyrimidine bases in double helices of nucleic acids. The coincidence of two factors is essential: (1) complementary positions of hydrogen-bond donor and acceptor sites and (2) proper geometry of the base pair which allows fitting into the backbone of the double helix. A=U(T) and G≡C occur regularly in RNA (DNA). I=C is shown as an example of possible substitution for G≡C which is found occasionally at some exceptional positions in catalytically active RNA molecules (tRNAs, ribosomal RNA). In principle, other purines like xanthine or diaminopurine may be incorporated, provided the appropriately substituted pyrimidines act as counterparts.

Polymerisation of activated adenylates, in particular adenosine 5'-phosphorimidazolide (ImpA) and various diadenylates, on polyuridylic acid, poly(U), as template has been studied by Lohrmann & Orgel (1979). They obtained appreciable amounts of oligo(A) with a degree of polymerisation up to five and more. In small yields they also found chain

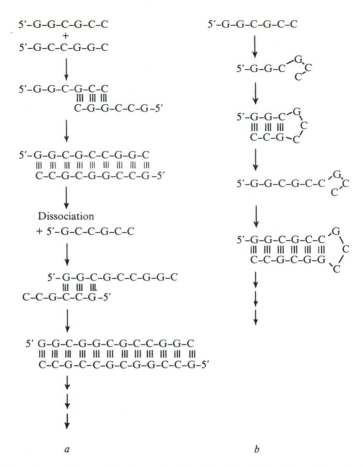

Fig. 2.8.   A possible prebiotic mechanism for the chain elongation of oligonucleotides due to template-induced polymerisation. *a*. Intermolecular, partial
double helix formation. *b*. Internal selfcopying via hairpin structures. For the sake
of simplicity the sequences were assumed to contain G and C exclusively.

lengths up to ten. The oligoadenylates obtained are not uniform with
respect to their sugar–phosphate backbones. The ratio of [2′–5′]- to
[3′–5′]-phosphate linkages varies, largely depending on the initial
compounds used. Analogous studies were performed also with poly(C)
and various heteropolymers as templates (Ninio & Orgel, 1978). In
general, Ninio & Orgel obtained similar results. The proper di- and
trinucleotides are incorporated more readily into oligomers than are
activated monomers. Two very recent findings seem to be very important
with respect to prebiotic evolution. Firstly, oligoadenylates are selec-

tively adsorbed on hydroxyapatite. The higher their molecular weight the stronger is the adsorption. Template-directed synthesis of oligoadenylates on poly (U) is hardly influenced by the mineral although the polymers stick tightly to the surface (Gibbs, Lohrmann & Orgel, 1980). This result opens a perspective for the prebiotic polynucleotide synthesis mentioned vaguely above. Oligomers of higher molecular weights and polymers are specifically protected against hydrolysis. Secondly, the influence of metal ions on template-induced enzyme-free polymerisation of activated nucleotides is as dramatic as in the case of the template-free reaction: Pb^{2+} ions have a specific catalytic effect. Polymerisation of ImpA on poly(U) templates (Sleeper, Lohrmann & Orgel, 1979) leads to oligoadenylates with chain lengths up to five and more. The catalytic influence of Pb^{2+} ions is remarkable: the yield in oligomers is much higher and, even more important, the product is predominantly [3'–5']-linked (75%) in contrast to the oligomer synthesised in Pb^{2+}-free media where one finds 90% or more [2'–5']-linkages. Polymerisation of ImpG on poly(C) gives an 85% yield of oligomers (L. E. Orgel, personal communication). The main chain lengths are around 15 nucleotides. Another interesting finding concerns the specificity of nucleotide incorporation. Starting from a mixture of equal amounts of ImpA and ImpG the correct base (G) is preferred to the wrong one (A) with a ratio of 10:1 in total, i.e. for all positions of the oligomer. Adenine accumulates at the ends and an accuracy of 20:1 is estimated for the incorporation of G into the internal positions of the chain. The recent results are very promising and give an idea of the efficiency and selectivity of specific catalysis and surface adsorption in prebiotic template-induced polymerisation.

Another interesting finding concerns the role of 2-deoxyribose, the sugar in contemporary genetic material. Attempts to use the deoxysugar instead of ribose gave only very poor results (L. E. Orgel, personal communication). Thus, at least under the conditions studied so far, the poly-ribonucleotides were much more likely to be the primary replicating molecules and deoxyribonucleic acid (DNA) is a latter product of evolution.

Now we shall try the opposite approach and attempt to look from present-day biochemistry towards the first organisms in a retrospective way. Transfer ribonucleic acids (tRNAs) are generally accepted to belong to the most ancient molecules in the cell. The primary sequences of about 130 tRNA molecules from about 20 organisms have been determined so far and are available for the purpose of alignment. The present collection of data, is thus far more complete than it was six years ago when

Holmquist, Jukes & Pangburn (1973) made a pioneering study on the evolution of tRNAs based on sequence data of 43 molecules. We shall discuss here only the results of the most recent comparative study in this field performed by Eigen & Winkler-Oswatitsch (1980).

Comparative studies of sequences, more frequently performed at the level of proteins than at that of nucleic acids – see for example the recent volume edited by Matsubara & Yamanaka (1978) – are based on the assumption that the evolutionary distance between two organisms is reflected by the number of differences between two analogous sequences. Phylogenetic trees may be reconstructed from these distances (in information theory they are usually called Hamming distances) provided a number of precautions are observed in order to avoid certain pitfalls. In the case of tRNAs we have to recall that the anticodon, the invariant and the almost invariant positions are fixed by the contemporary functions of these molecules and cannot be subjected to changes by mutation. The stem regions are flexible but have a variational constraint due to base pairing.

Eigen & Winkler-Oswatitsch (1980) find that a star-shaped phy-logenetic relationship (ideal bundle) fits the evolutionary distances obtained for tRNAs with different anticodons better than does a phy-logenetic tree (Fig. 2.9). It seems therefore that our present-day tRNAs are the result of a divergent evolution from a common ancestor. This ancestor may be either a single molecule or an ensemble of closely related sequences of a type called 'quasispecies' which will be discussed in the next section. Now, we check whether the sequences of tRNAs have been completely randomised in their variable positions or whether they are still carrying a 'memory' of their origin in the early Pre-Cambrian. For a given phylogenetic model such a proof is indeed possible by proper statistics of Hamming distances. Let us assume we start a process of independent evolution with N sequences, each v digits long. At the beginning $(t = 0)$ all sequences are identical with the initial sequence I_0. Due to mutations, differences are introduced independently into the individual sequences. At every instant (t) we can determine a master sequence $I_m(t)$, which is obtained by identifying the most frequently appearing symbol (A, U, G or C) for each position (Fig. 2.10). The Hamming distance between I_0 and $I_m(t)$ is denoted by $D_1(t)$. It will increase with time and reach a plateau value at complete randomisation. We define two average Hamming distances: we evaluate all individual distances between I_0 and the sequences $I_i(t)$ $(i = 1, 2, \ldots, N)$. The average of the N Hamming distances is defined as $D_2(t)$. Finally, we

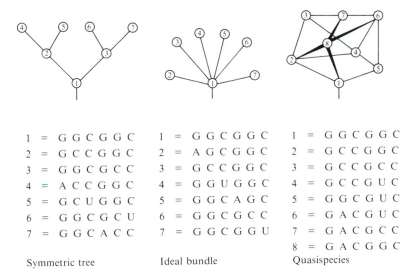

1 = G G C G G C	1 = G G C G G C	1 = G G C G G C
2 = G C C G G C	2 = A G C G G C	2 = G C C G G C
3 = G G C G C C	3 = G C C G G C	3 = G C C G C C
4 = A C C G G C	4 = G G U G G C	4 = G C C G U C
5 = G C U G G C	5 = G G C A G C	5 = G G C G U C
6 = G G C G C U	6 = G G C G C C	6 = G A C G U C
7 = G G C A C C	7 = G G C G G U	7 = G A C G C C
		8 = G A C G G C
Symmetric tree	Ideal bundle	Quasispecies

Fig. 2.9. Three models of divergence leading to different phylogenetic relations (Eigen & Winkler-Oswatitsch, 1980). Sequences connected by a straight line differ by single digit exchange. The symmetric tree and the bundle are idealised distributions. The quasispecies is a realistic distribution which consists of a master sequence and its most frequent mutants at stationary concentrations.

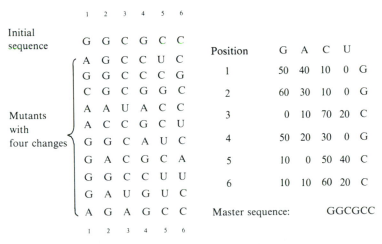

	1	2	3	4	5	6
Initial sequence	G	G	C	G	C	C

							Position	G	A	C	U	
	A	G	C	C	U	C	1	50	40	10	0	G
	G	G	C	C	C	G	2	60	30	10	0	G
	C	G	C	G	G	C						
	A	A	U	A	C	C	3	0	10	70	20	C
Mutants with four changes	A	C	C	G	C	U						
	G	G	C	A	U	C	4	50	20	30	0	G
	G	A	C	G	C	A	5	10	0	50	40	C
	G	G	C	C	U	U						
	G	A	U	G	U	C	6	10	10	60	20	C
	A	G	A	G	C	C	Master sequence:		GGCGCC			
	1	2	3	4	5	6						

Fig. 2.10. An example of the reconstruction of a master sequence from a set of ten four-error mutants. Base substitutions were performed in the sequences G→A→C→U and C→U→G→A respectively. The frequencies at which the digits G, A, C and U appear at the individual positions are given in percentages.

compare all possible pairs, $I_i(t)$ and $I_j(t)$ and obtain $\frac{1}{2}N(n-1)$ Hamming distances the average of which we denote by $D_3(t)$. Analytical expressions for all three functions can be given (P. Richter, personal communication).

$$D_1(t) = \frac{\nu}{2} erfc\left\{ \sqrt{\frac{N}{6[1+3\,\exp(-4t/3\nu)][1-\exp(-4t/3\nu)]}} \right.$$

$$\left. \times \left[3\,\exp\left(-\frac{4t}{3\nu}\right) - \sqrt{\frac{3}{2N}} \right] \right\} \tag{2.2}$$

$$D_2(t) = \frac{3}{4}\nu\left[1 - \exp\left(-\frac{4t}{3\nu}\right) \right] \tag{2.3}$$

$$D_3(t) = \frac{3}{4}\nu\left[1 - \exp\left(-\frac{8t}{3\nu}\right) \right] \tag{2.4}$$

By *erfc* we denote the complement of the error function and $\exp(\alpha)$ stands for e^α.

An example which comes close to the actual situation with tRNA molecules is shown in Fig. 2.11. $D_3(t)$ reaches the plateau values first, followed by $D_2(t)$. The Hamming distance between the initial and the master sequence, $D_1(t)$, increases very slowly during the first phase of evolution. Consequently, the master sequence can still be close to the initial one whereas the individual sequences have already accumulated many mutational changes. Eventually, they may even close to randomisation.

How far did the tRNA molecules come in their divergent evolution? To answer this question we have to look for organisms for which many tRNAs have been sequenced. *Escherichia coli* and *Saccharomyces cerevisiae* (yeast) are most appropriate: 26 sequences are known for *E. coli*, 20 for yeast. As we can see from Fig. 2.11 the tRNAs in yeast, the representative of eukaryotes, approach complete randomisation somewhat more closely than those in the bacterium *E. coli*. In both cases their individual sequences do not reveal much information about a common ancestor. The corresponding master sequence, however, should resemble the primordial polynucleotide rather closely.

Equipped with this confidence in a possible reconstruction of the 'primordial gene', Eigen & Winkler-Oswatitsch (1980) evaluated and discussed the master sequence from all available tRNA data. They arrive at the following three conclusions, which are relevant for the purpose of this essay.

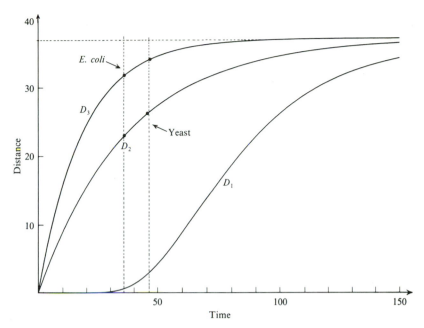

Fig. 2.11. Independent randomisation of $N = 63$ tRNA sequences consisting of $\nu = 50$ variable digits. The Hamming distances D_1, D_2 and D_3 were calculated according to eqns 2.2–2.4. The actual present-day values of D_2 and D_3 obtained from alignment studies of tRNAs in *E. coli* and yeast are shown by black dots (Eigen & Winkler-Oswatitsch, 1980). In their most significant analysis of sequences the whole procedure was based on a total of $\nu = 30$ variable positions. The conclusions derived, nevertheless, are the same as reported here.

(1) The primordial sequence is characterised by a high GC content. Actually, its GC content is higher than that of any individual contemporary tRNA.

(2) The sequence shows a significant preference for a repetitive pattern of the type —RNA—, particularly the triplets —GNC— are very common.*

(3) The ancient precursor of tRNAs was a highly symmetric clover-leaf molecule which could form a very stable hairpin structure as well (Fig. 2.12).

Experimental studies of the thermodynamics and kinetics of base pairing (Pörschke, 1977) have shown that high GC content makes the sequences 'sticky' and more resistant to hydrolysis. In the hairpin more

---

* Here and in the following sections we denote a purine base, G or A, by R, a pyrimidine base, C or U(T), by Y and an arbitrary base, G, A, C or U(T), by N.

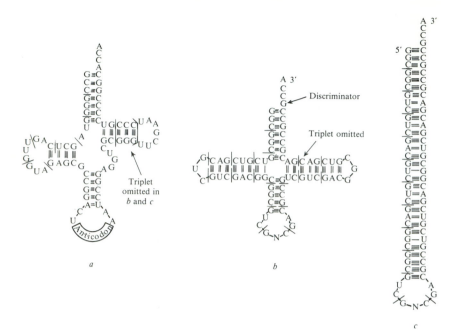

Fig. 2.12.   Reconstruction of the primordial gene according to Eigen & Winkler-Oswatitsch (1980). *a.* The master sequence of present-day tRNAs. *b.* Clover-leaf structure of the primordial gene after omission of a triplet indicated in *a* and introduction of a repetitive –GNC– pattern. *c.* Hairpin structure of *b*, the stem of the hairpin containing only one weak mismatch, a GU pair. Almost perfect complementary symmetry of the + and − strand is a consequence of this hairpin structure. The differences between the + and the − strand consist in the 3′ ACCG- end, the GU pair mentioned and the central position of the anticodon.

base pairs can be formed than in the clover leaf. The capability to form a hairpin is another feature which stabilises the molecule against degradation.

The repetitive pattern may be a remnant of a chain elongation mechanism as indicated in Fig. 2.8. A further consequence of this pattern and another independent hint on its origin will be discussed together with the origin of translation in a subsequent section (p. 65).

DYNAMICS OF SELFREPLICATION

In the previous sections we tried to collect arguments and experimental findings which suggest that the appearance of biopolymers on the prebiotic earth was a probable event. Moreover, it is very likely that polynucleotides capable of replication were among these polymers. The formation of biopolymers with regular structures consisting of a repetitive backbone and side chains introduced a previously unknown variability into the prebiotic scenario. The number of possible different sequences for only moderate chain lengths is already astronomic (Fig. 2.13). A huge number of these sequences will have identical brutto-formulas and thus can be understood as isomers of a certain class. They may also have very similar free energies, but, in general, there will be no

Number of digits (ν)	Sequences				N
1	G	A	C	U	4
2	GG	AG	CG	UG	16
	GA	AA	CA	UA	
	GC	AC	CC	UC	
	GU	AU	CU	UU	
3	GGG	AGG	CGG	UGG	64
	GGA	AGA	CGA	UGA	
	GGC	AGC	CGC	UGC	
	GGU	AGU	CGU	UGU	
	GAG	AAG	CAG	UAG	
	GAA	AAA	CAA	UAA	
	GAC	AAC	CAC	UAC	
	GAU	AAU	CAU	UAU	
	GCG	ACG	CCG	UCG	
	GCA	ACA	CCA	UCA	
	GCC	ACC	CCC	UCC	
	GCU	ACU	CCU	UCU	
	GUG	AUG	CUG	UUG	
	GUA	AUA	CUA	UUA	
	GUC	AUC	CUC	UUC	
	GUU	AUU	CUU	UUU	
4					256
10					1 048 576
100					1.607×10^{60}
ν					4^{ν}

Fig. 2.13. An illustrative example of the enormous variability of polynucleotide sequences. The number of possible different sequences ($N = 4^{\nu}$) soon reaches 'superastronomical' size with increasing ν.

equilibration between them, since many covalent bonds have to be broken and formed again along the reaction path leading from one sequence to another. What property was it, then, that could bring order into this vast and ample variety? We have to search for a kinetic mechanism which by itself steadily reduces the number of different sequences present. The answer is simple and hidden in the properties of polynucleotides: selfreplication is the most proper means to increase the number of certain sequences at the expense of others. Hence, we shall discuss the dynamics of selfreplication in this section.

Basic features of selfreplication

The capability of selfreplication introduces an otherwise unknown richness into the dynamics of chemical reactions. Oscillatory behaviour (Eyring & Henderson, 1978) or spatial pattern formation (Nicolis & Lefever, 1975) have never been observed in chemical systems which lacked autocatalytic steps.

Another important feature of selfreplicating systems is revealed by a simple stochastic model. Let us assume a system which contains N polynucleotides with different sequences at the time $t = 0$. Thus, we start with one copy of every polynucleotide. We assume further that all the molecules duplicate with the same probability. The probabilistic decay rate of the molecules is equal to the rate of duplication. A mathematical analysis of this stochastic process has been presented by Batholomay (1958). He treated this 'linear birth and death' system as a Markovian chain. His results may be applied directly to our problem (Eigen, 1971; Schuster, 1972) and can be summarised as follows (Fig. 2.14).

(1) The number of different sequences present in the system decreases steadily.
(2) The expectation value of the population numbers of the surviving species increases.
(3) After a very long time the system as a whole becomes extinct.

The properties (1) and (2) are slightly modified in cases where the probabilities for duplication and decay are not the same. This is not so with (3): a duplication rate which exceeds the decay rate leads to a non-zero probability for the survival of the system as a whole.

From this very simple example we learn that the capability of replication is a powerful means in reducing the enormous variability encountered with biopolymers. The choice of a certain sequence which is selected as a survivor in the system is a purely stochastic event. All

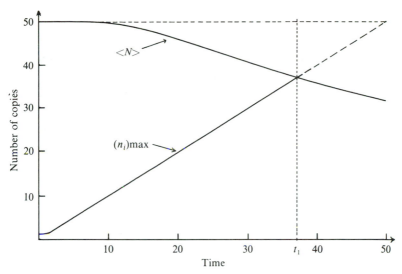

Fig. 2.14. Linear birth and death process applied to the replication of poly-nucleotides. Let us assume an initial situation with N_0 different sequences. Every sequence is present in a single copy only: $n_i(0) = 1$; $i = 1, 2, \ldots N_0$; $\Sigma n_i(0) = N_0$. The individual polynucleotides replicate and are hydrolysed independently but with equal rate constants: $I_i \overset{f}{\to} 2 I_i$ and $I_i \overset{d}{\to} 0$, $i = 1, \ldots, N_0$. Moreover we shall assume that the rate of replication is equal to the rate of hydrolysis: $f = d$. Under these conditions one might expect a constant total number of polynucleotides. An extension of the mathematical analysis presented by Batholomay (1958) to our problem shows, however, that the answer is not as simple. In order to illustrate what happens we consider two time-dependent quantities (1) the most probable value for the total number of molecules, $\langle N \rangle$, which is obtained by multiplying the probability of survival for a single polynucleotide by N_0 and (2) the most probable population number of an individual sequence in the system $(n_i)_{max}$. At $t = 0$ we have $\langle N \rangle = N_0$ and $(n_i)_{max} = 1$ since all sequences are present in the copy. Due to the replication process many sequences disappear and $(n_i)_{max}$ increases steadily. The value of $\langle N \rangle$ remains essentially constant at first but then starts to decrease slowly. The stochastic treatment shows that the system becomes extinct after long enough time, $\lim_{t \to \infty} \langle N \rangle = 0$. Long before that, there is a certain moment, $t = t_1$, at which the two functions cross, $\langle N \rangle_{t=t_1} = (n_i)_{max, t=t_1}$. From that point on, $t > t_1$, it is practically certain that we are dealing with a homogeneous population. In other words, the whole system contains a single sequence only present in $\langle N \rangle$ copies: $n_i(t_1) = 0$; $i = 1, 2, \ldots, N_0$, $i \neq k$ and $n_k(t_1) = \langle N \rangle$. For the concrete example shown in the diagram we have chosen $f = d = 1$ and $N_0 = 50$.

sequences are characterised by the same probabilities for duplication and for decay and hence have equal chances of survival. In this abstract case-study we observe an example for the mere tautology 'survival of the survivor': the outcome of the selection process can neither be predicted nor limited in advance because of fundamental lack of information.

The dynamics of polynucleotide replication, in general, follow a complicated many-step mechanism and are difficult to analyse. In order to concentrate on the basic characteristics of selfreplication we have to search for a model system which is likewise accessible to theoretical analysis *and* experimental verification. The primary goal of such a system is the creation of a constant environment, the parameters of which can be easily controlled. At the same time the number of variables such as pressure, temperature and concentrations of supplementary material has to be reduced to a minimum. The most simple and suitable system from the theoretician's point of view is the flow reactor shown in Fig. 2.15. The question of experimental verification is postponed for a moment. We shall simulate the whole replication process by an 'over-all' single-step reaction; the same is assumed for hydrolytic cleavage:

$$I_i + \sum_{\lambda=1}^{m} \nu_{i\lambda} A_\lambda \xrightarrow{k_i} 2I_i \tag{2.5}$$

$$I_i \xrightarrow{d_i} \sum_{\lambda=1}^{m} \nu_{i\lambda} B_\lambda; \qquad i = 1, 2, \ldots, N \tag{2.6}$$

By I_i we denote an individual polynucleotide. A_λ are the activated building blocks, e.g. the nucleoside triphosphates ATP, UTP, GTP and CTP. B_λ stands for the products of degradation, in particular for the nucleoside monophosphates AMP, UMP, GMP and CMP. Under the conditions of the flow reactor (Fig. 2.15) the kinetics of the reactions (2.5) and (2.6) is represented by the differential equations

$$\frac{dx_i}{dt} = \dot{x}_i = (k_i - d_i)x_i - \frac{x_i}{c}\phi; \qquad i = 1, 2, \ldots, N \tag{2.7}$$

The individual concentrations of polynucleotides are denoted by $x_i = [I_i]$, c is their total concentration ($c = \sum_i x_i$) and ϕ is a global dilution flux. Imposing the constraint of constant total concentration (Eigen, 1971)

$$\frac{dc}{dt} = \dot{c} = \sum_i x_i = 0; \qquad \phi = \sum_i (k_i - d_i)x_i \tag{2.8}$$

on the system we obtain the following fairₐy simple differential equations:

$$\dot{x}_i = (E_i - \bar{E})x_i, \qquad i = 1, 2, \ldots, N \tag{2.9}$$

Here we use $E_i = k_i - d_i$ for the 'excess productivity' of the individual sequence I_i and $\bar{E} = \sum_i (E_i x_i / c)$ for the corresponding mean value.

Solution curves for a concrete numerical example for (2.9) are shown in Fig. 2.16. The general result may be easily visualised by inspection of

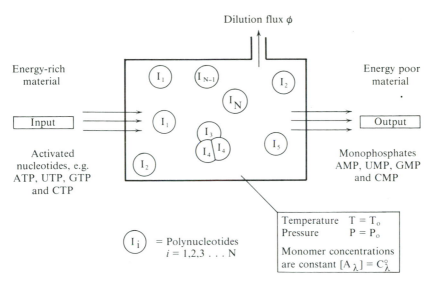

Fig. 2.15. The evolution reactor. This kind of flow reactor consists of a reaction vessel which allows for temperature and pressure control. Its walls are impermeable to polynucleotides. Energy-rich material is poured from the environment into the reactor. The degradation products are removed steadily. Material transport is adjusted in such a way that the concentration of monomers is constant in the reactor. A dilution flux ϕ is installed in order to remove the excess of polynucleotides produced by multiplication. Thus the sum of the concentrations $[I_1]+[I_2]+ \ldots +[I_N]=\sum_{i=1}^{N} x_i = c$, may be controlled by the flux ϕ. Under 'constant organisation' ϕ is adjusted such that the total concentration c is constant.

the differential equation. At every instant $\bar{E}$ has a certain value. The concentrations of all polynucleotides with an excess productivity lower than $\bar{E}$ will decrease in the next moment since we have $\dot{x}_i < 0$ for these sequences. Polynucleotides with $E_i > \bar{E}$, by contrast, are more productive than the average and their concentrations will increase. Thus, we find that selection for greater efficiency reflected in the net growth is built into the system. Consequently, $\bar{E}$ is increasing and the requirement for further net growth is set higher. This situation in a way resembles a high jump contest with $\bar{E}$ representing the bar which is put higher and higher, so that more and more jumpers are eliminated. The champion corresponds to the sequence with the largest value of E_i.

$\bar{E}(t)$ is a monotonically increasing function and represents the quantity which is optimised during the evolution of the reactor. Since the outcome of the selection process is completely determined and predictable we

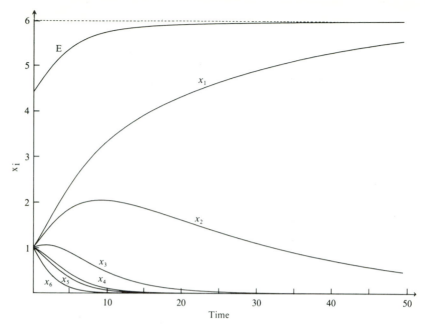

Fig. 2.16.   Solution curves for the differential equations $\dot{x}_i = (E_i - \bar{E})x_i, i = 1, \ldots,$ 6 with $E_1 = 1.0$, $E_2 = 0.95$, $E_3 = 0.8$, $E_4 = 0.65$, $E_5 = 0.6$ and $E_6 = 0.4$ and $\bar{E} = \sum_i x_i E_i / \sum_i x_i$. Initial conditions: $x_1(0) = x_2(0) = \ldots = x_6(0) = 1$. The mean excess production, $\bar{E}$, starts from an initial value of $\bar{E}(0) = 4.4$, increases steadily and approaches the maximum value $\lim_{t \to \infty} \bar{E}(t) = 6$ when the population becomes homogeneous: $x_1 = 6$, $x_2 = x_3 = \ldots = x_6 = 0$. One easily recognises the cause for the increase of $\bar{E}$: the less efficiently growing polynucleotides are eliminated. They disappear in the same sequence as their excess productivity E_i increases.

have a situation characterisable by 'survival of the fittest'. Fitness in physical terms is measured by the size of the excess productivity E_i.

Selfreplication with errors

No physical process proceeds with absolute accuracy, and we have to deal with certain non-zero frequencies of copying errors which give rise to mutant distributions. First we consider a stochastic aspect of the problem by means of a simple but illustrative model (Fig. 2.17; Schuster & Sigmund, 1980). Errors are assumed to occur independently and with equal probability for every digit. We define a single digit accuracy q which is the probability that a given digit is reproduced correctly in the daughter sequence. The probability of obtaining a correct copy of a polymer with ν

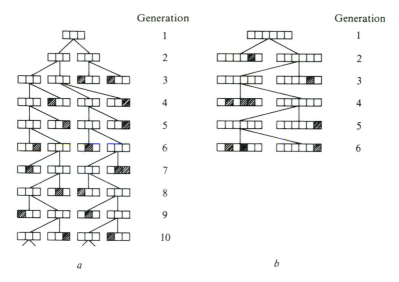

Fig. 2.17. A stochastic model for selfreplication with errors. A sequence of ν digits is represented by ν squares aligned in a row. Errors are indicated by shaded squares. A correct sequence produces a progeny of Σ copies per generation. All mutations are assumed to be lethal: error copies are not allowed to multiply. The accuracy of the replication process is characterised by a single digit accuracy q; q is the probability that a given digit is incorporated correctly into the growing chain. Two characteristic examples with different outcomes are shown. *a*. $q = 5/6$, $\nu = 3$ and $\Sigma = 2$, the system has a high probability of survival; *b*. $q = 5/6$, $\nu = 6$ and $\Sigma = 2$, the system dies out after a few generations.

segments, Q, then is of the form

$$Q = q^{\nu} \tag{2.10}$$

Every correct sequence is assumed to create a progeny of Σ copies per generation and to live for just one multiplication cycle. All error copies, i.e. mutants, in this model are lethal.

Under these assumptions the problem can be treated as a multiplicative chain (see Bartlett, 1978). The necessary and sufficient condition for certain extinction of the master copy is $m \leq 1$, where m represents the expectation value of reproductive success, i.e. the mean number of descendants per generation from a single copy. The number of offspring is a random variable X with binomial distribution $\beta(\Sigma, Q)$. This means that the probability

$$\text{Prob}\{X = k\} = \binom{\Sigma}{k} Q^k (1 - Q)^{\Sigma - k} \tag{2.11}$$

From the expectation we obtain

$$m = \Sigma Q = \Sigma q^\nu \qquad (2.12)$$

The condition for a non-zero probability of survival simply reads

$$Q\Sigma > 1 \quad \text{or} \quad Q > \Sigma^{-1} \qquad (2.13)$$

respectively. There is a sharply defined minimum accuracy of replication, $Q_{min} = \Sigma^{-1}$, below which the system goes certainly extinct. For a given value of q this minimum accuracy corresponds to a maximum chain length of the polymer

$$Q_{min} = q^{\nu_{max}} \quad \text{or} \quad \nu_{max} = -\frac{\ln \Sigma}{\ln q} \sim \frac{\ln \Sigma}{1-q}, \quad \text{if } q \sim 1 \qquad (2.14)$$

A given polymerisation mechanism is characterised by a certain q value which in turn sets a critical limit to the information content that can be transmitted from generation to generation by the copying mechanism. An increase in the information content transferred via replication requires an improvement of the replication machinery. We shall come back to this problem at the end of this section.

The stochastic model revealed one important feature of selfreplication with errors: the content of information to be transferred is limited by the accuracy of replication. In order to gain some more quantitative insight we have to drop the most restrictive (and unrealistic) assumption concerning lethality of mutants. Evolution can take place in an efficient way only if we allow for an ample mutant distribution with an almost continuous (or 'quasicontinuous') gradation of properties. The resulting mathematical problem is very hard to solve at the level of stochastic processes. As in the previous section we may try, nevertheless, to set up differential equations. In doing so we make two implicit assumptions, namely (1) that we are considering a sufficiently large ensemble of molecules and (2) that mutations occur frequently enough to reach stationary mutant distributions.

The deterministic approach to template-induced replication with errors has been studied extensively in the past (Eigen, 1971; Thompson & McBride, 1974; Jones, Enns & Ragnekar, 1976; Eigen & Schuster, 1977; Epstein & Eigen, 1979; Küppers, 1980). All these investigations start from conditions as encountered in the flow reactor (Fig. 2.15) which are mimicked by the differential equation

$$\dot{x}_i = \sum_{j=1}^{N} f_{ij} x_j - \frac{x_i}{c} \phi; \qquad i = 1, 2, \ldots, N \qquad (2.15)$$

Eqn (2.15) is a straightforward generalisation of eqn (2.7). In addition to correct selfreplication we allow also errors leading to mutants according to the reaction

$$I_j + \sum_{\lambda=1}^{m} \nu_{i\lambda} A_\lambda \xrightarrow{f_{ij}} I_i + I_j; \qquad i, j = 1, 2, \ldots, N \qquad (2.16)$$

The constant f_{ij} thus represents the rate at which the sequence I_i is obtained as an error copy of I_j. The 'diagonal' rate constants require further specification

$$f_{ii} = k_i Q_i - d_i \qquad (2.17)$$

Q_i is the quality factor for the replication of the sequence I_i. As before, it describes the probability for correct replication of the whole polymer. The rate constant k_i gives the number of copies, correct *and* erroneous, which are synthesised per time unit on the template I_i, d_i is the rate constant for hydrolysis of these templates as in eqn (2.6). The complete quantity f_{ii} has been named the 'value function' (Eigen, 1971) since it determines the selective value of the polynucleotide I_i. Again we define an excess productivity $E_i = k_i - d_i$.

From mass conservation we obtain the global relation

$$\sum_i k_i(1 - Q_i)x_i = \sum_i \sum_{j \neq i} f_{ij} x_j \qquad (2.18)$$

Accordingly, the mean excess productivity becomes identical to the dilution flux ϕ exactly as it was with error-free replication, except for a factor c^{-1}:

$$\bar{E} = \frac{1}{c} \sum_i E_i x_i = \frac{1}{c} \phi \qquad (2.19)$$

The remaining differential equation is

$$\dot{x}_i = [f_{ii} - \bar{E}(t)]x_i + \sum_{j \neq i} f_{ij} x_j; \qquad i, j = 1, 2, \ldots, N \qquad (2.20)$$

Eqn (2.20) can be analysed by second order perturbation theory (Jones *et al.*, 1976). Again selection takes place in the evolution reactor. By successive elimination of the slowly growing molecules the mean excess productivity, $\bar{E}(t)$, increases steadily until it reaches an optimal value

$$\lim_{t \to \infty} \bar{E}(t) = \lambda_{\max} = f_{mm} + \sum_{j \neq m} \frac{f_{mj} f_{jm}}{f_{mm} - f_{jj}} \qquad (2.21)$$

By I_m we denote the master sequence, i.e. the polynucleotide which has the largest value function $f_{mm} = \max(f_{ii}, i = 1, \ldots, N)$. The results of second-order perturbation theory are good approximations to the exact solutions provided $f_{im} \ll (f_{mm} - f_{ii})$ for all $i \neq m$.

What is the final outcome of the selection process? In order to answer this question we have to investigate the stationary state of the evolution reactor in case there exists one. From perturbation theory we obtain (stationary concentrations are indicated by bars, $\bar{x}_1, \bar{x}_2, \ldots, \bar{x}_N$):

$$\bar{x}_m = c\,\frac{f_{mm} - \bar{E}_{-m}}{E_m - \bar{E}_{-m}} = c\,\frac{Q_m - \sigma_m^{-1}}{1 - \sigma_m^{-1}} \tag{2.22}$$

and

$$\bar{x}_i = \bar{x}_m \frac{f_{im}}{f_{mm} - f_{ii}}; \qquad i = 1, 2, \ldots, N; i \neq m \tag{2.23}$$

For short notation we define two quantities in eqn (2.22)

$$\bar{E}_{-m} = \frac{\sum\limits_{i=1, i \neq m}^{N} E_i x_i}{c - x_m} \quad \text{and} \quad \sigma_m = \frac{k_m}{d_m + \bar{E}_{-m}} \tag{2.24}$$

The latter quantity will be named the superiority parameter σ_m. It is a measure for the kinetic superiority of the master sequence compared to the average of the other polynucleotides present in the evolution reactor. We distinguish two situations.

(1) $Q_m \leq \sigma_m^{-1}$: there is no stationary state of the evolution reactor. The accuracy of replication is too low and at every instant we find different sequences.

(2) $Q_m > \sigma_m^{-1}$: the master copy is accompanied by a mutant distribution at the stationary state. We call the whole ensemble of sequences a 'quasispecies' because of the similarity to an actual biological species as far as the existence of mutants present in the stationary population is concerned. The stationary concentration of a given mutant relative to the master sequence is determined by two quantities: the difference in value functions, $f_{mm} - f_{ii}$ and the off-diagonal term f_{im}, the probability of a mutation from I_m to I_i. A mutant sequence I_i will thus be present at higher stationary concentration, $\bar{x}_i$, if either the value function f_{ii} is larger and consequently $f_{mm} - f_{ii}$ is smaller, or if the mutation rate f_{im} is higher. What factors outside the accuracy of replication determine mutation rates? Clearly, the relation, more precisely the Hamming distance, between the sequences is important. Roughly speaking, a given single-error copy will

occur more frequently than a two-error copy, which in turn has a higher probability than a three-error copy etc. In other words a sequence will be the more abundant the more closely it is related to the master copy.

So far we have excluded systems in which two or more sequences have identical or almost identical value functions. In these cases the simple second-order formalism of perturbation theory is not applicable. An approximate solution may be obtained, however, by the perturbation formalism for quasi-degenerate states which is well known in quantum mechanics (Courant & Hilbert, 1953).

As with the stochastic approach we obtain a lower limit for the accuracy of the copying process below which the transfer of information breaks down:

$$Q_m > Q_{\min} = \sigma_m^{-1} \tag{2.25}$$

We introduce an average single digit accuracy $\bar{q}_m$ for the replication of a given polynucleotide with a chain length of ν_m:

$$Q_m = q_1 \times q_2 \times \ldots \times q_{\nu_m} = \bar{q}_m^{\nu_m} \tag{2.26}$$

q_i represents the single digit accuracy for the incorporation of the complementary nucleotide at a given position (i) along the chain. The quality factors q_i will depend on the nature of the nucleotide (A, U, G or C) and on its neighbourhood in the sequence. For sequences of sufficient length, with similar nucleotide contents, in general the values for $\bar{q}$ are expected to be similar.

The lower limit of the quality factor implies an upper limit for the chain length of the polynucleotide

$$\nu_m < \nu_{\max} = -\frac{\ln \sigma_m}{\ln \bar{q}_m} \sim \frac{\ln \sigma_m}{1 - \bar{q}_m} \tag{2.27}$$

The approximation in the last part of eqn (2.27) will be fulfilled in almost every case since $\bar{q}_m$ can be expected to be close to 1.

It seems worth-while to illustrate the meaning of the superiority σ_m: according to eqn (2.24) the master sequence can be superior to the others by replicating faster, $k_m > \bar{k}_{-m}$, or by being degraded more slowly, $d_m < \bar{d}_{-m}$. The averages of the rest are defined as before

$$\bar{k}_{-m} = \frac{\sum\limits_{i=1, i \neq m}^{N} k_i x_i}{c - x_m} \quad \text{and} \quad \bar{d}_{-m} = \frac{\sum\limits_{i=1, i \neq m}^{N} d_i x_i}{c - x_m}$$

Let us assume identical rate constants for degradation, $d_1 = d_2 = \ldots = d_N$.

Then, the superiority is determined only by the replication rate constants

$$\sigma_m = k_m/\bar{k}_{-m}$$

Complete lethality of mutants in the deterministic approach ($\bar{k}_{-m} = 0$) leads to an interesting and important result: σ_m becomes infinite and the quality factor Q_m may take any positive value, no matter how small it may be, to fulfil eqn (2.25). Hence, according to eqn (2.27) chains of any length can exist in stationary states in cases where there is no competition with mutants. Still, we would have to consider the restrictions of the stochastic aspects as given by eqn (2.14). These are of minor importance only, when we are dealing with large numbers of molecules. Then a very long time will elapse, on average, before the system becomes extinct. This aspect deserves further attention and will be discussed in some detail in a forthcoming booklet (P. Schuster & K. Sigmund, in preparation).

The error threshold defined by eqn (2.27) strongly depends on the average single digit accuracy of the replication process. Due to the logarithmic function the influence of the superiority on the maximum chain length is not as pronounced. Some characteristic numerical values are summarised in Table 2.2. The top row gives some numbers typical for prebiotic, template-induced, polynucleotide replication. The value taken for the single digit accuracy, $\bar{q} = 0.95$, corresponds very well to the results of Orgel and coworkers discussed in the previous section. Originally, values between $\bar{q} = 0.90$ and $\bar{q} = 0.99$, the extremes corresponding to AU

Table 2.2. *Single digit accuracy, superiority and error threshold in some chemical and biological systems*

Single digit accuracy, $\bar{q}_m$	Error rate per digit, $1 - \bar{q}_m$	Superiority, σ_m	Maximum digit content, ν_{max}	Biological examples
0.95	5×10^{-2}	2	14	Enzyme-free RNA replication
		20	60	
		200	106	tRNA precursor, $\nu = 80$
0.9995	5×10^{-4}	2	1386	Single-stranded RNA replication
		20	5991	via specific replicases
		200	10597	Phage $Q\beta$, $\nu = 4500$
0.999999	1×10^{-6}	2	0.7×10^6	DNA replication via polymerases
		20	3.0×10^6	and proof-reading
		200	5.3×10^6	*E. coli*, $\nu = 4 \times 10^6$

rich and GC rich sequences respectively, have been derived from thermodynamic considerations (Eigen, 1971; Pörschke, 1977). Under prebiotic conditions the length of polynucleotides with defined sequences was limited to molecules up to the size of present day tRNAs. Longer polynucleotides could form but their sequences were changing permanently.

The accuracy of replication and the maximum chain length in viral and bacterial replication will be discussed in the next section.

Some experiments on RNA-replication

The possibilities of experimental realisation of precellular evolution have been reviewed recently by Küppers (1979). He describes an evolution machine which comes close to the idealised flow reactor shown in Fig. 2.15 and which allows the study of selfreplication of polynucleotides. An interesting experiment in this direction was reported by Schneider, Neuser & Heinrichs (1979): they studied the system poly(A)-poly(U) in a stirred flow reactor. The enzyme, RNA-polymerase, as well as excess ATP and UTP were streamed through the system in which the polymer is synthesised. RNA polymerisation roughly follows a mechanism consisting of nucleation and polymerisation steps (Fig. 2.18). Under the conditions of the flow reactor the overall reaction kinetics are represented by the growth law for complementary replication originally discussed by Eigen (1971). At the steady state the $(\pm)$ ensemble grows according to the simple differential eqn (2.7). The replication rate constant for the ensemble is obtained as the geometric mean of the individual rate constants, $k_\pm = \sqrt{k_+ \cdot k_-}$. Hence, efficient replication requires that both strands are well recognised by the RNA-polymerase. This fact favours structures with complementary symmetry, as discussed on p. 53. At least both ends of the two strands $(\pm)$ have to be complementary.

The $Q\beta$-bacteriophage system has been studied most extensively by in-vitro experiments. Evolution experiments on this system were performed by Spiegelman (1971). He mimicked the flow reactor by serial transfer experiments and found a characteristic response of the system to selection constraints: under these ideal growth conditions the phage RNA lost its infectivity but the rate of polymerisation increased many times. The faster growing pieces of RNA had smaller molecular weights. This change in properties was interpreted initially as partial loss of the viral genome. One of the smallest pieces of RNA isolated in the serial transfer experiments was 218 nucleotides long. It was called the

P. Schuster

$$\text{poly (A)} + E \; \overset{k_1}{\underset{k_2}{\rightleftharpoons}} \; E \cdot \text{poly (A)}$$

$$\text{poly (U)} + E \; \overset{k_1}{\underset{k_3}{\rightleftharpoons}} \; E \cdot \text{poly (U)}$$

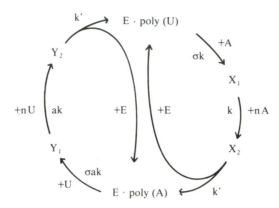

Fig. 2.18. A cyclic mechanism of enzyme catalysed polynucleotide replication according to Schneider, Neuser & Heinrichs (1979). The symbols E, A and U are used for free RNA polymerase, ATP and UTP respectively. X_1 and Y_1 are initiation complexes (enzyme·poly(U)·A_m or enzyme·poly(A)·U_m, m being a small number). X_2 and Y_2 are reaction intermediates of the type enzyme·poly(U)·poly(A) or enzyme·poly(A)·poly(U). k, k', k_1, k_2, k_3 are rate constants, σ is the co-operativity parameter and a is a factor expressing the ratio of the polymerisation rates of U and A.

midivariant and its sequence was completely determined (Kramer, Mills, Cole, Nishihara & Spiegelman, 1974). The molecule shows a characteristic structure with complementary symmetry in the sense mentioned before (Fig. 2.19), so that both strands are efficient templates for the enzyme. Later experiments led to the conclusion that the situation is not as simple: the $Q\beta$-polymerase is capable of de-novo synthesis of RNA in the absence of template (Sumper & Luce, 1975). The outcome of these experiments was polynucleotides closely related to the midivariants in size and fingerprint. It is important to stress the fact that a present-day RNA-polymerase is able to produce its own template directly from a solution of nucleoside triphosphates although we do not know yet whether this finding is relevant for prebiotic conditions where the proteins could be only very primitive catalysts. Küppers & Sumper (1975)

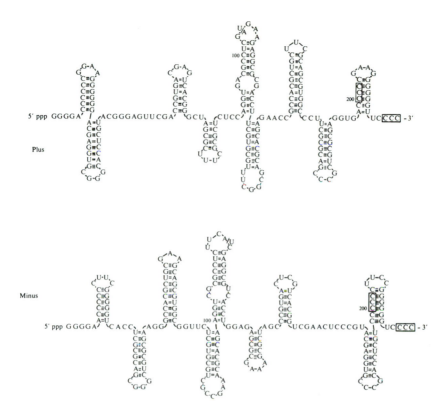

Fig. 2.19. The sequence and a possible secondary structure for both strands (±) of the midivariant of bacteriophage $Q\beta$ (Kramer *et al.*, 1974). The recognition site for the enzyme, the $Q\beta$ specific RNA polymerase, consists in two CCC triplets at the 3' end of the RNA molecules as indicated in the figure. The two triplets in the correct relative steric position are the minimum requirement without which the RNA molecules are not accepted as templates for polymerisation by the enzyme. The replication process thus starts at the 5' end of the new or daughter strand and proceeds always in the direction 5' → 3'. Complementary symmetry leading to similar or even identical secondary structures of the two strands (±) is an important advantage for complementary replication: both strands are replicated with similar or even the same rate constants and the overall growth rate for the (±) ensemble takes an optimum value ($k_{\pm} = \sqrt{k_{+}k_{-}}$).

succeeded in identifying the recognition site of the enzyme. This recognition site consists of two CCC-triplets and represents the minimum requirement of a polynucleotide for being accepted as a template for replication. Further, it can be shown (Eigen & Schuster, 1978*b*) that 169 of the 218 positions of the midivariant sequenced by Kramer *et al.* (1974) are identical with those of an artificial sequence consisting of CCC, CCCC, UUCG blocks and their complements (GGG, GGGG and CGAA respectively). The enzyme has recognition sites for CCC triplets as we have seen above, and also for UUCG blocks. The latter recognition site is reminiscent of the $T\Psi$ loop of tRNA molecules where we find a sequence $T\Psi CG$ which is equivalent to UUCG in unmodified bases. Indeed, $Q\beta$-polymerase consists of one virus-specific subunit and three constituents from the host cell which are involved in ribosomal protein synthesis as well, namely the s_1-protein and the elongation factors EFTu and EFTs. One of the three host proteins most probably carries the recognition site for the $T\Psi$ loop.

A more recent, elegant series of in-vivo and in-vitro experiments on $Q\beta$ bacteriophage replication performed by Domingo, Flavell & Weissmann (1976) provides a possibility for checking the relevance of eqn (2.27) for RNA replication. An error copy of the phage genome was produced *in vitro* by site-directed mutagenesis. The base change was extracistronic and the mutant was infectious in in-vivo experiments, i.e. in the host bacteria. Making use of a model from population genetics for revertant mutants in a growing phage population Batschelet, Domingo & Weissmann (1976) were able to determine the accuracy of replication. An analysis of their data (Eigen & Schuster, 1977) yields a selective advantage for the master copy ($f_{\text{wildtype}}/f_{\text{mutant}} \sim 2$ to 4) and a single digit accuracy of $q \sim 0.9997$. Inserting these values into eqn (2.27) we obtain a maximum chain length for $Q\beta$-RNA which is very close to the actual value $\nu \sim 4500$ (Table 2.2). The bacteriophage $Q\beta$ thus approaches the upper limit of the information content as closely as possible. This idea is in general agreement with results on the simple DNA phage ϕX 174, for which the complete sequence of the genome is available (Fiddes, 1977). The constraint on information storage is so strong that the genes overlap in the genome, i.e. the same sequence of polynucleotides is used twice or even three times for coding but with different frames.

Moreover, Weissmann and coworkers provided direct proof for the quasispecies-like nature of $Q\beta$-bacteriophage. The standard sequence, i.e. the master copy, is present as a relatively small fraction only in the wild-type distribution. The majority of sequences found are those of frequently occurring mutants. According to eqn (2.22) such a situation is characteristic of an ensemble which grows closely to the error threshold $(Q_m \sim \sigma_m^{-1})$. Thus, this finding agrees well with the fact that ν is close to ν_{max}.

The success in the description of RNA bacteriophage replication by the model system of eqn (2.15) encourages the extrapolation to pro-karyotic DNA replication (Eigen & Schuster, 1977). For this purpose we recall the semi-conservative nature of the process (Fig. 2.20) and the fact that two enzymes, roughly speaking a polymerase and a repair enzyme, are responsible for the average single digit accuracy $\bar{q}$ of the replication process as a whole. We may try an off the cuff estimate by assuming that both enzymes operate with similar accuracy as $Q\beta$ replicase (Table 2.2). Thereby, one obtains a maximum chain length of $1 \times 10^6 < \nu_{\max} < 10 \times 10^6$ base pairs, in excellent agreement with the length of known bacterial genomes. We have to be very careful with this speculation, however, since the process of DNA replication as far as we know at present is very complicated (see e.g. Kornberg, 1974; Bernardi & Ninio, 1978; Staudenbauer, 1978). Mutations in the polymerase may cause an increase and also a decrease in the mutation rates (Bessmann, Muzyczka, Goodman & Schnaar, 1974). The enzyme thus does not appear to operate at the optimal accuracy. An estimate of the error threshold of DNA replication based on safer grounds would require more accurate quantitative data on the replication process.

Co-operation between selfreplicative elements

So far, we have seen that selfreplicative elements compete and selection causes the essential narrowing in the enormous variability of poly-nucleotide sequences. A 'quasispecies', i.e. an ensemble consisting of a master sequence and its most frequent mutants, is the ultimate outcome of such a selection process. The information content and the catalytic properties that can be stored in a quasispecies are limited by the error threshold. The evolution of independent competitors thus comes to a 'dead end' sooner or later. Therefore, we have to ask another fundamental question: how can we force co-operation between otherwise competitive polynucleotides?

The answer can be given in very general terms (Eigen & Schuster, 1978a; Epstein, 1979a). Introducing the next higher order, in particular second-order terms of mass action, into the kinetic eqns (2.7) describing selfreplication in a flow reactor we obtain the general differential equation:

$$\dot{x}_i = (k_i - d_i)x_i + \sum_{j=1}^{N} k_{ij}x_ix_j - \frac{x_i}{c}\phi; \qquad i, j = 1, 2, \ldots, N \qquad (2.28)$$

P. Schuster

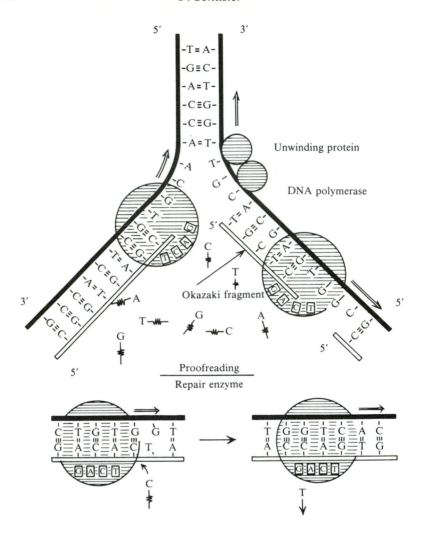

Fig. 2.20. A schematic illustration of the mechanism of prokaryotic DNA replication. The replication of double stranded DNA is a highly sophisticated process including many individual steps of reaction and control (Kornberg, 1974). The process as a whole is of semi-conservative nature, which means that each daughter molecule contains one parental and one new strand (indicated as black and white lines). The most important steps of DNA polymerisation are shown in a highly simplified manner. Both daughter strands are synthesised in the 5′ → 3′ direction. The parental double helix is unwound by an enzyme, the unwinding protein. RNA primers initiate the polymerisation process. At least two different polymerases are involved in the primary process of DNA synthesis (polymerase-I and polymerase-III). Chain elongation up to fragments with 1000 to 2000 nucleotides seems to be effected mainly by the polymerase-III complex. These

For the sake of simplicity we have omitted the terms accounting for the mutations. Eqn (2.28) describes the two reactions (2.5) and (2.8) together with the reaction

$$I_i + \sum_{\lambda=1}^{m} \nu_{i\lambda} A_\lambda + I_j \xrightarrow{k_{ij}} 2I_i + I_j \qquad (2.29)$$

In reaction (2.29) the polynucleotide I_j acts as a catalyst in the replication of I_i.

The first-order terms in eqn (2.28) lead to competition, as we know. In order to search for co-operation we can restrict ourselves to the general system with second-order terms only:

$$\dot{x}_i = \sum_{j=1}^{N} k_{ij} x_i x_j - \frac{x_i}{c} \phi; \qquad i, j = 1, 2, \ldots, N \qquad (2.30)$$

This general differential equation with dimension up to $N = 4$ was studied under the constraint of constant total concentration (Hofbauer, Schuster, Sigmund & Wolff, 1980a): co-operation occurs where the catalytic terms form a closed positive feedback loop (Fig. 2.21).

The differential equation considering the feedback loop alone

$$\dot{x}_i = k_i x_i x_j - \frac{x_i}{c} \phi; i = 1, 2, \ldots, N; j = i - 1 + N\delta_{i1} \qquad (2.31)$$

has been studied extensively (Eigen, 1971; Eigen & Schuster, 1978a; Schuster, Sigmund & Wolff, 1978). The underlying dynamic principle has been named 'hypercycle'. A hypercycle is the simplest system ensuring a coexistence of several selfreplicative units. However, it is the logical extension of the series: chemical reaction → catalyst = a cycle of chemical reactions → autocatalyst = a cycle of catalysts → hypercycle = a cycle of autocatalysts (Eigen & Schuster, 1977).

---

nascent so-called Okazaki fragments are linked together by a ligase. A sophisticated repair mechanism, presumably involving several enzymes, corrects mistakes made during the primary polymerisation process. It seems to operate on the 3′ end of the fragments. Mismatched bases are excised on the daughter strand. The repair enzyme presumably recognises the daughter strand due to the lack of methyl groups on the adenyl residues of newly synthesised DNA (Glickman, 1979). The parental strands are methylated. Methylation of the daughter strands seems to be a kind of finish performed after repair. Mutants of *E. coli* which are deficient in DNA adenine methylation are, at the same time, ineffective in the post-replicative error-avoiding mechanism and show a substantial increase in spontaneous mutagenesis.

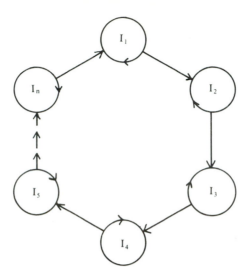

Fig. 2.21. Hypercyclic coupling. The closed loops indicate selfinduced replication; the straight arrows stand for catalytic activity. The formation of a closed
loop with positive feedback introduces new properties into the system.

Co-operativity in hypercycles with arbitrary numbers of elements has
been proved by analytical techniques (Schuster, Sigmund & Wolff, 1979).
The details of the dynamics depend on the number of elements in the
cycle. Hypercycles of dimension 2, 3 and 4 have stable stationary states.
In these states all members of the cycle are present at defined constant
concentrations. Oscillations, i.e. periodic changes in concentrations, are
obtained for systems with five and more members.

An important generalisation of this theorem of co-operation has been
made recently (Hofbauer, Schuster, Sigmund & Wolff, 1980*b*): the fact
that the selfreplicating elements co-operate does not depend on the
particular algebraic form of the coupling term in the rate equations. The
mass-action kinetics may be replaced by more complicated mechanisms
as long as they fall into the class of the following differential equation:

$$\dot{x}_i = x_i x_j F_i - \frac{x_i}{c}\phi; \qquad i = 1, 2, \ldots, N; \qquad j = i - 1 + N\delta_{i1} \qquad (2.32)$$

The only restriction is that the function F_i has to be continuous and has to
have a positive lower bound:

$$0 < a_i \leq F_i < \infty \qquad (2.33)$$

We shall make use of this very general property in the next section. It expresses in mathematical terms that any closed feedback loop with positive catalytic action will lead to co-operation.

RNA-phage infection of a bacterial cell can be understood as a simple hypercyclic process. The non-linearity in the growth rate is brought about by translation of the (+) strand. The protein obtained is a factor of the phage-specific RNA-replicase and its concentration determines the rate of phage RNA replication (Fig. 2.22).

Apart from the previous simple example, direct experimental verification of hypercyclic processes involving several polynucleotides is not easy to perform *in vitro*. Most in-vivo systems are not sufficiently well known for our purposes. A promising example is represented by influenza virus (Palese, 1977). The genome of this animal virus is segmented into eight pieces of single-stranded RNA which replicate in the host cell. The eight polynucleotides differ in length by factors up to four.

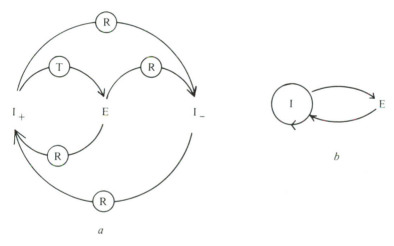

a

Fig. 2.22. *a*. RNA-phage infection of a bacterial cell as a simple hypercyclic process (Eigen & Schuster, 1977). Using the translation machinery (T) of the host cell, the infectious plus strand (I_+) first instructs the synthesis of a protein subunit (E) which associates itself with other host proteins to form a phage-specific RNA-replicase. This replicase complex (R) exclusively recognises the phenotypic features of the phage-RNA which are exhibited by both $(\pm)$ strands due to the complementary symmetry in special regions of the RNA chain (Fig. 2.19). The result is a burst of phage production which follows a hyperbolic growth law (Eigen & Schuster, 1978a) until one of the intermediates becomes saturated or the metabolic supply of the host cell is exhausted. *b* shows that it is sufficient if one of the intermediates possesses autocatalytic or selfinstructive function provided other partners feed back on it via a closed cyclic link.

It would be interesting to know how the replication of these independent genes is controlled.

Catalytic linkage via translation products presents a general possibility to build up hypercycles. It will be considered in the next section.

It seems worth-while mentioning briefly that the theorem of co-operation plays a certain role in the theory of genetically determined, social behaviour of animals. Maynard-Smith (1974) developed a mathematical model for animal contests by means of game theory. His model centres on the concept of evolutionary stable strategies. This approach can be translated into differential equations (Taylor & Jonker, 1978; Hofbauer, Schuster & Sigmund, 1979; Zeeman, 1979). Interestingly, one obtains precisely the differential eqn (2.30) and hence the theorem of co-operation can be understood as an extension of evolutionary stable strategies.

Hypercycles are characterised by another relevant property: due to the non-linear growth term (see eqn 2.31) hypercycles as individuals compete much more strongly than do ordinary selfreplicative elements with linear growth terms (eqn 2.7) (Eigen, 1971; Eigen & Schuster, 1978a; Hofbauer et al., 1980b). In a system containing several hypercycles competition soon leads to elimination of all but one. There is 'once and for all' selection: once the concentration of one hypercycle has exceeded a certain threshold value practically no second system has a chance to start, even if the latter is more efficient. Bad luck for the latecomer! Individual sequences, nevertheless, may be incorporated into an already selected system by hypercycle extension (see e.g. Fig. 2.28). The opposite process, elimination of members by constriction of the cycle, is possible as well.

DESIGNING CATALYSTS – THE ORIGIN OF TRANSLATION

The appearance of selfreplicating molecules on the primordial scene solved one of the problems of chemical evolution: the copying process brought order into the enormous variability of possible polynucleotide sequences. At the same time, however, selfreplication created a new problem. Due to the error, threshold molecules replicating with defined sequences could not exceed a certain length. The short-term conservation* of primary sequences is a more serious prerequisite of prebiotic evolution than it might look at a superficial glance. We know, from X-ray

---

* We distinguish short-term development over some ten, twenty or hundred generations from long term evolution.

diffraction studies (Kim *et al.*, 1974; Ladner *et al.*, 1975) and various indirect methods, that single-stranded RNA molecules have a rich terti- ary structure, the persistence of which depends on a stable primary sequence. The tertiary structure is responsible for the properties of the molecule which must not be changed too much in a continuous develop- ment. As we saw in the preceding section the maximum chain lengths for enzyme-free replication lie around 70–80 nucleotides. This size cor- responds to that of the present-day tRNA molecules. This limit is set by the accuracy of enzyme-free mononucleotide-template recognition. The recent experiments by Orgel and coworkers reported on p. 33 indicated that metal ions and crystal surfaces were the primordial catalysts which made such an accuracy feasible. In order to improve further the accuracy of replication, more specific catalysts are necessary. Preferentially, these catalysts should be adjustable, i.e. changeable and gradually improveable by the system itself. The most direct approach to reach such a goal would use the tRNA molecules themselves: by means of their tertiary structures they could help to improve the accuracy of replication. Mutations would provide a highly variable and flexible reservoir of different tertiary structures. Are there any hints that tRNA molecules can act as catalysts? Indeed, there are such hints and they point towards the primary aim of this catalytic action. We encounter two classes of ribonucleotides as catalysts in present-day biochemistry.

(1) The tRNA molecules act as transmitters of amino acids in ribosomal protein synthesis. Some tRNAs which are not recognised by a codon transmit amino acids in glyco-peptide synthesis.

(2) Ribosomal RNA occurs as a structural factor in both subunits of the ribosome: one RNA molecule (16S in bacteria, 18S in eukaryotic organisms) occurs in the small subunit and two (5S and 23S in bacteria) or three (5S, 5.8S and 28S in eukaryotes) RNA molecules are building blocks of the large subunit. The ribosomal ribonucleic acids are considered to be very old molecules. The smallest of them (5S) has a length of about 120 nucleotides in prokaryotes and was used for alignment studies (Hori & Osawa, 1979). Besides these two classes of molecules cellular DNA and RNA is used exclusively for coding proteins or, in the intercistronic regions, for incompletely understood regulatory purposes.

Thus, all catalytically active polynucleotides known so far are involved in the formation of peptide bonds, in particular in ribosomal protein synthesis. This appears to be a remnant of their ancient history. After the system had learned to design its catalysts by coding a message in a polynucleotide sequence and translating it into a polypeptide, there was no further need to make use of new potential catalytic properties of polynucleotides: RNA synthesis is more expensive than protein synthesis and polypeptides are more efficient catalysts. The problem of how to improve the accuracy of polynucleotide replication is shifted towards another open question: the origin of translation and the genetic code.

The origin of the translation machinery is a particularly hard problem which all comprehensive speculations on prebiotic evolution have to face and to tackle in one way or another. The experimental facts available are very few, and the 'intellectual fossils' allow various interpretations. No wonder many different and contradictory models have been proposed in this field! Some appear more, some less, plausible. We shall give a few examples chosen more or less arbitrarily from the ample collection to be found in the literature. Then we shall try to concentrate on those ideas which seem to be compatible or even to support each other.

Let us start by mentioning some relevant facts.

(1) There is little specific interaction between individual amino acids and mono- or oligonucleotides. One striking exception is the strong interaction between the carboxylate group and guanine (Lancelot & Hélène, 1977). The association complex formed by guanine and acetate (Fig. 2.23) is even stronger than the corresponding guanine-cytosine complex.

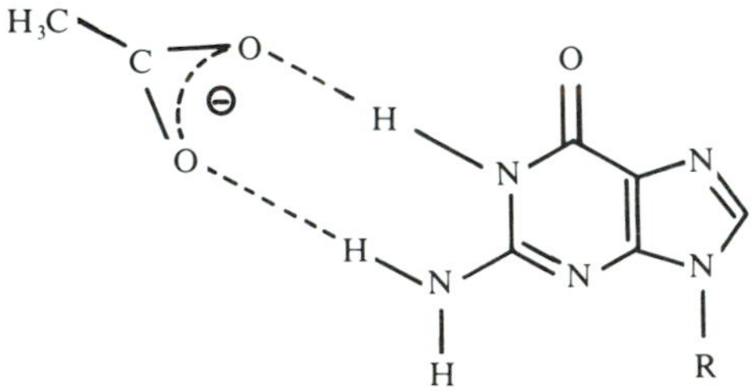

Fig. 2.23. The structure of the acetate-guanine complex. It is worth noticing that the stability of this complex is higher than that of the Watson–Crick base pair G≡C shown in Fig. 2.7 (Lancelot & Hélène, 1977).

(2) The fourth base from the 3′ end of the tRNAs, i.e. the base following the 3′pACC triplet and usually called discriminator, correlates to some extent to the nature of the amino acid which is transmitted by this tRNA (Crothers, Seno & Söll, 1972). To give an example, the discriminator is G for asp and glu in agreement with strong carboxylate–guanine interaction mentioned above.

(3) The genetic code (Fig. 2.24) appears to be universal or almost universal in the biosphere. The only deviations known so far occur in mitochondria (Hall, 1979): there the termination codon UGA codes for trp. Further it was established that, at least in a few cases, CUA codes for thr instead of leu and AUA for met instead of ile. On the whole these differences seem to be marginal variations rather than a stab with a dagger into the heart of the central dogma of a universal genetic code.

(4) There is some apparent structure in the code table (Fig. 2.24). The most obvious feature concerns the flexibility in the third base which has been explained elegantly by Crick's (1966) wobble hypothesis. Other less pronounced regularities may be deduced from our

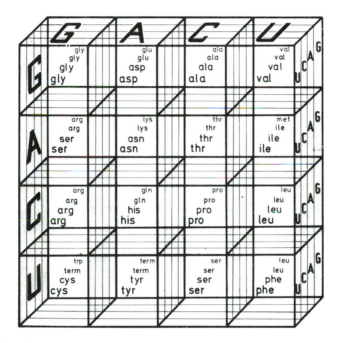

Fig. 2.24. The genetic code is the universal key for translation of genetic information from the legislative language of nucleic acids into the executive language of proteins. In this representation the three co-ordinates of geometric space are assigned to the positions of letters in the triplet codewords. 1, base from top to bottom; 2, base from left to right; 3, base from back to front. The four letters of the nucleic acid alphabet are arranged in such a way that the codewords for the most abundant amino acids appear in the top layer of the cube. (Reproduced from Eigen & Schuster, 1978*b*, by courtesy of Springer-Verlag, Heidelberg.)

diagram: U as second base of the codon determines a hydrophobic amino acid, GAN is translated into an amino acid with a carboxylate group in the side chain, etc.

Starting out from Cairns-Smith's (1966, 1975) ideas of replicating clays as alien ancestors of our present genetic system Hartman (1975a, b) constructed a model for the origin of translation and metabolism. A general feature of this concept is that the code evolved gradually and that the initial start was an interaction of clays, oligonucleotides and oligopeptides. At the level of polynucleotides the storage of information changed in Hartman's model from a single base-doublet code (GG = gly) to a two base-doublet code (GG = gly, GC = ala, CG = arg and CC = pro) and finally to the present-day triplet code. The last step definitely would have been the most crucial one. By changing the number of letters per codon, all previously stored messages would have become nonsense and such a change necessarily would have been lethal. One way to overcome this difficulty does not sound very plausible: ordered spacers are postulated to have existed in the doublet code. Another difficulty of this model lies in the early incorporation of two amino acids, namely arg and pro, which were rare in the primordial soup (Table 2.1).

One of the early suggestions on the origin of translation was to start from a triplet code and to make restrictions for simplicity on the level of the amino acids. A few amino acids only might have been coded but almost all the codons would have to be used in order to avoid too many nonsense triplets (Crick, 1968; Orgel, 1968). Following this idea, the ancient translation would have had more redundancy which would have been gradually reduced by making the code more sophisticated, in particular by incorporation of more amino acids.

The next questions obviously are: which were the first amino acids in the primitive code? Were there any plausible or even evident restrictions on the level of polynucleotides as well, in the sense that there existed regularities in sequences which excluded certain triplets? A first suggestion concerning the second question was made by Orgel (1972). He proposed alternating sequences of purines and pyrimidines to be a good starting point: ... RYRYRYRYRYR. To read such a message Orgel proposed two classes of tRNA-like molecules with RYR and YRY anticodons which could transmit hydrophilic and hydrophobic amino acids respectively.

A more recent, detailed model for protein synthesis without ribosomes has been suggested by Crick, Brenner, Klug & Pieczenik (1976). It centres on the idea that three base pairs are insufficient for strong enough binding of two nucleotides. In particular, the ancient messenger and the ancient tRNA would not stick together long enough to allow for the formation of a peptide bond. They made use of a two-conformation model of assisted codon-anticodon interaction (Fuller & Hodgson, 1967; Woese, 1970; see Fig. 2.25) which had also been proposed for prebiotic polypeptide synthesis by Woese (1967, 1973). The postulated conformational change is not likely to occur at contemporary ribosomes (Matzke, Barta & Küchler, 1980): a covalently linked complex of tRNA and messenger is translocated from the aminoacyl to the peptidyl binding site although the geometrical restrictions do not allow a hf-FH transition (Fig. 2.25). Nevertheless, the model might be relevant under prebiotic

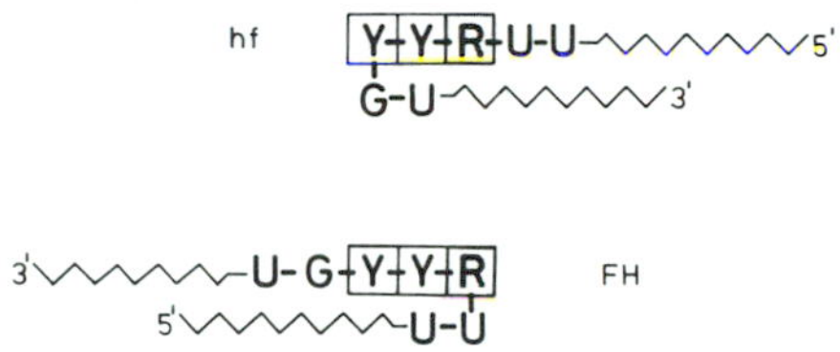

Fig. 2.25. Two possible configurations of the anticodon loop of tRNAs: FH according to Fuller & Hodgson (1967) and hf according to Woese (1970). The anticodon pattern (framed) refers to the model of Crick *et al.* (1976). (Reproduced from Eigen & Schuster, 1978*b* by courtesy of Springer-Verlag, Heidelberg.)

conditions. Polypeptide synthesis would have followed the steps shown schematically in Fig. 2.26. In order to reduce sliding of the tRNAs along the messenger, primitive translation would require a frame consisting of a repetitive pattern. This frame necessarily imposed a restriction on acceptable polynucleotides. Crick *et al.* (1976) suggest a sequence ... RRYRRY ... for the messenger which implies RYY to be the anticodon of the first tRNAs. Four amino acids are predicted to be the first in this model, namely (codons are given in parentheses): asn (AAU & AAC), ser (AGU & AGC), asp (GAU & GAC) and gly (GGU & GGC).

Finally, we would like to present a more recent model (Eigen & Schuster, 1978*b*) which is essentially based on the preceding one by Crick *et al.* (1976) but adopts certain modifications which will turn out to be of relevance as far as the plausibility arguments are concerned. The primary repetitive pattern (RRY) is replaced by a more general RNYRNY ... frame, which releases the constraint on the central nucleotide. Accordingly, messengers and tRNAs may fall into the same class of molecules – the anticodon is also of the type RNY. At the

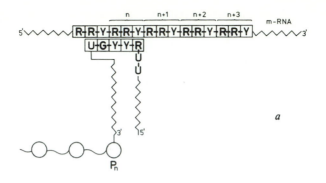

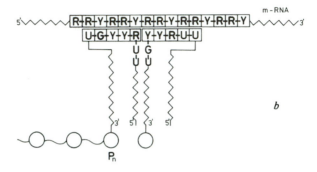

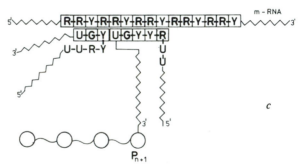

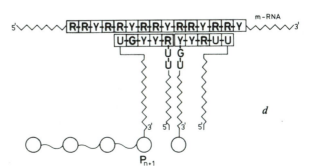

beginning the primordial tRNAs may well have been the primordial messenger or genes as well. Later on, genetic message and catalytic efficiency could easily be decoupled by a kind of 'gene duplication' (see also below). Thermodynamic and kinetic considerations based on the experimental data of Pörschke (1977) make high GC content of the sequences a '*conditio sine qua non*' for prebiotic peptide synthesis. AU-rich polynucleotides are not 'sticky' enough. A base-paired region of five nucleotides would not stick together sufficiently long to allow for enzyme-free formation of a peptide bond. Thus, it was the next logical step to propose a GNC frame which would be sticky enough. A certain degree of A and U (incorporated as 'N') would allow for sufficiently fast dissociation of double helices (Eigen & Winkler-Oswatitsch, 1980; see also Fig. 2.8 caption). Based on the present-day code the model leads to the following prediction concerning the first four amino acids:

gly (GGC), ala (GCC), asp (GAC) and val (GUC).

Let us now put together some independently derived facts which support the model proposed and make it more plausible.

(1) Alignment studies of almost all known tRNA sequences (Eigen & Winkler-Oswatitsch, 1980) revealed that a common ancestor or quasispecies distribution was rich in RNY triplets and had an extraordinarily high GC content (Fig. 2.12).

(2) The ancestor had a highly symmetric structure which allowed both strands ($+ \& -$) to replicate with equal efficiency.

(3) The adaptors of the four GNC codons, tRNAs which transmit gly, ala, asp and val, are the four most closely related tRNA molecules (in *E. coli* and in yeast at least). They may well be descendants of the same quasispecies.

Fig. 2.26. The primitive translation mechanism requires 'sticky' interactions between the messenger and the peptidyl-tRNA. It thereby allows the growing peptide chain to remain in contact with the message until translation is completed. According to Crick *at al.* (1976) the transport is effected by a flip mechanism involving conformational changes of the tRNA (FH$\rightleftharpoons$hf). The nascent peptide chain is always connected with the messengers via five base pairs with some additional stabilisation by the adjacent aminoacyl-tRNA. The partial overlap of base pairing guarantees a consistent reading of a message encoded in base triplets. The letters *a*, *b* and *c* denote consecutive steps in the process of elongation of the polypeptide chain by one amino-acid residue. (Reproduced from Eigen & Schuster, 1978*b* by courtesy of Springer-Verlag, Heidelberg.)

(4) The four amino acids gly, ala, asp and val are the most abundant representatives of this class of molecules in prebiotic simulation experiments and in meteorites (Table 2.1). All other amino acids appear at substantially smaller concentrations. It should be noted that none of the other models mentioned here predicted the same to be the first four.

(5) Jukes (1973) made an attempt to extrapolate from present-day codon–anticodon pairing to a preceding form of the genetic code. All four codons mentioned above are members of his ancestral code for ten amino acids.

(6) Rossmann, Moras & Olsen (1974) made an alignment study of various dehydrogenases and other nucleotide binding proteins in order to reconstruct a possible common ancestor protein. Walker (1977) analysed the 21-site amino-acid sequences of the nucleotide-binding surface and tried to identify the precursor amino acids coded into the ancestral surface which is assumed to be 3×10^9 years old (Rossmann *et al.*, 1974). Actually, Walker concludes that '. . . valine and one or both of the aspartic acid-glutamic acid group would be likely precursors with valine predominating'. Both constituents of the ancient recognition site are members of our ancestral 'gang of four'.

So far we have collected some arguments stressing the plausibility of the proposed model for the start of translation. How could the system undergo further development and incorporate more amino acids in the code? There is little evidence. The other tRNA sequences are not as closely related as the first four. We can, nevertheless, try to conceive a theoretical model. In order to avoid competition between the individual tRNAs we have to invoke some kind of hypercyclic coupling. Idealised systems involving replication, translation and specific coupling have been proposed and discussed previously (Eigen, 1971; Eigen & Schuster, 1978*b*; Epstein, 1979*b*; Eigen, Schuster, Sigmund & Wolff, 1980). There are various mechanisms for such a catalytic action: they may vary from protection against hydrolysis through specific complex formation to catalysis of replication by specific polymerisation factors. Because of our present lack of information on primordial polypeptides and their properties it is very hard to make a meaningful speculation on the detailed mechanism at this stage. An example of such a catalytic system is shown in Fig. 2.27. The generalised theorem of co-operation is valid for this and also for many other more complicated mechanisms. Starting from two genes or two quasispecies and their translation products the system may be enlarged stepwise and thus incorporate more and more amino acids

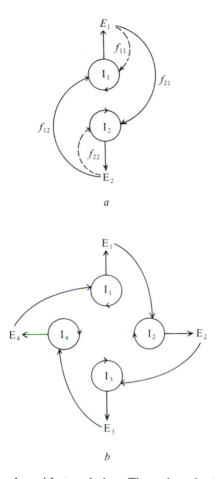

Fig. 2.27. Hypercycles with translation. The polynucleotides I_i exhibit two kinds of instruction, one for their own reproduction due to their capability to act as template and the other for the translation into proteins E_i which are effective potential catalysts. *a*. Co-operation between two selfreplicative elements I_1 and I_2 requires that altruistic catalysis is more efficient than catalytic selfenhancement, $f_{12} > f_{22}$ and $f_{21} > f_{11}$. *b*. The graph for a four-membered hypercycle with translation.

(Fig. 2.28). This development comes to an end when the incorporation of a new amino acid does not lead to sufficiently strong improvement of the catalytically active translation products. Later on or already during this process, the role of the messengers had to be separated from the role of the adaptors and the system as a whole became a highly complicated network of reactions and catalytic activities.

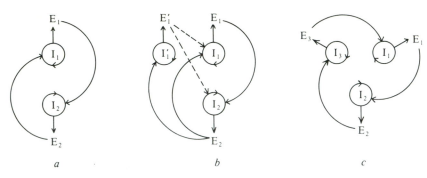

Fig. 2.28. A mutation-incorporation mechanism for stepwise extension of hypercycles. *a*. A two-membered hypercycle with translation. *b*. A mutant of I_1 denoted by I_1' appears. It has a certain catalytic effect on the replication of I_1 and I_2. *c*. Extension of the hypercycle by one member occurs if the mutant has the following two properties: (1) I_1' ($= I_3$) is recognised better by E_2 than is I_1 and (2) the translation product of I_1' ($= I_3$), the protein E_1' ($= E_3$), is a better catalyst for the replication of I_1 than is E_2. In other words, the new catalytic coupling terms have to fulfil a criterion as mentioned in Fig. 2.27: mutual enhancement along the cycle $1 \to 2 \to 3 \to 1$ has to prevail over selfenhancement.

Among others, there are at least three urgent classes of problems which are not yet understood and thus wait for forthcoming investigations.

(1) The interaction between primitive polypeptides and polynucleotides is an ample field for experimental studies. We have more or less precise thermodynamic and kinetic data as far as some of the present-day polynucleotides and proteins are concerned. Our knowledge of the catalytic properties of their ancient precursors or of polynucleotides and polypeptides in general is rather rudimentary. An attempt to synthesise the translation product of the primordial gene, the possible ancient ancestor of present-day tRNAs, has been mentioned recently (Eigen & Winkler-Oswatitsch, 1980).

(2) The role of aminoacyl-tRNA synthases is largely unknown. Some authors suggest a start of translation without enzymes, based on polynucleotide-amino-acid recognition alone, while others postulate the existence of primitive precursors of the synthetases. Systematic studies on the relationship between the present-day enzymes, which appear to be grouped into a few families (Wetzel, 1978) and their evolutionary history may shed more light on this question.

(3) In the previous section we discussed the problem of error propagation in polynucleotide replication. Protein synthesis through

polynucleotide translation, in principle, is subjected to a very similar error problem. The quantitative analysis, however, is much more involved since the whole feedback cycle of error propagation in the cell or in a prebiotic system involves many different steps; transcription from DNA to messenger RNA, translation of the messenger into proteins and the catalytic action of the enzymes. Orgel (1963, 1973) discussed this error problem with respect to the problem of ageing in clones of mammalian cells. In addition, two papers (Hoffmann, 1974; Goel & Islam, 1977) representative of others, aim at the derivation of a mathematical model. We should keep in mind that such an analysis is also necessary for the prebiotic start of primitive translation.

Hypercyclic coupling provides a straightforward explanation for the uniqueness or almost uniqueness of the genetic code: different hypercycles are incompatible and mutually exclusive (see the preceding section). An established system can only be extended (or, in principle, also reduced) by the 'mutation–incorporation' mechanism shown in Fig. 2.28. There are 'once and for all' decisions and established assignments are very unlikely to be changed. The behaviour of the system becomes different only when sets of molecules are enclosed in compartments and thereby lose their permanent contacts as we shall discuss in the next section.

COMPARTMENTS AND INDIVIDUALS

We have now considered prebiotic evolution in the vapour phase, in homogeneous solution, on catalytically active surfaces and in the solid state. To sum up so far: the primordial system makes use of all types of reactions in the three phases and on interfaces which were possible under realistic conditions. Probably, we obtain an organised mixture of polynucleotides and proteins. The mixture as a whole is multiplying and proliferating over the whole accessible space as long as the supply of building material is sufficient. Presumably, the system developed a genetic code by gradual improvement. If the start succeeded more than once, all systems that came into contact with it would have been extinguished by strong competition except the first one. The capability to encode proteins allows the system to design catalysts for its various needs. The final situation at this stage may be best characterised as a molecular replication and translation machinery without organised metabolism and without spatial separation.

Is the system described capable of a precellular evolution of Darwinian type? The underlying questions have been discussed briefly in previous papers (Eigen, 1971; Eigen & Schuster, 1978b; Maynard-Smith, 1979). We shall try to be somewhat more detailed here and shall consider two idealised classes of mutations which we call 'phenotypic' and 'genotypic' (Fig. 2.29). A phenotypic mutation in a polynucleotide sequence causes a

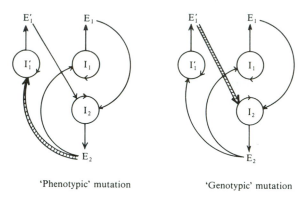

'Phenotypic' mutation   'Genotypic' mutation

Fig. 2.29.   Two idealised classes of mutations in primitive replication-translation systems. The 'phenotypic' mutations lead to mutants (I_1') which are better targets for the specific replicase, whereas the properties of their translation products (E_1') are about the same as in the wild type (E_1). The 'genotypic' mutant, on the contrary, is characterised by a better translation product but roughly unchanged recognition by the replicase.

major change in the tertiary structure of the molecule. Accordingly, the polynucleotide becomes a better or a worse target for replication. Selection immediately sets in and the phenotypic properties are gradually improved until an optimal replication rate is reached. The properties of the translation product are changed very little in the case of a typical phenotypic mutation. In the language of molecular genetics an extracistronic mutation would be phenotypic. Genotypic mutations behave in the opposite manner in this respect: they lead to a change in the translation product without substantial modification of the tertiary structure of the polynucleotide. Let us assume that the modified translation product is more efficient than its precursor. In homogeneous solution this advantageous protein does not specifically favour the production of the mutant gene. A general enhancement of polynucleotide and/or polypeptide synthesis depending on the catalytic activity of the translation product will take place. The advantageous genotypic mutant thus favours the whole volume element where it appeared. It will lead to a local increase of synthesis and also of concentrations of macromolecules provided diffusion is sufficiently slow. At the steady state we shall find a concentration gradient and all molecules from the more prosperous volume element will spread slowly over a larger area together, of course, with the favourable mutant gene. This reflection suggests a possible solution to the problem: in spatial isolation the favoured volume element would have the most pronounced advantage.

It seems worth-while to consider the opposite effect for a moment, since general arguments of this kind have been used by mistake against the possibility of development of selfreplicating systems in homogeneous solution. Let us assume a catastrophically deleterious mutant the translation product of which is a perfect ribonuclease, i.e. an enzyme which hydrolyses RNA. All polynucleotides present in the volume element where the mutant appeared will be degraded and together with them also the evil gene. The ribonuclease molecules can do harm only as long as they live in solution. Eventually they will be hydrolysed and the mischief has gone. Molecules from the surroundings will invade the polynucleotide-free volume element. We note that the limits of diffusion and the finite life-times of macromolecules together yield a finite spatial range for spreading a deleterious mutant and the 'catastrophe' may well be of a local nature only.

'Phenotypic' and 'genotypic' mutations, of course, are abstractions, which we used here for the purpose of illustration. In practice, every change in a sequence which codes for a protein will alter both the tertiary structure of the polynucleotide and the properties of the translation product. Usually, one effect will be dominating and then the classification is meaningful.

As we saw, slow diffusion limits spatial spreading and has a weak isolating effect. Nevertheless, it cannot effectively support selection of favourable genotypic mutations. The actual solution to the problem reveals a new principle of selforganisation: formation of compartments. In a compartment the fate of the mutant gene is strictly coupled to the efficiency of its translation product, because selection acts on the compartment as a whole.

Isolation of genes in compartments, however, creates a number of problems that do not exist in homogeneous solution.

(1) Material transport has to be organised in some way. The growing system requires an input of the activated building blocks of biopolymers. The products of hydrolysis have to be removed. One may

postulate either walls of the compartments which are permeable to low-molecular-weight materials or some transport mechanism opening and closing specific gates.

(2) Formation and division of compartments lead to packaging problems. A complete set of molecules is necessary to make a compartment viable. We are confronted therefore with a combinatorial problem as well.

(3) The synthesis of macromolecules and the rate of formation or division of compartments have to be synchronised in order to guarantee stationary concentrations of polymers inside the compartment.

(4) The compartment as a whole is subject to an error propagation problem.

It follows from points (1) to (4) that growth in compartments requires a rather elaborate molecular organisation. At present many details are unknown and it is very difficult to suggest a plausible model for this stage of prebiotic evolution. An appropriate kinetic formalism to describe selfreplication in compartments will be the subject of a forthcoming paper (M. Eigen, W. C. Gardiner & P. Schuster, in preparation).

When do compartments appear in prebiotic evolution? Based on the arguments mentioned above we are likely to suggest: as soon as necessary but as late as possible.

How did compartments originate and from which materials were they built? A number of experiments reporting the formation of spherical, compartment-like particles were published in the literature (Fox, Haranda & Kendrick, 1959; Fox & Dose, 1972; Dose & Rauchfuss, 1975). Commonly, these particles of different size are obtained by heating a mixture of amino acids for some time to temperatures above 100°C and dissolving the material obtained in hot water. The molecular structures of these 'protenoids' are not completely known. In the case of amino acids with suitable functional groups in the side chains, hydroxy-, amino- or carboxygroups, other linkages besides the ordinary peptide bonds between α-amino- and α-carboxygroups were found. The most remarkable feature of 'protenoids' may be that they have catalytic activity. They may well have played a role as catalysts in primordial chemistry as did clays or other inorganic materials. They may have hindered diffusion by their presence and thereby could have speeded up evolution. Protenoid particles, however, are not likely to solve our problem of compartmentalisation. They form under conditions under which most biopolymers are not stable. Moreover, it is hard to imagine how the system would get control of protenoid formation which is a necessary prerequisite for the development of a mechanism for cell division.

The following suggestion is more or less based on speculation and should not be taken too seriously. It is guided by the idea of a smooth transition from prebiotic reactions to present-day biochemistry. A recent review on the assembly of cell membranes (Lodish & Rothman, 1979) may be useful in this context. Much has been learned from studies on animal viruses. The virus particles are formed from the host-cell membrane, viral RNA, viral proteins, viral glycoproteins and a specific matrix protein which again is encoded by the viral RNA. Thus, by means of a few proteins, a membrane-coated particle is formed from the comparatively large pool of lipids in the host-cell membranes. Let us assume that lipids or somewhat related ampholytes were present in the primordial soup. Major constituents of lipids are straight-chain mono-carboxylic acids. A very recent analysis of the organic material in the Murchison meteorite (Lawless & Yuen, 1979) showed that they are at least present up to a chain length of C_8. The replication–translation machinery might succeed in producing a hydrophobic or ampholytic polypeptide which is incorporated into one of the lipid bilayers in the surroundings. Some of these proteins may induce vesicle formation. Then under favourable conditions the whole machinery could itself be enclosed by the membrane. From the very beginning the membrane is asymmetric since the protein is always synthesised on one side. Further protein production may lead to growth of the membrane, and eventually to division of the system which already has control over part of the process of membrane formation. These compartments would already have the characteristic features of individuals and may be considered as a kind of protocell.

After the replication machinery became accurate enough the independent genes, eventually, were combined to a single, large RNA molecule.

FOUR PREREQUISITES FOR DARWINIAN EVOLUTION

The Darwinian principle of biological evolution found its verification on the microscopic level through the spectacular results of molecular genetics. Nevertheless, Darwin's ideas are steadily attacked and questioned by some biologists who tend to belittle Darwin's concept and to reduce it to a mere tautology: survival of the survivor (see Gould (1976) and his arguments against this tendency). On the other hand, sequence alignment studies mainly performed on the level of proteins revealed that neutral mutations and genetic drift play a non-negligible role in the evolutionary changes of proteins (King & Jukes, 1969). They called this drift non-Darwinian evolution. Aiming only at the enumeration of some prerequisites of biological evolution we do not want to get involved in this scientific dispute. Hence we shall try to understand Darwinian evolution in more general terms as an evolutionary strategy combining random mutations with an optimisation process. Neutral mutations are included in this concept as errors occurring at loci which are not subjected to selection at the moment.

In this sense we are able to suggest four necessary prerequisites of Darwinian evolution.

(1) *Formation of polymers with a regular and repetitive building principle.* Polymers with regular backbone and variable side chains introduce an otherwise unknown variability into chemistry. Biochemistry would not exist without this principle. This enormous variability is the primary condition for the development of a biosphere. All the variety of biological species makes use of a negligible part of the possible sequences only. In aqueous solution biopolymers, in particular polynucleotides and proteins, are not stable against hydrolytic degradation. Thus, all reactions involving biopolymers at some stationary concentration are necessarily bound to subsidiary processes which provide energy-rich material for the synthesis of macromolecules. On the earth these secondary processes ultimately have to use sunlight as the source of energy. A primitive type of metabolism, therefore, already existed for the first biopolymers.

(2) *Template-induced replication.* The capability to act as a template for replication is bound to specific molecular properties which centre on the requirement of complementarity in intermolecular interaction. Replication brings order into the enormous variability of biopolymers through competition and selection. Replication, on the other hand, implies errors. The errors are the source of a limited variability as it is found in a mutant distribution. Erroneous replication thus provides the reservoir from which evolution chooses the appropriate candidates for further development. The accuracy of replication sets a limit to the chain length of polynucleotides. If this limit is exceeded the master sequence cannot compete successfully with its own mutants. The mechanism of replication thus determines the content of information that can be stored in polynucleotides.

(3) *Co-operation between selfreplicative elements.* Selfreplicating elements usually compete. Competition can be suppressed by a certain type of catalytic coupling which consists of a positive feedback loop. The corresponding kinetic system is called a hypercycle. The hypercycle is a principle of organisation which allows for integration of information stored in structurally unlinked selfreplicating elements. Hypercyclic coupling seems to play a role in symbiosis, coexistence of species in ecosystems, social behaviour of animals and in various other fields of biology. As far as prebiotic evolution is concerned, we see a need for hypercycles at this period which led to the development of the translation machinery and to the fixation of the genetic code. The information stored in a single polynucleotide of the same size as present-day tRNAs or in a quasispecies is not sufficient to encode the catalysts for the whole system. More information carriers had to co-operate. For that purpose they had to use a kind of hypercyclic organisation. Enzymes specifically catalysing polynucleotide replication were developed and represent one class of catalytic linkages between the replicating molecules. The first replication–translation machinery will readily optimise polynucleotide–replicase interactions. The system selects favourable phenotypic mutations and thereby optimises the targets of the RNA polymerases.

(4) *Formation of compartments.* In order to be able to select efficiently for better translation products of polynucleotides the system has to use an additional principle of organisation which consists essentially of spatial isolation. The use of preformed compartments is one step in this direction. In order to make compartmentation effective the system has to have active control on the formation or division process. Equipped with such a mechanism of self-controlled protocell division the system reached the level of an individual in the biological sense.

RESPONSES TO ENVIRONMENTAL CHANGES

Environmental changes, as we know very well, are extremely important driving forces of biological evolution. They introduce weighting factors into the ensemble of previously degenerate neutral mutants. They represent the challenge for evolutionary change. Did the changing environment drive prebiotic evolution as well? The answer as we see it today is yes. Changes were important from the very beginning of prebiotic chemistry. Periodic drying and wetting, to give just one example, appears to be an almost necessary condition for the primordial synthesis of nucleotides and derivatives. We can imagine many other situations for which a proper change in the environment is an important factor speeding up evolution.

On the other hand we have seen that, at least in principle, molecular self-organisation may take place also in a constant environment. An extreme case is represented by the evolution reactor when it is run under conditions under which the concentrations of polynucleotides are the only variables. The same device may be applied to study the response of the system to various changes from outside. Very little has been done so far in this direction and we have to wait for further results in theory and experiment.

The major problem, as it appears now, is to make reliable predictions on the details of a prebiotic environment and its possible changes. A model consisting of sequences of processes coupled to the environment has been proposed by Kuhn (1972). Although his model considerations on prebiotic evolution are very detailed, they seem to be rather vulnerable: the whole sequence becomes improbable if only one of the steps postulated is shown to lack plausibility.

CONCLUDING REMARKS

A possible sequence of steps in prebiotic evolution is shown schematically in Fig. 2.30. It summarises the lines we have followed throughout this review. From the work of many scientists we have rather clear ideas on some of the important steps in prebiotic chemistry, but there still remains a dusky core around the origin of the genetic code and the formation of the protocells. We have tried to make a few speculations, which we hope will be subjected to further experimental tests. In the future some of these concepts might undergo the important transition from being plausible to being probable. Actually, many plausible models can be proposed, but it is very hard to decide between them or even to suggest relevant experiments to prove or disprove them.

One of the difficulties of reconstruction may be visualised by an illustrative metaphor. Let us imagine people who see a Gothic cathedral or an old Roman bridge without having had any architectural or engineering education in their lives. They would have no idea how the pieces of stone could have been put together, because they have never seen the scaffolding that was there when the building was constructed. It may be that some molecular devices which were important during the critical phase of evolution leading to the first cells no longer exist in present-day biochemistry. One hint that nature works in this direction

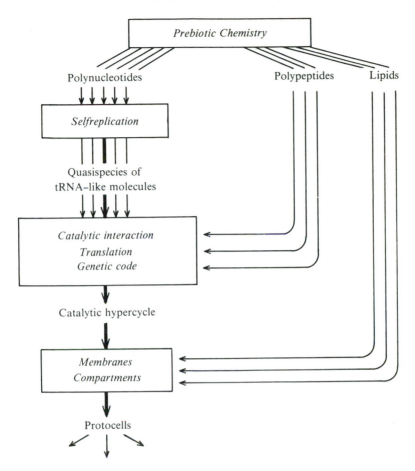

Fig. 2.30. A scheme for the logical sequence of steps in prebiotic evolution
leading from unorganised macromolecules to protocells.

has been discovered recently in case of the selfassembly of the coat of T4
virus (Wood & King, 1979). There are intermediate proteins which keep
the subunits together for some critical period and fall off afterwards.
These proteins do not appear in the intact coat, but they are necessary for
the process of selfassembly. When they are not there, no intact virus
capsule is formed.

There is one problem about prebiotic evolution which is often discussed, but which we have not mentioned here: the origin of chirality in
the biosphere. There are several possible explanations. We mention one
representative example: reactions catalysed by solid-state surfaces
inevitably introduce chirality into the system since they cannot proceed

efficiently with both enantiomers. The result is bistability of the dynamics (Decker, 1973a, b). The choice of one particular type of antipodes like L-α-amino acids may well have occurred by chance.

It is sometimes easier to search for principles than to develop concrete models; in particular when the facts are uncertain. We suggested four principles for prebiotic evolution in the last-but-one section (p. 76). The main driving force for biological evolution, nevertheless, is competition and this seems always to have been the case since the first replicating molecules appeared on the earth. The way in which competition is transformed into steady improvement of the properties on which selection acts is nicely illustrated by the Red Queen's Hypothesis of Van Valen (1973, 1974). He took the idea from *Alice in Wonderland*: '. . . as the Red Queen said to Alice: "now here, you see, it takes all the running you can do, to keep in the same place".' Competition allows no rest but requires continuous struggle and improvement. Many new principles have had to be used since the early days in the Pre-Cambrian, many new techniques have had to be invented in order to eliminate the rivals in the great contest of nature (Schwarz & Dayhoff, 1978; Maynard-Smith, 1978). How far those first polynucleotides were driven by the permanent challenge through their neighbours is magnificently documented in the biosphere that surrounds us.

Many stimulating discussions with Dr Ruthild Winkler-Oswatitsch, Professors Manfred Eigen, William C. Gardiner, Ernst Küchler, Leslie Orgel and Karl Sigmund are gratefully acknowledged. M. Eigen, E. Küchler, L. Orgel and A. W. Schwartz made unpublished material available and Dr R. Wolff provided several computer plots. Financial support was provided by the Austrian Fonds zur Förderung der Wissenschaftlichen Forschung, Project No. 3502. Figs. 2.24, 2.25 and 2.26 are reproduced from Eigen & Schuster (1978b) by courtesy of Springer-Verlag, Heidelberg. Last but not least I would like to thank Dr B. Schreiber for typing the manuscript and Mr J. Schuster for drawing the figures.

REFERENCES

Anders, E., Hayatsu, R. & Studier, M. H. (1974). Catalytic reactions in the solar nebula: Implications for interstellar molecules and organic compounds in meteorites. *Origins of life*, **5**, 57–67.
Bartlett, M. S. (1978). *An introduction to stochastic processes.* 3rd edn pp. 41–8. Cambridge University Press.
Batholomay, A. F. (1958). On the linear birth and death processes of biology as Markov chains. *Bulletin of mathematical biophysics*, **20**, 97–118.

Batschelet, E., Domingo, E. & Weissmann, C. (1976). The proportion of revertant and mutant phage in a growing population, as a function of mutation and growth rate. *Gene*, **1**, 27–32.

Bernardi, F. & Ninio, J. (1978). The accuracy of DNA replication. *Biochemie*, **60**, 1083–95.

Bessmann, M. J., Muzyczka, N., Goodman, M. F. & Schnaar, R. L. (1974). Studies on the biochemical basis of spontaneous mutation II. The incorporation of a base and its analogue into DNA by wild-type, mutator and antimutator DNA polymerases. *Journal of molecular biology*, **88**, 409–21.

Cairns-Smith, A. G. (1966). The origin of life and the nature and the primitive gene. *Journal of theoretical biology*, **10**, 53–88.

Cairns-Smith, A. G. (1975). A case for an alien ancestry. *Proceedings of the royal society of London, series B*, **189**, 249–74.

Courant, R. & Hilbert, D. (1953). *Methods of mathematical physics*, Vol. 1. New York: Interscience.

Crick, F. C. H. (1966). Codon-anticodon pairing: the wobble hypothesis. *Journal of molecular biology*, **19**, 548–55.

Crick, F. H. C. (1968). The origin of the genetic code. *Journal of molecular biology*, **38**, 367–79.

Crick, F. H. C., Brenner, S., Klug, A. & Pieczenik, G. (1976). A speculation on the origin of protein synthesis. *Origins of life*, **7**, 389–97.

Crothers, D. M., Seno, T. & Söll, D. G. (192). Is there a discriminator site in transfer RNA. *Proceedings of the national academy of sciences of the USA*, **69**, 3063–7.

Decker, P. (1973*a*). Evolution in open systems: bistability and the origin of molecular asymmetry. *Nature*, **241**, 72–4.

Decker, P. (1973*b*). Possible resolution of racemic mixtures by bistability in 'biods', open systems which can exist in several steady states. *Journal of molecular evolution*, **2**, 137–43.

Domingo, E., Flavell, R. A. & Weissmann, C. (1976). In vitro site-directed mutagenesis: generation and properties of an infectious extracistronic mutant of bacteriophage $Q\beta$. *Gene*, **1**, 3–25.

Dose, K. & Rauchfuss, H. (1975). *Chemische Evolution und der Ursprung lebender Systeme*. Stuttgart: Wisschenschaftliche Versags Ges.m.b.H.

Eigen, M. (1971). Selforganization of matter and the evolution of biological macromolecules. *Naturwissenschaften*, **58**, 465–526.

Eigen, M. & Schuster, P. (1977). The hypercycle. A principle of natural self-organization. A: Emergence of the hypercycle. *Naturwissenschaften*, **64**, 541–65.

Eigen, M. & Schuster, P. (1978*a*). The hypercycle. A principle of natural self-organization B: The abstract hypercycle. *Naturwissenschaften*, **65**, 7–41.

Eigen, M. & Schuster, P. (1978*b*). The hypercycle. A principle of natural self-organization C: The realistic hypercycle. *Naturwissenschaften*, **65**, 341–69.

Eigen, M., Schuster, P., Sigmund, K. & Wolff, R. (1980). Elementary step dynamics of catalytic hypercycles. *Biosystems*, **13**, 1–22.

Eigen, M. & Winkler-Oswatitsch, R. (1980). Transfer-RNA an early adaptor. *Science*, in the press.

Epps, D. E., Sherwood, E., Eichberg, J. & Oró, J. (1978). Cyanamide mediated syntheses under plausible primitive earth conditions V. The synthesis of phosphatidic acids. *Journal of molecular evolution*, **11**, 279–92.

Epstein,I. R. (1979a) Coexistence, competition and hypercyclic interaction in some systems of biological interest. *Biophysical Chemistry*, **9**, 245–50.

Epstein, I. R. (1979b). Competitive coexistence of self-reproducing macromolecules. *Journal of theoretical biology*, **8**, 271–98.

Epstein, I. R. & Eigen, M. (1979). Selection and elf-organisation o self-reproducing macromolecules under the constraint of constant flux. *Biophysical Chemistry*, **10**, 153–160.

Eyring, H. & Henderson, D. (ed.) (1978). *Theoretical chemistry: advances and perspectives: periodicities in chemistry and biology*, **4**. New York: Academic Press.

Ferris, J. P., Joshi, P. C., Edelson, E. H. & Lawless, J. G. (1978). HCN: a plausible source of purines, pyrimidines and amino acids on the primitive earth. *Journal of molecular evolution*, **11**, 293–311.

Ferris, J. P., Kuder, J. E. & Catalano, A. W. (1969). Photochemical reactions and the chemical evolution of purines and nicotinamide derivatives. *Science*, **166**, 765–6.

Fiddes, J. C. (1977). The nucleotide sequence of a viral DNA. *Scientific American*, **237**, No. 6, 54–67.

Fox, S. W. & Dose, K. (1972). *Molecular evolution and the origin of life*. San Francisco: Freeman.

Fox, S. W., Haranda, K. & Kendrick, J. (1959). Production of sperules from synthetic protenoid and hot water. *Science*, **129**, 1221–3.

Fuller, W. & Hodgson, A. (1976). Conformation of the anticodon loop in t-RNA. *Nature*, **216**, 817–21.

Fuller, W. D., Sanchez, R. A. & Orgel, L. E. (1972). Solid-state synthesis of purine nucleosides. *Journal of molecular evolution*, **1**, 249–57.

Gabel, N. W. & Ponnamperuma, C. (1967). Model for origin of monosaccharides. *Nature*, **216**, 453–5.

Gibbs, D., Lohrmann, R. & Orgel, L. E. (1980). Template-directed synthesis and selective adsorption of oligo-adenylates on hydroxyapatite. *Journal of molecular evolution*, **15**, 347–54.

Glickman, B. W. (1979). Spontaneous mutagenesis in *Escherichia coli* strains lacking 6-methyladenine residues in their DNA. *Mutation Research*, **61**, 153–62.

Goel, N. S. & Islam, S. (1977). Error catastrophe in and the evolution of the protein synthesizing machinery. *Journal of theoretical biology*, **68**, 167–82.

Gould, S. J. (1976). Darwin's untimely burial. *Natural History*, **35**, No. 8, 24–32.

Hall, B. D. (1979). Mitochondria spring surprises. *Nature*, **282**, 129–30.

Hartman, H. (1975a). Speculations on the evolution of the genetic code. *Origins of life*, **6**, 423–7.

Hartman, H. (1975b). Speculations on the origin and evolution of metabolism. *Journal of molecular evolution*, **4**, 359–70.

Hofbauer, J., Schuster, P. & Sigmund, K. (1979). A note on evolutionary stable strategies and game dynamics. *Journal of theoretical biology*, **81**, 609–12.

Hofbauer, J., Schuster, P., Sigmund, K. & Wolff, R. (1980*a*). Dynamical systems under constant organization II: Homogeneous growth functions of degree p=2, *SIAM J. Appl. Math. C*, **38**, 282–304.

Hofbauer, J., Schuster, P., Sigmund, K. & Wolff, R. (1980*b*). Competition and cooperation in catalytic selfreplication. *Journal of mathematical biology*, in the press.

Hoffmann, G. W. (1974). On the origin of the genetic code and the stability of the translation apparatus. *Journal of molecular biology*, **86**, 349–62.

Holland, H. D. (1962). Model for the evolution of the earth's atmosphere. In *Petrologic Studies – A volume to honor A. F. Buddington*, ed. A. E. J. Engel, H. L. James & B. F. Leonard, pp. 447–77. Boulder, Colorado: Geological Society of America.

Holland, H. D. (1978). *The chemistry of the atmosphere and oceans*. New York: Wiley-Interscience.

Holmquist, R., Jukes, T. H. & Pangburn, S. (1973). Evolution of transfer RNA. *Journal of molecular biology*, **78**, 92–116.

Hori, H. & Osawa, S. (1979). Evolutionary change in 5S RNA secondary structure and a phylogenetic tree of 54 5S RNA species. *Proceedings of the national academy of sciences of the USA*, **76**, 381–5.

Jacob, B. (1977). Evolution and tinkering. *Science*, **196**, 1161–6.

Jones, B. L., Enns, R. H. & Ragnekar, S. S. (1976). On the theory of selection of coupled macromolecular systems. *Bulletin of mathematical biology*, **38**, 15–28.

Jukes, T. H. (1973). Possibilities for the evolution of the genetic code from a preceding form. *Nature*, **246**, 22–6.

Katchalsky, A. (1973). Prebiotic synthesis of biopolymers on inorganic templates. *Naturwissenschaften*, **60**, 215–20.

Kim, S. H., Suddath, F. L., Quigley, G. J., McPherson, A., Sussman, J. L., Wang, A. H., Seeman, N. C. & Rich, A. (1974). Three-dimensional tertiary structure of yeast phenylalanine transfer RNA. *Science*, **185**, 435–40.

King, J. L. & Jukes, T. H. (1969). Non-Darwinian evolution. *Science*, **164**, 788–98.

Kornberg, A. (1974). *DNA synthesis*. San Francisco: Freeman & Co.

Kramer, F. R., Mills, D. R., Cole, P. E., Nishihara, T. & Spiegelman, S. (1974). Evolution in vitro: Sequence and phenotype of a mutant RNA resistant to ethidium bromide. *Journal of molecular biology*, **89**, 719–36.

Küppers, B. O. (1979). Towards an experimental analysis of molecular self-organization and precellular Darwinian evolution. *Naturwissenschaften*, **66**, 228–43.

Küppers, B. O. (1980). Some remarks on the dynamics of molecular self-organization. *Bulletin of mathematical biology*, **41**, 803–12.

Küppers, B. & Sumper, M. (1975). Minimal requirement fr template recognition by bacteriophage *Qβ* replicase: approach to a general RNA-dependent RNA synthesis. *Proceedings of the national academy of sciences of the USA*, **72**, 2640–3.

Kuhn, H. (1972). Selbstorganisation molekularer Systeme und die Evolution des genetischen Apparates. *Angewandte Chemie*, **84**, 838–61.

Kvenvolden, K. A., Lawless, J. G. & Ponnamperuma, C. (1971). Nonprotein Amino Acids in the Murchison Meteorite. *Proceedings of the national academy of sciences of the USA*, **68**, 486–90.

Kvenvolden, K. A. (1974). Natural evidence for chemical and early biological evolution. *Origins of Life*, **5**, 71–86.

Ladner, J. E., Jack, A., Robertus, J. D., Brown, R. S., Rhodes, D., Clark, B. F. C. & Klug, A. (1975). Structure of yeast phenylalanine transfer RNA at 2.5 Å resolution. *Proceedings of the national academy of sciences of the USA*, **72**, 4414–18.

Lancelot, G. & Hélène, C. (1977). Selective recognition of nucleic acids by proteins: The specificity of guanine interaction with carboxylate ions. *Proceedings of the national academy of sciences of the USA*, **74**, 4872.

Lawless, J. G. & Yuen, G. U. (1979). Quantification of monocarboxylic acids in the Murchinson carbonoceous meteorite. *Nature*, **282**, 396–8.

Lewinsohn, R., Paecht-Horowitz, M. & Katchalsky, A. (1967). Polycondensation of amino acid phosphoanhydrides III. Polycondensation of alanyladenylate. *Biochimica biophysica acta*, **140**, 24–36.

Lodish, H. F. & Rothman, J. E. (1979). The assembly of cell membranes. *Scientific American*, **240**, No. 1, 38–53.

Lohrmann, R. & Orgel, L. E. (1973). Prebiotic activation processes. *Nature*, **244**, 418–20.

Lohrmann, R. & Orgel, L. E. (1979). Studies of oligo-adenylate formation on a poly (U) template. *Journal of molecular evolution*, **12**, 237–57.

Matsubara, H. & Yamanaka, T. (1978). *Evolution of protein molecules*. Tokyo: Japan Scientific Societies Press.

Matzke, A., Barta, A. & Küchler, E. (1980). On the mechanism of translocation: relative arrangement of t-RNA and m-RNA on the ribosome. *Proceedings of the national academy of sciences of the USA*. In press.

Maynard-Smith, J. (1974). The theory of games and the evolution of animal conflicts. *Journal of theoretical biology*, **47**, 209–21.

Maynard-Smith, J. (1978). *The evolution of sex*. Cambridge University Press.

Maynard-Smith, J. (1979). Hypercycles and the origin of life. *Nature*, **280**, 445–6.

Miller, S. L. & Orgel, L. E. (1974). *The origins of life on the earth*. Engelwood Cliffs, New Jersey: Prentice Hall, Inc.

Miller, S. L., Urey, H. C. & Oró, J. (1976). Origin of organic compounds on the primitive earth and in meteorites. *Journal of molecular evolution*, **9**, 59–72.

Moorbath, S., O'Nions, R. K. & Pankhurst, R. J. (1973). Early Archean Age for the Isua Iron Formation, West Greenland. *Nature*, **245**, 138–9.

Morowitz, H. J. (1977). Perspectives on thermodynamics and the origin of life. *Advances in biological and medical physics*, **16**, 151–63.

Nicolis, G. & Lefever, R. (1975). *Membranes, dissipative structures and evolution*. *Advances in chemical physics*, **29**. New York: Interscience, John Wiley & Sons.

Ninio, J. & Orgel, L. E. (1978). Heteropolynucleotides as templates for nonenzymatic polymerizations. *Journal of molecular evolution*, **12**, 91–9.

Orgel, L. E. (1963). The maintenance of the accuracy of protein synthesis and its relevance to ageing. *Biochemistry*, **49**, 517–21.

Orgel, L. E. (1968). Evolution of the genetic apparatus. *Journal of molecular biology*, **38**, 381–93.

Orgel, L. E. (1972). *A possible step in the origin of the genetic code. Gerhard Schmidt Memorial Volume.* Jerusalem: The Israel Chemical Society.

Orgel, L. E. (1973). Ageing of clones of mammalian cells. *Nature,* **243,** 441–5.

Orgel, L. E. & Lohrmann, R. (1974). Prebiotic chemistry and nucleic acid replication. *Accounts of chemical research,* **7,** 368–77.

Oró, J. (1961). The mechanism of synthesis of adenine from hydrogen cyanide under possible primitive earth conditions. *Nature,* **191,** 1193–4.

Osterberg, R. & Orgel, L. E. (1972). Polyphosphate and trimeta-phosphate formation under potentially prebiotic conditions. *Journal of molecular evolution,* **1,** 241–8.

Paecht-Horowitz, M. (1978). The influence of various cations on the catalytic properties of clays. *Journal of molecular evolution,* **11,** 101–7.

Paecht-Horowitz, M., Berger, J. & Katchalsky, A. (1970). Prebiotic synthesis of polypeptides by heterogeneous poly-condensation of amino acid adenylates. *Nature,* **228,** 636–9.

Palese, P. (1977). The genes of influenca virus. *Cell,* **10,** 1–10.

Pörschke, D. (1977). Elementary steps of base recognition and helix-coil transitions in nucleic acids. In *Chemical relaxation in molecular biology,* ed. I. Pecht & R. Rigler, Molecular biology, biochemistry and biophysics, **24,** pp. 191–218. Heidelberg: Springer-Verlag.

Raulin, F. & Toupance, G. (1975). Formation of prebiochemical compounds in models of the primitive earth's atmosphere II: CH_4–H_2S atmospheres. *Origins of life,* **6,** 91–7.

Reid, C. & Orgel, L. E. (1967). Synthesis of sugars in potentially prebiotic conditions. *Nature,* **216,** 455.

Rossmann, M. G., Moras, D. & Olsen, K. W. (1974). Chemical and biological evolution of a nucleotide-binding protein. *Nature,* **250,** 194–9.

Rutten, M. G. (1971). *The origin of life by natural causes,* pp. 185–252. Amsterdam: Elsevier.

Saffhill, R. (1970). Selective phosphorylation of the cis-2′, 3′-diol of unprotected ribonucleosides with tri-metaphosphate in aqueous solution. *Journal of organic chemistry,* **35,** 2881–3.

Schopf, J. W. (1978). The evolution of the earliest cells. *Scientific American,* **239,** No. 3, 84–102.

Schneider, F. W., Neuser, D. & Heinrichs, M. (1979). Hysteretic behaviour in poly (A)-poly (U) synthesis in a stirred flow reactor. In *Molecular Mechanisms of Biological Recognition,* ed. M. Balaban, pp. 241–52. Amsterdam: Elsevier-North Holland Biochemical Press.

Schuster, P. (1972). Vom Makromolekül zur primitiven Zelle-die Entstehung biologischer Funktionen. *Chemie in unserer Zeit,* **6,** 1–16.

Schuster, P. & Sigmund, K. (1980). Self-organization of biological macromolecules and evolutionary stable strategies. In *Dynamics of synergetic systems,* ed. H. Haken, pp. 156–69. Berlin: Springer-Verlag.

Schuster, P., Sigmund, K. & Wolff, R. (1978). Dynamical systems under constant organization I. Topological analysis of a family of non-linear differential equations. *Bulletin of mathematical biology,* **40,** 743–69.

Schuster, P., Sigmund, K. & Wolff, R. (1979). Dynamical systems under constant organization III. Cooperative and competitive behaviour in hypercycles. *Journal of differential Equations*, **32**, 357–68.

Schwartz, A. W. (1969). Specific phosphorylation of the 2'- and 3' positions in ribonucleosides. *Chemical communications*, 1393.

Schwartz, A. W. (1974). An evolutionary model for prebiotic phosphorylation. In *The origin of life and evolutionary biochemistry*, ed. K. Dose, S. W. Fox, G. A. Deborin & T. E. Pavlovskaya, pp. 435–43. New York: Plenum Press.

Schwartz, A. W. (1979). Chemical evolution – the genesis of the first organic compounds. In *Marine organic chemistry*, ed. E. K. Duursma & R. Dawson, pp. 7–30. Amsterdam: Elsevier.

Schwartz, R. M. & Dayhoff, M. O. (1978). Origins of prokaryotes, eucaryotes, mitochondria and chloroplasts. *Science*, **199**, 395–403.

Sleeper, H. L., Lohrmann, R. & Orgel, L. E. (1979). Template-directed synthesis of oligoadenylates catalyzed by Pb^{2+} ions. *Journal of molecular evolution*, **13**, 203–14.

Sleeper, H. L. & Orgel, L. E. (1979). The catalysis of nucleotide polymerization by compounds of divalent lead. *Journal of molecular evolution*, **12**, 357–64.

Spiegelman, S. (1971). An approach to the experimental analysis of precellular evolution. *Quarterly review of biophysics*, **4**, 213–53.

Staudenbauer, W. (1978). Structure and replication of the colicin E1 plasmid. *Current topics in microbiology and immunology*, **83**, 93–156.

Sumper, M. & Luce, R. (1975). Evidence for a de novo production of self-replicating and environmentally adapted RNA structures by bacteriophage $Q\beta$ replicase. *Proceedings of the national academy of sciences of the USA*, **72**, 162–6.

Taylor, P. & Jonker, L. (1978). Evolutionary stable strategies and game dynamics. *Mathematical biosciences*, **40**, 145–56.

Thompson, C. J. & McBride, J. L. (1974). On Eigen's theory of self-organization of matter and the evolution of biological macromolecules. *Mathematical biosciences*, **21**, 127–42.

Toupance, G., Raulin, F. & Buvet, R. (1971). Primary transformation processes under the influence of energy for models of primordial atmospheres in thermodynamic equilibrium. In *Chemical evolution and the origin of life*, ed. R. Buvet & C. Ponnamperuma, pp. 83–95. Amsterdam: North-Holland Publishing Company.

Toupance, G., Raulin, F. & Buvet, R. (1975). Formation of prebiochemical compounds in models of the primitive earth's atmosphere I: CH_4–NH_3 and CH_4–N_2 atmospheres. *Origins of life*, **6**, 83–90.

Van Valen, L. (1973). A new evolutionary law. *Evolutionary theory*, **1**, 1–30.

Van Valen, L. (1974). Molecular evolution as predicted by natural selection. *Journal of molecular evolution*, **3**, 89–101.

Walker, G. W. R. (1977). Nucleotide-binding site data and the origin of the genetic code. *Biosystems*, **9**, 139–50.

Weber, A. L. & Orgel, L. E. (1979). The formation of peptides from glycine thioesters. *Journal of molecular evolution*, in the press.

Weiss, A. (1975). Plenary lecture presented at the 'Jahrestagung der Gesellschaft Deutscher Chemiker', Köln.

Wetzel, R. (1978). Aminoacyl-t-RNA synthetase families and their significance to the origin of the genetic code. *Origins of life*, **9**, 39–50.

Winnewisser, G., Churchwell, E. & Walmsley, C. M. (1979). Astro-physics of interstellar molecules. In *Modern aspects of microwave spectroscopy*, ed. G. W. Chantry, pp. 311–503. London: Academic Press.

Woese, C. R. (1967). *The genetic code*. New York: Harper and Row.

Woese, C. R. (1970). Molecular mechanics of translation: a reciprocating ratchet mechanism. *Nature*, **226**, 817–20.

Woese, C. R. (1973). Evolution of the genetic code. *Naturwissenschaften*, **60**, 447–59.

Wood, W. B. & King, J. (1979). Genetic control of complex bacteriophage assembly. In *Comprehensive Virology*, ed. H. Fraenkel-Conrat & R. R. Wagner, **13**, 581–633. New York: Plenum Press.

Zeeman, E. C. (1979). Population dynamics from game theory. In *Proceedings of an International conference on global theory of dynamic systems*. Evanston: North-Western University.

3

Data handling for phylogenetic trees

DEREK PEACOCK

DEPARTMENT OF BIOCHEMISTRY, ADRIAN BUILDING,
THE UNIVERSITY, LEICESTER, UK

The possibility that evolutionary history could be clearly seen from comparisons of protein sequences was recognised soon after such data started to accumulate. Statements by Crick (1958) and Zuckerkandl & Pauling (1965) led to an initially optimistic view of the usefulness of protein data in reconstructing phylogenetic trees, but there were several reasons for this optimism.

Firstly the amino-acid sequence of the protein is a direct translation of the genetic message. In comparison, morphological features have the added difficulty that their genetic basis is usually unknown. Thus two morphological characters may be the product of a widely differing number of genes some of which may be common to both characters.

Secondly, it is easier to establish homology. Two characters are homologous if they have arisen from a common ancestor by divergent evolution. To reconstruct a phylogeny it is essential that similarity in any two characters can be identified as having arisen by descent from a common ancestral character rather than by convergent evolution from two distinct ancestors. Proteins have very specific biochemically identifiable functions in the cell which make it relatively easy to determine the functional homology of a set of proteins. Also many of the amino-acid positions in a protein are invariant even between such widely separated species as man and yeast.

Finally, proteins are ideally suited to objective numerical methods, as all their characters (amino-acid positions) are equivalent in type, i.e. they are all multistate characters. There were therefore reasonable theoretical arguments for predicting that molecular sequence data might offer a whole new source of information for the construction of phylogenetic trees and that such trees might be freer of ambiguity and distortion than those based on other data.

Some of the first phylogenetic trees constructed from protein-sequence data showed considerable agreement with trees constructed from classical morphological characters and fossil evidence, in particular those constructed from cytochrome *c* (Fitch & Margoliash, 1967). However as more data have become available an increasing number of cases where the molecular phylogeny disagrees in part with accepted biological or palaeontological phylogenies have been found (Goodman, 1976; Beintema, Gaastra, Lenstra, Welling & Fitch, 1977; De Jong, Gleaves & Boulter, 1977; Romero-Herrera, Lehmann, Joysey & Friday, 1978; Lyddiat, Peacock & Boulter, 1978). It is now clear that molecular data are not free from one major distorting factor, the occurrence of parallel evolutionary events. In fact the 'best' estimates (i.e. the lowest), of the levels of parallelism common in protein data sets suggest that parallel events are at least as common as divergent events (see Table 3.1).

The common occurrence of such high levels of evolutionary 'noise' places heavy burdens on all the methods for reconstructing phylogenetic trees. There are however several different approaches upon which methods are based, and in the following sections I shall discuss just some of the methods that are particularly suited to the reconstruction of phylogenies from amino-acid or nucleotide sequence data. I shall

Table 3.1. *The proportion of parallel substitutions in molecular data sets. The number of parallel substitutions is estimated by the computer program LABEL using 16 species data sets. Parallelism is expressed as the number of parallel substitutions as a proportion of the total number of non-singular substitutions. The data used are published in* The Atlas of Protein Sequence and Structure *(Dayhoff, 1978)*

Protein	Species Group	Parallelism (%)
Cytochrome *c*	Flowering plants	53
β-Haemoglobin	Vertebrates	50
Ferredoxin	Plants	49
Plastocyanin	Flowering plants	49
Myoglobin	Mammals	47
α-Haemoglobin	Vertebrates	40
α-Crystallin	Vertebrates	37
Cytochrome *c*	Vertebrates	29
Insulin	Vertebrates	29
Plastocyanin	Legumes	14[a]
Cytochrome *c*	Mammals	12

[a] There are only 10 species in this data set.

concentrate on those methods which have been commonly used in practice and for which, therefore, computer programs are available. In the last two sections I shall discuss some of the practical difficulties inherent in this kind of analysis, and examine some of the evidence to show how well these methods can reconstruct phylogeny.

HOMOLOGY

Although, intuitively, few would doubt that the similarities between the amino-acid sequences of cytochrome *c* from man and yeast are caused by their sharing a common ancestral cytochrome *c* gene, such a belief needs substantiating. More generally before any set of protein sequences can be used to reconstruct phylogeny, their homology needs to be established. In some cases it is not sufficient to show that proteins are homologous, it is also necessary to show that they are orthologous rather than paralogous. Lysozyme, for example, has been shown to have two genetic loci one of which is expressed in the eggs of a duck, and the other in the eggs of a goose (Arnheim & Steller, 1970). Two proteins may appear similar because they descend divergently from a common ancestral gene (homologous) or because they descend convergently from separate ancestral genes (analogous). Thus two proteins not sharing a common ancestral gene may converge to a similar chemical structure simply as the result of restrictions imposed by sharing a similar biological function. Ideally any test for homology should prove that the similarities between two proteins are greater than could be expected both by chance and by sharing a common function (i.e. by analogy).

The methods for establishing homology can be broken up into several separate problems. (1) A method to search for sections of one protein that match a given section in the other. (2) A method for scoring the degree of similarity between any two amino acids. (3) A method for testing how significant any overall similarity between two proteins is in comparison to the similarity expected by chance. (4) A method to distinguish any established similarity as being due to homology rather than analogy. (5) Finally, before the set of homologous sequences can be used in any tree building method, the locations of any insertions or deletions have to be found.

The first systematic procedure for testing homology was that of Fitch (1966). In this procedure fragments of one protein sequence are compared with those of the other. The length of the fragments compared was somewhat arbitrary but Fitch gave examples based upon 30 residues.

Every possible fragment of 30 consecutive amino acids from one sequence was compared with every possible fragment of that length in the other. In every comparison there were 30 pairs of amino acids, and for each pair a mutation value was assigned which was the minimum number of nucleotide substitutions needed to convert one amino acid to the other. These mutation values were summed for the 30 amino acids in the fragments. In comparing all possible pairs of fragments for two proteins, a collection of 5000–30 000 or more mutation values is obtained. If the two proteins being tested for homology are unrelated, then a probit plot of cumulative distribution (a plot of the relative frequency of a given mutation value against the mutation values) will be a straight line. If the two proteins are homologous and not too many mutations have occurred since their common ancestral gene, then those comparisons in the collection which pair truly homologous fragments should increase the number of low mutation values above that expected by chance alone. This will cause the probit plot of cumulative distribution of mutation values to depart from linearity.

Fitch (1970a) went on to improve the sensitivity of this method by showing that for most proteins a restriction factor can be applied to the shorter of the two sequences. For example in comparing haemoglobin, with 153 amino acids, with myoglobin with 148, a restriction factor of 0.75 would mean that the first 30 N-terminal residues of haemoglobin would be compared with 30 residue fragments of myoglobin up to $148 \times (1 - 0.75) = 37$ residues from the N-terminus of myoglobin. This avoids making very unlikely comparisons such as the first 30 N-terminal residues of one protein with the last 30 C-terminal residues of the other. The method can estimate the overall probability that any result obtained is due to chance as, knowing the amino-acid composition of the proteins, one can find the individual probabilities that two randomly selected amino acids will have codons which differ by i or more nucleotides.

This method proved to be a sensitive one for detecting similarity between two proteins, but leaves unresolved the two problems of location of gaps and the distinction between homology and analogy. Fitch (1969) suggests that gap location can be achieved by plotting the mutation value for each pair of 30 amino-acid segments, against the mid-point of one segment. Valleys in the curve which fall below a critical level indicate homologous regions.

Fitch (1970b) provides the only test for distinguishing between similarity caused by analogy from similarity due to homology. This test requires that two groups of sequences are available. In the example used by Fitch (1970b) there were five fungal and 19 metazoan cytochromes c. These were tested to show that the fungal-type cytochrome c was indeed homologous rather than analogous to the metazoan cytochrome c, with the probability that this result was due to chance of 10^{-9}. The method relies on demonstrating that the minimum nucleotide differences between the reconstructed ancestral sequences of the two groups (using maximum parsimony methods), are much smaller than the average distance between the original sequences of the two groups. This method cannot however distinguish between 24 cytochrome c sequences that have diverged from a single common ancestor, and 24 independently arising cytochromes c that are converging on a single future descendant form. It therefore does not contradict a basic philosophical principle that given information on a system at one point in time, one cannot tell in which direction the system is moving with respect to time. So although the method cannot distinguish between 24 independent origins and one it can (assuming divergence from ancestors) distinguish between one and two origins for the two groups.

Although Fitch has pioneered much of the early work in this field, many other authors have suggested alternatives. Fitch (1973) has reviewed much of the early work in the field so I shall concentrate on describing those alternative methods which have proved at least as sensitive as the early procedure of Fitch.

The procedure of Needleman & Wunsch (1970) offers a way of testing for significant similarity between a pair of proteins while at the same time allowing gaps in all possible places in both sequences. This method therefore has the advantage of not only testing for similarity but also of locating the positions of gaps in the sequences at the same time. The amino acids of each sequence are numbered starting at the N-terminus. The sequences form the sides of a two-dimensional array with one column for each amino acid in one sequence, and one row for each amino acid in the other (see Fig. 3.1). Thus all amino-acid matchings are represented as cells in this array, and each cell can be given a score which is a measure of the similarity of the pair of amino acids. In the simple example in Fig. 3.1 a value of 0 has been given for non-identical amino acids, and a value of 1 for identical amino acids. The method allows any scoring to be used and easily accommodates values of 0, 1, 2, 3 for the minimum nucleotide substitutions of Fitch (1966). By transforming the values in this array it is possible to find a pathway through the array which shows the maximum matching of one sequence with the other. If both sequences are aligned, and need no gaps to obtain the maximum match,

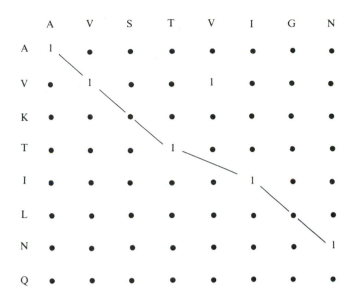

Fig. 3.1. An illustration of part of the procedure of Needleman & Wunsch (1970).

then the pathway will simply be the diagonal. If however gaps are needed to get a maximum match then the pathway will deviate from the diagonal.

To obtain realistic alignments it is however necessary to assign a penalty factor for each gap introduced. Thus a gap will only be introduced if the extra matchings exceed the penalty value. Unfortunately the value for this penalty factor is somewhat arbitrary. The significance of any matching found is assessed by Monte Carlo trials with randomly constructed sequences with the same overall composition as the proteins being tested. This unfortunately adds considerably to the computing time required, as the entire procedure has to be repeated for each new protein pair compared. Nevertheless this method is one of the most sensitive; for example the probability that the similarity between beta haemoglobin and myoglobin was due to chance comes to less than 10^{-16}.

McLachlan (1971) describes a method for comparing proteins which is similar to the methods of both Fitch (1966) and Needleman & Wunsch (1970). McLachlan uses a two-dimensional array to find the stretches of the two proteins that are similar (i.e. stretches of unbroken diagonal paths). One of the main differences in this method lies in the method of scoring. The degree of similarity between two amino acids is assessed on a scale which is based upon the observed frequencies with which one amino acid is substituted for by another. One such table of substitution frequencies is in Dayhoff (1972). The rationale behind the use of these frequencies is that the frequency of a given type of amino-acid substitution will depend not only on the genetic code which controls the rate of mutation but also on the chemical nature of the amino acids upon which the process of natural selection may operate. Thus the substitution of phenylalanine to tryptophan is very common, whereas arginine to glycine is very uncommon. Yet both have a minimum of one base change. A scoring system based on the frequencies of observed substitutions differs significantly from one based on the genetic code. The choice between them depends upon which more accurately reflects the overall rate of amino-acid substitution.

Miyata, Miyazawa & Yasunaga (1978) have demonstrated that the frequency of amino-acid substitutions relative to the frequency expected by chance decreases linearly with increasing physico-chemical difference between the amino acids. However this relationship does not exist for the amino-acid substitutions found in the abnormal human haemoglobins. These substitutions are mutational events most of which will not become fixed in the population as a whole. The relationship is also not followed by certain amino-acid residues in normal proteins which are known to be evolving at very fast rates. Thus it is likely that in most proteins the type of amino-acid substitutions occurring may be controlled largely by physico-chemical similarity in one part of the protein, and by the genetic code in other, more rapidly evolving parts. Tables of observed substitution frequencies, such as Dayhoff (1972), will of course reflect this mixture of controlling influences, and for this reason might be expected to provide a more accurate measure of amino-acid similarity. However it must be remembered that such a table is constructed from 'closely related homologous proteins, and their reconstructed ancestral sequences'. The accuracy of this table is therefore governed by the methods used to construct it. There are 400 entries in the table, and to obtain an accurate estimate of the relative substitution frequency requires many observed substitutions per entry. The original table of Dayhoff (1972) was based upon 814 substitutions, or on average about two substitutions per entry. Many entries are however empty. It is therefore a pity that this table has not been updated with the many more proteins that are now available.

Various procedures have been published to overcome the problem of using an arbitrary penalty for the introduction of gaps. Sankoff & Cedergren (1973) introduced one test to quantify the significance of a given number of deletion/insertion events in nucleotide sequences. The method is based upon Monte Carlo sampling of random sequences with the same base composition to compute the probability distributions for a test of significance.

Reichert, Cohen & Wong (1973) applied information theory to develop a novel method. Three types of information were defined and summed: (1) the location, (2) the process (e.g. insertion, deletion or replacement), (3) the nucleotide. The method however requires values for the relative frequencies of insertions, deletions, and substitutions, which are not usually available from independent measures. It is also difficult to see any biological significance in their location factor.

Elleman (1978) has extended the procedure of Needleman & Wunsch (1970) so that gaps can be successively introduced into both sequences and objectively assessed by Monte Carlo tests on random sequences. As the optimal location of each gap is found the two dimensional array is transformed in a suitable way to allow the next gap to be tried in all locations. Application of this method improves the sensitivity of the Needleman and Wunsch (1970) method particularly for those sequences which either require only a few gaps, or which differ markedly in length.

Methods for establishing similarity between proteins are not only an essential first step in constructing phylogenetic trees, but are also extremely useful in providing information on the evolution of proteins and protein families. A useful summary of the evolution of protein families is to be found in Dayhoff (1978). Many of the methods described can be used to examine a single protein for internal duplication. Barker, Kecham & Dayhoff (1978) have recently examined 163 proteins and found evidence for duplicated regions in 18 proteins, and evidence for repetitive duplications in three proteins. The duplication of genes, and parts of genes seems to play a large role in the evolution of structural proteins.

PHYLOGENIES FROM DISTANCE MEASURES

Relationships constructed on the basis of overall similarity (or dis-similarity), are phenetic (Sneath & Sokal, 1973). Such relationships do not necessarily reflect the ancestral pathways by which the characters (such as amino-acid positions in a sequence) arose. However many workers have used phenetic relationships as guides to the possible phylogeny of the organisms. This difference between phenetic and phy-logenetic relationships can be illustrated by a simple example.

D. Peacock

Characters	Species		
1	(1, 2, 3)	(4, 5)	
2	(1, 2)	(3, 4, 5)	
3	(1, 5)	(2, 3, 4)	
4	(1)	(2, 3) (4)	(5)

In this example species 1, 2 and 3 all share a common character state for character one, e.g., they may have alanine at position 4 in the sequence of cytochrome *c*, whereas species 4 and 5 share an alternative character state for the same character.

When all characters are examined it is clear that species 2 and 3 share a common character state in 3 of the 4 characters, whereas species 1 and 2 only share a common character state in 2 of the 4 characters. The full matrix of differences is shown in the bottom half of Fig. 3.2.

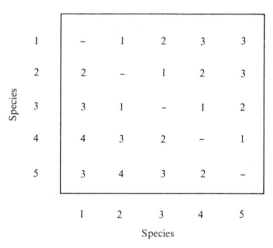

Fig. 3.2. Pair-wise differences between species. The top half of the matrix shows the number of non-singular differences between species pairs. The bottom half shows the total number of differences.

There is little doubt phenetically that species 2 is more closely related to species 3 than to any other species, and these species will be linked directly by most distance methods, giving the tree illustrated in Fig. 3.3*a*. However parsimony and compatibility methods will link species 1 and 2 to give the tree illustrated in Fig. 3.3*b*. The reason for the difference lies in the way the two types of methods deal with singularities, and parallelism (or homoplasy). Both parsimony and compatibility would identify a single parallel substitution between species 1 and 5 found in character 3. This parallel substitution distorts the distance matrix.

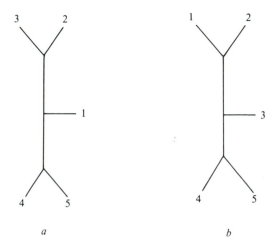

Fig. 3.3. The trees reconstructed using distance methods. (*a*) The tree reconstructed by phenetic distance methods from the distances given in the bottom half
of the matrix in Fig. 3.2. (*b*) The tree similarly constructed from the top half of the
matrix of Fig. 3.2.

The second difference is that character 4 contains no phylogenetic
information, it only contains information about the lengths of certain
branches in the tree. It does not influence the choice of the order of
branching in the tree. The species 1, 4 and 5 all have character states for
character 4 that are singular (Fitch, 1977). Only one group of species
share a common character state, and when this character state is assigned
to all ancestors in the tree, it makes no difference where species 1, 4 and 5
are placed on the tree, only three substitutions are required to account for
the evolution of these character states. If singularities are omitted when
counting differences between species, the upper half of the matrix shown
in Fig. 3.2 is obtained. When a straight phenetic method (such as
described by Fitch & Margoliash, 1967) is applied to this corrected
difference table, there is no longer any disagreement between the
different types of method.

The book by Sneath & Sokal (1973) discusses in detail many of the
standard numerical taxonomic methods which can be used with distance
measures. Two phenetic methods that have been used with proteins are
the Unweighted Pair Group Method (UPGMA) of Sokal & Michener
(1958), and a similar method by Fitch & Margoliash (1967). Both
methods successively pair species, or species clusters, on the basis of pairs
with the smallest average differences. The difference measure can be the

number of amino-acid differences, or the estimated minimum number of nucleotide differences. Both measures can be corrected for the estimated number of multiple substitutions at the same site (Dayhoff, 1972). The trees produced can be evaluated on how well the reconstructed distances fit the original difference matrix (Fitch & Margoliash, 1967) or alternatively by a cophenetic correlation coefficient (Sokal & Rohlf, 1962).

The methods can produce alternative trees if they find equally close pairs to join, however the UPGMA method as commonly implemented only selects one alternative set of pairs. Fitch & Margoliash (1967), on the other hand, suggest that alternative pairings be deliberately followed even if they are slightly less close than others. This strategy does in fact lead to trees which are 'better' (i.e. produce branch lengths which better reflect the original difference matrix) than ones built just from closest pairings. These methods do however assume that the rate of evolution is relatively constant, and will only produce correct cladograms when rates are not too unequal, the branches are not too close, and there is little convergence (Jardine, Rijsbergen & Jardine, 1969; Colless, 1970). How constant the rate of molecular evolution has been is the subject of much debate, which has been recently reviewed by Wilson, Carlson & White (1977). Although there is no doubt that different proteins evolve at markedly different rates, it is less clear how constant the rate of evolution is for one protein in different lines of descent. There is however sufficient evidence to suggest that although molecular evolution may behave like a clock over sufficiently long time periods, it is not accurate enough to use as a central assumption in any phylogenetic tree building method (see, for example, Fitch & Langley, 1976).

One distance method that is not so heavily reliant upon the assumption of equal rates of evolution is that of Farris (1970). This method was originally produced for quantitative data and constructs Wagner trees by successively adding a species to a growing tree in such a way as to minimise the overall length of the tree. A modification was suggested by Farris (1972) to allow the method to operate upon distance matrices. However, when used on sequence data the method can no longer guarantee to calculate lengths accurately in all cases. To reduce the overall error in estimating branch lengths a second error function is also evaluated before the next species is added. This leads to alternative procedures in that the choice of the next species to add is made by trying to minimise two functions at once, the length function and the error on the lengths. Which has priority is not made clear by Farris (1972).

Nevertheless this represents an important method for constructing phylogenetic trees from distance measures.

According to the *Concise Oxford Dictionary* the law of parsimony states that no more causes or forces should be assumed than are necessary to account for the facts. This axiom underlies all scientific theory and is fundamental to the construction of mathematical or logical models of observed phenomena. The facts required to be accounted for by such a model are those currently known, and so the law has the corollary that new discoveries may lead to the adoption of new theories.

Contrary to the allegations of those who oppose its use in building phylogenetic trees, parsimony does not assume that nature follows the most direct path in changing from one character state to another. It simply states that the number of parallel and reverse character state changes be kept to the minimum necessary to account for all the present day character states. Put another way by Sneath (1974) 'without a general belief that evolution has followed the shortest pathways there is no constraint on the wildest of postulated pathways'. It is as well to remember however that parsimony is a heuristic principle and not a statistic or a proven mathematical theorem.

The problem of finding the set of most parsimonious trees for a given data set can be broken down into two parts. The first problem is to count the minimum number of character state changes for a given tree topology (the 'Little' parsimony problem (Moore, 1976)). The second problem is to find all the topologies that have the minimum number of state changes. The first problem has been solved, the second can only at present be solved by an exhaustive examination of all possible topologies. It may in fact be that there is no other solution for this second problem.

Camin & Sokal (1965) suggested a procedure for computing parsimonious trees, which was not directly applicable to amino-acid data as it required cladistic characters (characters for which the order of evolution of the character states is known). Dayhoff & Eck (1966) produced the first algorithm for finding parsimonious trees specifically for amino-acid sequence data. The Dayhoff & Eck (1966) procedure, although correctly calculating the minimum number of substitutions in the majority of cases, overestimated them in a significant minority of cases (Peacock, unpublished results). A subsequent modification by Dayhoff (1972) improved the accuracy of counting at the cost of a considerable increase in computing time.

The method is however a good example of a search procedure for finding parsimonious trees. All possible four species trees are first examined. One of the most parsimonious four species trees is selected and a fifth species is tried in all possible positions on this tree. One of the most parsimonious five species trees is then selected, and a sixth species tried in all possible positions, and so on. When all species have been added the best tree is shuffled around by breaking every possible branch in the tree, and trying it in all other positions on the tree. Any better tree found is then used as the starting point of another round of shuffling. In this way a wide range of alternatives is tested, although it is difficult to program the method in a way which efficiently prevents repeated examination of the same tree topology formed by another route.

A solution to the little parsimony problem was found by Fitch (1971) and was later proved mathematically by Hartigan (1973). The procedure works by considering one character (position in the amino-acid or nucleotide sequence) at a time.

As an example, consider a part of a given tree in Fig. 3.4 which shows five species each having a single character state A, B, or C. Ancestors are assigned to each node of the tree starting at the tips and working towards the centre. An ancestor is formed by the following rules.

All character states in common along the two joining arms are assigned to the ancestor. If there are no character states in common from the two

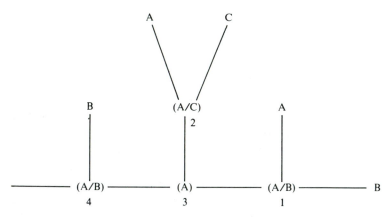

Fig. 3.4. An illustration of the parsimony counting procedure of Fitch (1971). The figure shows part of a tree for a single character, with ancestral characters in brackets.

joining arms, then all character states are assigned to the ancestor. More mathematically, each ancestor is assigned the intersection of the joining character state sets if this intersection is non-empty, otherwise the union of those sets is assigned. The number of times a union is needed by this procedure is the smallest number of substitutions needed in the given tree.

Thus in Fig. 3.4, starting at the rightmost tip of the tree, the joining arms have character states A and B. There is therefore nothing in common so that the ancestor is assigned A/B, and the count of substitutions is incremented by one. This does not mean that the ancestor is A/B but that at this stage no decision can be reached between these alternatives. Node 2 has to be assigned next, as at this stage it is the only node with two joining arms with assigned character states. Node 2 is assigned A/C as there is nothing in common, and the substitution count becomes two. Node 3 can now be assigned to A as this state is common to node 1 and node 2.

As this procedure copes with ambiguity in the assigned ancestors, the method will also cope with ambiguity in the tips or present-day sequences. Such ambiguity is sometimes found when proteins are sequenced from a sample of organisms taken from a population, and also when multiple forms of the protein are present in the same individual. Plant cytochrome *c*s are relatively free from ambiguity, whereas in plant plastocyanin ambiguity was commonly found, particularly in some plant families (Boulter, Gleaves, Haslett, Peacock & Jensen, 1978). Moore, Barnabas & Goodman (1973) and Fitch & Farris (1974) have developed similar though multi-pass procedures which will count the minimum number of nucleotide substitutions in a given tree starting with amino-acid sequences.

Solutions therefore exist for counting accurately the minimum number of amino-acid, nucleotide, or inferred nucleotide substitutions. The outlook for a solution to the remaining parsimony problem is, however, not good. Fitch (1977) explores many aspects to this problem, and presents several ways of reducing its size. We are however left with only one certain way of finding the set of most parsimonious trees for a given data set, and that is by exhaustively searching all possible alternative trees. This approach is only feasible for small data sets (less than 16 species), with even the fastest available computer.

The alternatives for larger data sets all involve a systematic shuffling of an initial starting tree, such as Dayhoff & Eck (1966), or Goodman, Moore & Barnabas (1973). There are several ways of arriving at a starting tree, including any biologically feasible tree, the parsimonious building of Dayhoff & Eck, or the use of another tree-building method. Whichever way is used there is no certainty of finding the set of most parsimonious trees. It is not uncommon to find that only one computer run out of 20–30 trials contains the minimum tree so far found (Peacock, unpublished results). Who is to say whether a further 200–300 runs would not do better, or 2000–3000 even better still?

PHYLOGENIES USING COMPATIBILITY

The concept of compatibility, or rather its antithesis incompatibility, stems from the realisation that different data sets can lead to different proposed phylogenies. Indeed incompatibilities can also exist between different sets of characters within the same data set. The phenomenon is not independent of the method used to reconstruct the phylogeny, and Farris (1971) recognises two types which he terms mosaic incongruence and incompatibility. Mosaic incongruence is caused by varying rates of evolution, which produce different distances between organisms which are separated by the same time period. Thus mosaic incongruence affects phenetic methods only, and will not be considered further.

Le Quesne (1969) suggested a method for detecting the incompatibility between two state characters. This method forms the basis of subsequent methods, which extended the idea to multistate characters such as amino-acid or nucleotide sequences (Sneath, Sackin & Ambler, 1975; Estabrook & Landrum, 1975; Fitch, 1975). The test for incompatibility is made between characters in pairs. A co-occurrence table is drawn up which has as rows the character states for one character, and as columns the character states for the other character. Each entry in the table is a list of those species which have the particular character states for that row and column. For example in Fig. 3.5*a* species 1 and 2 both have an alanine (A) for character 1, and a tryptophan (W) for character 2. The two characters are incompatible if the co-occurrence table contains a circuit (or cycle; Fitch, 1975). As Fig. 3.5*a* contains a circuit, any tree describing the evolution of those species must contain at least one parallel substitution. A formal proof of this is provided by Estabrook & McMorris (1977). The two characters in Fig. 3.5*b*, on the other hand, are compatible.

Estabrook, Johnson & McMorris (1976) provided a formal proof for cladistic characters (characters for which the order of evolution of the character states is known), and have shown that if all possible pairs of

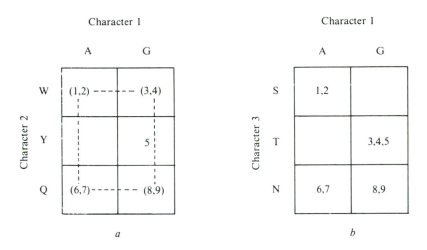

Fig. 3.5. An example of a co-occurrence table. *a.* Two characters which are incompatible as species (1, 2), (3, 4), (8, 9) and (6, 7) form the four corners of a circuit. *b.* Two compatible characters.

characters are compatible then the data set as a whole is compatible. A similar proof can be given for two state characters (McMorris, 1977). Unfortunately for multistate characters there can be incompatibilities in the data set as a whole which cannot be found by pairwise testing (Fitch, 1975). Nevertheless for most protein data sets examined less than 3% of all incompatibilities cannot be detected by pairwise testing (Guise, Peacock & Gleaves, unpublished). Fitch (1975) suggests ways of testing characters in threes for these 'hidden' incompatibilities, and implies that all such incompatibilities could be found for nucleotide sequences by testing in groups of three. Although this may be true for nucleotide sequences with only four character states, it is definitely not true for amino-acid sequences which have more than four amino acids at any site (Guise, unpublished).

The concept of compatibility has been used as the basis for two tree-building methods. Estabrook, Strauch & Fiala (1977) have demonstrated the use of compatibility in reconstructing phylogeny using the previously published cladistic data from the Orthopteroid insects. Using the pairwise test for compatibility they found all the mutually compatible cliques of characters. Thus out of the 91 characters, there were two cliques of 23 characters each of which was mutually compatible. These two cliques differed in only two characters. They took the 21 characters common to both cliques, and because all 21 characters are mutually compatible a tree diagram can be easily derived.

Those areas of the tree where the species are not separated can be further refined by repeating the process on just the unseparated species groups one by one. The method can be applied to amino-acid or unambiguous nucleotide sequences. With some data sets there is little doubt of the significance of the largest clique of compatible characters found. However, when applied to a complex data set such as higher plant plastocyanins it produces many small equal-sized cliques. The method has recently been developed further (Estabrook, 1979) to include a statistical test for the probability that a given clique is due to chance. With data sets that do not contain too much parallelism this method is elegant, quick and simple to use.

Compatibility has been used as the basis for a second method which can be applied to more complex data sets (Guise, Peacock & Gleaves, unpublished). Compatibility is used to identify the locations of parallel substitutions in the most parsimonious tree (Fitch, 1975). For each character all the co-occurrence tables are examined and a list is made of all the species sets that form part of a circuit. For example in Fig. 3.5a the species sets (1, 2), (3, 4), (6, 7), and (8, 9) would be counted, but not (5). The count would be summed for all pairwise comparisons of character 1 with other characters, it may be that the species set (1, 2) is found more often as part of a circuit than any other set. The procedure is repeated for all other characters. It is assumed that this is the most likely location of a parallel substitution, and it is labelled as such by changing, say, alanine A to A*. This will reduce the number of circuits. The process is repeated until all the circuits are accounted for. When there is a choice of labelling which would reduce the number of circuits by the same amount the alternatives are followed recursively in the computer program LABEL.

This method can be applied to large data sets (see, for example, Boulter, Peacock, Guise, Gleaves & Estabrook, 1979). The accuracy of the method has been evaluated using simulated and real data, and it is found that the number of labels added to a data set corresponds to the number of parallel substitutions in the data set to within 2% even when the proportion of parallel substitutions in the data set reaches 70% of all non-singular substitutions. This method was used to make the comparisons in Table 3.1 which show the relative amounts of parallelism in different real protein and nucleotide sequences.

SOME PRACTICAL CONSIDERATIONS

Ideal data would allow most methods to reconstruct a unique phylogeny. With data sets which contain high levels of parallelism it is very unlikely that there is only one 'best' tree. Many alternatives may also be found with a data set not because of parallelism, but because the method is restricted to constructing bifurcating trees when the data best fit a trifurcating or higher order of branching.

Thus with most data sets it is expected that alternative trees will be found which are equally good. All the methods discussed, except Estabrook's Clique Analysis, can in the process of building a tree be faced with equally good choices, and therefore these methods must follow all the alternative pathways they find. Felsenstein (1978) has highlighted how many alternative trees there are for a given number of species. Taking unrooted bifurcating trees as an example, for eight species there are approximately 10^4 trees, for 16 species there are 10^{14} and for 22 species 10^{22} trees. Thus the overall efficiency of any tree-building method is a function of the efficiency with which it selects the 'best' trees from the multitude of alternatives, and the efficiency with which it builds and assesses each tree it does consider. There are now well over 1000 protein and nucleotide sequences collected together in the latest *Atlas of Protein Sequence and Structure* (Dayhoff, 1978). With the increasing use of DNA sequencing the number of available sequences is likely to rise dramatically again. Already several large data sets exist (90 cytochrome cs, 91 partial plastocyanins). The number of possible trees for 90 species is too large a number for conventional representation in most computers. With the increasing size of data sets, the practical restraints in constructing trees become more and more acute.

Most workers will have access to a mainframe timesharing system which usually places a time limit on jobs of something like 20 min. The CDC Cyber 73 at Leicester University is no exception, and Table 3.2 shows some sample timings for various tree-building methods. It must be emphasised that these timings are only a guide to an order of magnitude, as timings vary enormously with the complexity of the data set, and the efficiency with which each algorithm has been translated into a computer program. It does however give some indication of the relative practical limits of the different methods when applied to a data set with high levels of parallelism. It is however not possible to say how many of the total number of equally good trees are found by each procedure, apart from the total search.

Table 3.2. *A comparison of timings for various tree building methods. These timings are from a CDC Cyber 73 computer operating upon plasto-cyanin data sets. Figures in brackets are estimated finishing times*

Method	Execution time (s)			No. of species No. of alternative trees considered		
	16	32	64	16	32	64
UPGMA	10^1	10^2	10^2	10^0	10^0	10^0
Fitch & Margoliash	10^9	–	–	10^9	–	–
Distance Wagner	(10^4)	(10^5)	–	10^4	10^5	–
Dayhoff parsimony search	10^1	10^2	10^3	10^2	10^3	10^4
Total parsimony search[a]	10^3	–	–	10^7	–	–
Estabrook Clique Analysis	10^1	10^2	10^2	10^0	10^0	10^0

[a] There were only 11 species in this data set.

A total search of all trees does guarantee that all of the most parsi-monious (or best, on whatever criterion) trees for a given data set are found. It is however the method with the lowest practical limit of around 13 species for the less complicated data sets. To extend its range from 13 to, say, 16 species requires that the efficiency of computation is raised by 15 000 fold. The efficiency of this method has recently been improved (Gleaves, Peacock & Guise, unpublished) based upon the ideas of Fitch (1977). The new implementation can consider over 10 000 trees per second, and is on that basis one of the most efficient algorithms available. The procedure of Estabrook differs from all others in that it is only applicable to those data sets which yield a suitably large clique of compatible characters, or a small number of equally good cliques. All other methods will, with complex data, generate alternative choices, and even if all alternatives are considered there will be no guarantee that all the 'best' trees will be found. The UPGMA method can cope with several hundred species in a practicable time. However, it does so by choosing only one arbitrary set of pairings when there are equal alternatives. If other methods ignore the question of equally good alternatives, they too would then be practicable for the largest of data sets. The UPGMA method can in principle be modified to consider all equal alternatives, but it is certain that this will greatly reduce the number of species the method can handle, without any surety that all the trees with the best cophenetic correlation will be found. The method of Fitch & Margoliash (1967), which is similar to UPGMA but which does consider likely alternatives clearly shows how many alternative, equally good trees exist when using phenetic methods.

Thus all methods when applied to complex data sets, and when they consider all equally good alternatives, are severely limited in the number of species (and characters) they can consider within practicable time constraints. Any attempt to increase the limits on the number of species requires an increase in computing power of many orders of magnitude.

There are a number of ways of extending the practical limits. The timings given in Table 3.2 were for the CDC Cyber 73, which is a slow machine. For those with access to a machine like the CDC 7600, a further tenfold increase in computing power is available. Another approach is to use a dedicated computer which was, up to a few years ago, an expensive alternative. However the wide range of microcomputers available at around £1500 now make this a very attractive alternative. A machine like the Research Machines 380Z offers speeds comparable to a CDC Cyber 73. However by using computing times of the order of one week per data set, a microcomputer offers an increase of 500 fold in computing time. The rate of development in microtechnology is extremely rapid, and it would seem reasonable to expect an increase of around tenfold in speed within the next ten years.

With these possible increases in computing power some may believe that there is no practical problem that time and money cannot overcome. However even if we accept that a 5000 fold increase in computing power will shortly become available through the microtechnology this is still unlikely to be sufficient to allow all equally good alternative trees to be found for data sets above 100 species. It does not take much extrapolation to foresee that data sets of several hundred species may become available shortly. It would seem a pity if phylogenetic insights cannot be obtained from such a wealth of molecular information due to limitation in both computing power and algorithm efficiency. Much more effort is needed to increase the efficiency of the implementation of the existing algorithms as well as to develop new and more powerful methods.

THE ACCURACY OF RECONSTRUCTED PHYLOGENIES

There are several factors that influence the accuracy of the reconstructed phylogenetic tree. Firstly, there must be a sufficient number of characters to partition the species into the number of phylogenetic groups required. Most tree-building methods build bifurcating trees (this does not preclude higher order branching as some branch lengths may be zero).

Take for example higher plant cytochrome *c*, where there are 25 sequences available (Boulter, 1974), each with 113 amino-acid residues.

However of the 113 residues, there are only 25 residues which contain phylogenetic information. To define a bifurcating tree of 25 species requires a minimum of 22 character state changes, each of which defines a different partition of the species. In the remaining 25 potentially useful characters there are 40 non-singular character state changes, at least 50% of which are parallel. It can be seen that this data set does not contain enough information to specify a fully bifurcating tree in spite of being a protein 113 residues long. It can however specify a tree that is largely bifurcating.

This is not atypical for many other protein data sets. Thus one important requirement for being able to construct an accurate phylogeny is that sufficient phylogenetically useful characters are available. If one protein has insufficient characters then it should not be used to construct trees on its own but should be pooled with the useful residues from other proteins.

The second factor influencing overall accuracy is the rate of molecular evolution. If the rate is too slow then the protein or nucleotide sequence will have too few useful characters. On the other hand if the rate is too fast then a number of substitutions will be superimposed and will not be seen in the present-day sequences. Corrections for these superimposed substitutions can be made for distance methods by procedures such as those of Dayhoff (1972) and, for parsimony methods by procedures described by Moore (1977), Holmquist (1978a) and Moore, Goodman, Callahan, Holmquist & Moise (1976). Although such corrections may be reasonably accurate, they must lead to errors in some cases. Holmquist (1978b) suggests a method for evaluating a reconstructed tree to identify the dense and less dense areas of the tree. In a fully dense tree there is no uncertainty concerning the ancestral sequences. The measure of denseness can give values for each node in the tree. Once the less dense nodes have been identified they can be used as a guide for choosing which species to add next to the tree in order to reduce the uncertainty which surrounds sparse areas of the tree. This method quite clearly shows that published phylogenetic trees vary considerably in their average denseness.

The evolutionary rate of proteins varies enormously (Dayhoff, 1972; Wilson et al., 1977). Proteins with high evolutionary rates are best used to delineate the phylogenetic relationships between species and genera, and slowly evolving proteins can be best used with higher orders. Plastocyanin is a protein with a relatively high evolutionary rate (Boulter, Haslett, Peacock, Ramshaw & Scawen, 1977) which can be used to delineate some of the relationships between higher plant families (Boulter *et al.*, 1978). However the protein gives more acceptable results when restricted within a family (Boulter *et al.*, 1979).

The last main factor which must affect the overall accuracy of reconstructed trees is the proportion of parallelism in the data set. From Table 3.1 it can be seen that none of the molecular data sets is free of parallelism, and that in most 50% of the phylogenetically useful substitutions are parallel. This is a very high figure, and as yet there are no studies to show how well any of the available methods can cope in the face of such evolutionary 'noise'. I believe this very high level of parallelism has caused the increasing disappointment with the phylogenetic trees constructed from molecular data. However most of the morphological and other data sets contain as much parallelism. Thus molecular sets are probably not any worse than the more classical characters; however, they are also probably no better in this respect.

There are, however, significant differences between proteins. Cytochrome *c*, when restricted to mammals and birds, has only 12% parallelism. Similarly, plastocyanin, when restricted to the family Leguminosae, has only 14% parallelism in an 11 species set. As only a small proportion of proteins and nucleotide sequences have been examined in large enough numbers to estimate the proportion of parallelism there is hope that other proteins will be found with small amounts of parallelism.

There are very few studies which have attempted to evaluate the relative merits of the different tree-building methods. Most workers have been heavily influenced in their choice of method by how well constructed trees fit with established phylogenetic schemes. While this kind of evaluation has some validity, it assumes that the classical phylogenetic schemes were constructed from data free of parallelism by methods that are completely accurate and objective. In most cases however these conditions are not met.

An alternative method of evaluation has been used by Peacock & Boulter (1975). They used the technique of computer simulation to produce simulated amino-acid sequences whose evolutionary history is known. Using these sequences they have compared the parsimony method of Dayhoff & Eck (1966) with a distance method of Moore, Goodman & Barnabas (1973). This study showed that the parsimony method was most accurate at small interspecies distances (and correspondingly low parallelism) and became increasingly less accurate as the interspecies distances were increased. The distance method on the other hand was less accurate at small interspecies distances, and became more accurate as the interspecies distances were increased. This study did, however, only compare these two methods, and did not study the effect of changing separately the proportion of parallelism and the rate of evolution.

Schwartz & Dayhoff (1978) have done a similar study using computer simulation to compare their own distance method (Dayhoff, 1976) which is a development of the work of Fitch & Margoliash (1967) with their parsimony method (Dayhoff & Eck, 1966). The mean distances between species were very much larger than those used by Peacock & Boulter (1975); however, their results agree in that both studies conclude that at these large interspecies distances (less than 20 amino-acid substitutions per 100 residues) a distance-based method is more accurate.

Prager & Wilson (1978) have compared three distance methods, the UPGMA method, Fitch & Margoliash (1967) and Farris (1972). Their method of comparison was however very different from the previous studies. Firstly they used several real data sets, including DNA hybridisation, immunological and amino-acid sequence studies. They constructed trees from these data sets using the three methods, and evaluated the trees by comparing the goodness of fit between the original distance matrix and the reconstructed distance matrix. They concluded that in most cases the method of Fitch & Margoliash gave the better fit. However this study can be criticised on two counts. Firstly the method of Farris (1972) does not aim to minimise the difference between the reconstructed matrix and the original. It attempts to find the tree with the minimum overall length. It is hardly fair to criticise this method for not doing what it was not designed to do. Secondly, and more importantly, evaluating the goodness of fit of the original and reconstructed matrices is not a test of the goodness of the reconstructed phylogeny. It is a test of phenetic reconstruction, and not of phylogeny which is concerned primarily with the order of branching in the tree, and only secondarily with the lengths of the branches.

Mickevich (1978) has applied another type of test to eight methods, only two of which have been discussed in this present article, UPGMA, and parsimony limited search (with an initial tree built using the distance Wagner method). Mickevich estimated the degree of congruence (similarity in groupings) between the trees constructed for the same species, from different sets of characters. Nine paired sets of characters were used, including paired morphological and allelic character sets, as well as two paired proteins. The methods that gave similar groupings from the different character sets were termed stable. The results obtained with the two protein data sets for the two tree-building methods showed no difference in stability.

Felsenstein (1978) compared several phylogenetic methods to determine conditions under which the methods fail to converge on the true phylogeny as more and more data are accumulated. Using very simple examples with binary data he showed that the Camin & Sokal (1965) and the Farris (1970) parsimony methods, as well as Estabrook's clique analysis method all failed to be statistically consistent under certain circumstances. The conditions for the failure were that parallel changes exceeded informative changes. A maximum likelihood estimate, on the other hand, was statistically consistent under the same circumstances. However a maximum likelihood approach requires a probabilistic model of character evolution that can be believed. Such a model is not yet available.

With so few studies, it is not possible at this stage to reach any definite conclusions concerning the accuracies of the various tree-building methods. Though there is much theoretical discussion to take into account, there are few experimental tests of the methods themselves. The few tests that have been completed suggest clearly that different methods are better for different data sets. Therefore any future testing should be done upon data sets whose characteristics have been clearly defined. It also seems important that the criteria for evaluating the methods should be equally applicable to all the methods tested, and should be a measure of the accuracy of the phylogenetic relationships constructed by the different methods.

CONCLUSIONS

There is an immense and as yet little tapped store of phylogenetic information contained within the nucleic-acid sequence of the genome. Such information has certain advantages when it comes to trying to reconstruct phylogeny. The characters (base or amino-acid positions) are directly comparable without prior weighting, and there are many characters in common throughout the entire range of eukaryotic organisms. However, there are many problems associated with their use. They are not free of parallel evolution, and the methods used to reconstruct phylogenies from them have serious practical limits, even when the most powerful computing resources are available; also, the accuracy of the various methods has yet to be established. However, in spite of these difficulties, the phylogenetic study of protein and nucleic-acid sequences has provided many new insights into the process of evolution at the molecular and organismal level.

REFERENCES

Arnheim, N. & Steller, R. (1970). Multiple genes for lysozyme in birds. *Archives of biochemistry and biophysics*, **141**, 656–61.

Barker, W. C., Kecham, L. K. & Dayhoff M. O. (1978). A comprehensive examination of protein sequences for evidence of internal Gene Duplication. *Journal of molecular biology*, **10**, 265–81.

Beintema, J. J., Gaastra, W., Lenstra, J. A. & Fitch, W. M. (1977). The molecular evolution of pancreatic ribonuclease. *Journal of molecular evolution*, **10**, 49–71.

Boulter, D. (1974). The evolution of plant proteins with special reference to higher plant cytochrome *c*. *Current advances in plant science*, **8**, 1–16.

Boulter, D., Haslett, B. G., Peacock, D., Ramshaw, J. A. M. & Scawen, M. D. (1977). Chemistry, function and evolution of plastocyanin. *Plant Biochemistry*, II, **13**, 1–40.

Boulter, D., Gleaves, J. T., Haslett, B. G., Peacock, D. & Jensen, U. (1978). The relationships of eight tribes of the compositae as suggested by plastocyanin amino acid sequence data. *Phytochemistry*, **17**, 1585–9.

Boulter, D., Peacock, D., Guise, A., Gleaves, J. T. & Estabrook, G. (1979). Relationships between the partial amino acid sequences of plastocyanin from members of ten families of flowering plants. *Phytochemistry*, **18**, 603–8.

Camin, J. H. & Sokal, R. R. (1965). A method for deducing branching sequences in phylogeny. *Evolution*, **19**, 311–26.

Colless, D. H. (1970). The phenogram as an estimate of phylogeny. *Systematic zoology*, **19**, 352–62.

Crick, F. H. C. (1958). On protein synthesis. In *The biological replication of macromolecules, 12th symposium of the society for experimental biology*, ed. F. K. Sanders, pp. 138–63. Cambridge University Press.

Dayhoff, M. O. & Eck, R. V. (1966). *Atlas of protein sequence and structure*, **2**. Washington, DC: National Biomedical Research Foundation.

Dayhoff, M. O. (1972). *Atlas of protein sequence and structure*, **5**. Washington, DC: National Biomedical Research Foundation.

Dayhoff, M. O. (1976). *Atlas of protein sequence and structure*, **5**, suppl. 2. Washington, DC: National Biomedical Research Foundation.

Dayhoff, M. O. (1978). *Atlas of protein sequence and structure*, **5**, suppl. 3. Washington, DC: National Biomedical Research Foundation.

De Jong, W. W., Gleaves, J. T. & Boulter, D. (1977). Evolutionary changes of alpha crystallin and the phylogeny of mammalian orders. *Journal of molecular evolution*, **10**, 123–35.

Elleman, T. C. (1978). A method for detecting distant evolutionary relationships between protein or nucleic acid sequence in the presence of deletions or insertions. *Journal of molecular evolution*, **11**, 143–61.

Estabrook, G. F. (1979). *Character compatibilities.* Proceedings of the 2nd International symposium on chemosystematics, principles and practice.

Estabrook, G. F. & McMorris, F. R. (1977). When are two qualitative characters compatible? *Journal of mathematical biology,* **4**, 195–200.

Estabrook, G. F., Strauch, J. G. Jr. & Fiala, K. L. (1977). An application of compatibility analysis to the Blackiths' data on othopteroid insects. *Systematic zoology,* **26**, 269–76.

Estabrook, G. F. & Landrum, L. (1975). A simple test for the possible simultaneous evolutionary divergence of two amino acid positions. *Taxon,* **24**, 609–13.

Estabrook, G. F., Johnson, C. S. Jr. & McMorris, F. R. (1976). A mathematical foundation for the analysis of cladistic character compatibility. *Mathematical biosciences,* **29**, 181–7.

Farris, J. S. (1970). Methods for computing Wagner trees. *Systematic zoology,* **19**, 83–92.

Farris, J. S. (1971). The hypothesis of nonspecificity and taxonomic congruence. *Annual review of systematics,* **2**, 277–302.

Farris, J. S. (1972). Estimating phylogenetic trees from distance matrices. *American Naturalist,* **106**, 645–68.

Felsenstein, J. (1978a). The number of evolutionary trees. *Systematic zoology,* **27**, 27–33.

Felsenstein, J. (1978b). Cases in which parsimony or compatibility methods will be positively misleading. *Systematic zoology,* **27**, 401–10.

Fitch, W. M. (1966). An improved method for detecting evolutionary homology. *Journal of molecular biology,* **16**, 9–16.

Fitch, W. M. (1969). Locating gaps in amino acid sequences to optimise the homology between two proteins. *Biochemical genetics,* **3**, 99–108.

Fitch, W. M. (1970a). Further improvements in the method of testing for evolutionary homology among proteins. *Journal of molecular biology,* **49**, 1–14.

Fitch, W. M. (1970b). Distinguishing homologous from analogous proteins. *Systematic zoology,* **19**, 99–113.

Fitch, W. M. (1971). Toward defining the course of evolution: minimum change for a specific tree topology. *Systematic zoology,* **20**, 406–16.

Fitch, W. M. (1973). Aspects of molecular evolution. *Annual review of genetics,* **7**, 343–80.

Fitch, W. M. (1974). Evolutionary trees with minimum nucleotide replacements from amino acid sequences. *Journal of molecular evolution,* **3**, 263–78.

Fitch, W. M. (1975). *Toward finding the tree of maximum parsimony, Proceedings of the eighth international conference on numerical taxonomy,* pp. 189–230. San Francisco, USA: W. H. Freeman & Company.

Fitch, W. M. (1977). On the problem of discovering the most parsimonious tree. *American Naturalist,* **111**, 223–57.

Fitch, W. M. & Farris, J. S. (1974). Evolutionary trees with minimum nucleotide replacements from amino acid sequences. *Journal of molecular evolution,* **3**, 263–78.

Fitch, W. M. & Langley, C. H. (1976). Protein evolution and the molecular clock. *Federation Proceedings,* **35**, 2092–7.

Fitch, W. M. & Margoliash, E. (1967). Construction of phylogenetic trees. *Science*, **155**, 279–84.

Goodman, M. (1976). Toward a genealogical description of the primates. In *Molecular Anthropology*, ed. M. Goodman & R. E. Tashian, pp. 321–53. New York: Plenum Press.

Hartigan, J. A. (1973). Minimum mutation fits to a given tree. *Biometrics*, **29**, 53–65.

Holmquist, R. (1978a). The augmentation algorithm and molecular phylogenetic trees. *Journal of molecular evolution*, **12**, 17–24.

Holmquist, R. (1978b). A measure of the denseness of a phylogenetic network. *Journal of molecular evolution*, **11**, 225–31.

Jardine, N., Rijsbergen, C. J. & Jardine, C. J. (1969). Evolutionary rates and the inference of evolutionary tree forms. *Nature*, **224**, 185.

Le Quesne, W. J. (1969). A method of selection of characters in numerical taxonomy. *Systematic zoology*, **18**, 201–5.

Lyddiat, A., Peacock, D. & Boulter, D. (1978). Evolutionary change in invertebrate cytochrome *c*. *Journal of molecular evolution*, **11**, 35–45.

Margoliash, E. (1963). Primary structure and evolution of cytochrome *c*. *Proceedings of the national academy of sciences of the USA*, **50**, 672.

McLachlan, A. D. (1971). Tests for comparing related amino acid sequences, cytochrome *c* and cytochrome *c*551. *Journal of molecular biology*, **61**, 409–24.

McMorris, F. R. (1977). On the compatibility of binary qualitative taxonomic characters. *Bulletin of mathematical biology*, **39**, 133–8.

Mickevich, M. F. (1978). Taxonomic congruence. *Systematic zoology*, **27**, 143–58.

Miyata, T., Miyazawa, S. & Yasunaga, T. (1978). Two types of amino acid substitutions in protein evolution. *Journal of molecular evolution*, **12**, 219–36.

Moore, G. W., Barnabas, J. & Goodman, M. (1973). A method for constructing maximum parsimony ancestral amino acid sequences on a given network. *Journal of theoretical biology*, **38**, 459–85.

Moore, G. W. (1976). Proof of the maximum parsimony (Red King) algorithm. In *Molecular Anthropology*, ed. M. Goodman & R. E. Tashian, pp. 117–37. New York: Plenum Press.

Moore, G. W. (1977). Proof of the populus path algorithm for missing mutations in parsimonious trees. *Journal of theoretical biology*, **66**, 95–106.

Moore, G. W., Goodman, M. & Barnabas, J. (1973). An iterative approach from the standpoints of the additive hypothesis to the dendrogram problem posed by molecular data sets. *Journal of theoretical biology*, **38**, 423–57.

Moore, G. W., Goodman, M., Callahan, C. & Holmquist, R. (1976). Stochastic versus augmented maximum parsimony method for estimating superimposed mutations in the divergent evolution of protein sequences: methods tested on cytochrome *c* amino acid sequences. *Journal of molecular biology*, **105**, 15–38.

Needleman, S. B. & Wunsch, C. D. (1970). A general method applicable to the search for similarities in the amino acid sequence of two proteins. *Journal of molecular biology*, **48**, 443–53.

Peacock, D. & Boulter, D. (1975). Use of amino acid sequence data in phylogeny and evaluation of methods using computer simulation. *Journal of molecular biology*, **95**, 513–27.

Prager, E. M. & Wilson, E. C. (1978). Construction of phylogenetic trees for proteins and nucleic acids: empirical evaluation of alternative matrix methods. *Journal of molecular evolution*, **11**, 129–42.

Reichert, T. A., Cohen, D. N. & Wong, A. K. (1973). An application of information theory to genetic mutations and the matching of polypeptide sequences. *Journal of molecular biology*, **42**, 245–61.

Romero-Herrera, A. E., Lehmann, H., Joysey, K. A. & Friday, A. E. (1978). On the evolution of myoglobin. *Philosophical transactions of the Royal society of London*, **283**, 61–163.

Sankoff, D. & Cedergren, R. J. (1973). A test for nucleotide sequence homology. *Journal of molecular biology*, **77**, 159–64.

Schwartz, R. M. & Dayhoff, M. O. (1978). Origins of prokaryotes, eukaryotes, mitochondria and chloroplasts. *Science*, **199**, 395–403.

Sneath, P. H. A. (1974). Phylogeny of micro-organisms. In *Evolution in the microbial world, 24th symposium of the Society for General Microbiology*, ed. M. J. Carlile & J. J. Skehel, pp. 1–39. Cambridge University Press.

Sneath, P. H. A., Sackin, M. J. & Ambler, R. P. (1975). Detecting evolutionary incompatibilities from protein sequences. *Systematic zoology*, **24**, 311–32.

Sneath, P. H. A. & Sokal, R. R. (1973). *Numerical Taxonomy*. San Francisco, USA: W. H. Freeman and Company.

Sokal, R. R. & Michener, C. D. (1958). A statistical method for evaluating systematic relationships. *University of Kansas Science Bulletin*, **38**, 1409–38.

Sokal, R. R. & Rohlf, F. J. (1962). The comparison of dendrograms by objective methods. *Taxon*, **11**, 33–40.

Wilson, A. C., Carlson, S. S. & White, T. J. (1977). Biochemical Evolution. *Annual reviews of Biochemistry*, **46**, 573–639.

Zuckerkandl, E. & Pauling, L. (1965). Molecules as documents of evolutionary history. *Journal of theoretical biology*, **8**, 357–66.

4

Enzymes in bacterial populations

PATRICIA H. CLARKE

DEPARTMENT OF BIOCHEMISTRY,
UNIVERSITY COLLEGE LONDON, LONDON WC1E 6BT, UK

INTRODUCTION

Evolution of bacteria

Bacteria are the most ancient of all living creatures. They are ubiquitous in the world today and have been so at all times since life first appeared on the earth over three thousand million years ago. We do not know what the very first organisms looked like, but recent studies in paleobiology have revealed fossils looking remarkably like present-day bacteria. Traces of what may be spherical or rod-shaped bacteria have been seen in the Onverwacht Shale about 3.5×10^9 years old. More distinct microfossils have been described in the Fig Tree chert, also in South Africa, about 3.2×10^9 years old. The Gunflint chert of Ontario is about 2×10^9 years old and fossil traces here have been identified as bacterial remains. The rocks of the Bitter Springs region in Central Australia are particularly rich in microfossils. From their morphological appearances they have been classified into 30 or so different species and it is suggested that some of these microfossil structures are very similar to bacteria and green algae that exist today (Schopf & Barghoorn, 1967).

One remarkable fossil found in the Gunflint chert was named *Kakabekia umbellata* because it looked like an umbrella or parachute. Later, organisms with a very similar appearance were found in soil around Harlech Castle in North Wales and also in Alaska and Iceland (Siegel & Siegel, 1970). In these places *Kakabekia* was found in soil rich in ammonia and this might suggest that this ancient family of bacteria had retained a taste for ammonia that had persisted from the chemical conditions of the early earth. Extrapolations such as this are fraught with problems. A close similarity in morphology of contemporary bacteria is no indication of physiological relatedness and we have no way of examining the biochemical characteristics of fossil remains. The modern

Kakabekia are fermentative organisms but, unlike the strict anaerobes, they are able to live in air. The conditions on earth when life began to evolve were anaerobic (Schuster, this volume) and oxygen toxicity may have presented a major problem overcome at some period by the *Kakabekia*. Although we cannot recognise the stages in biochemical evolution reached by the fossil bacteria it is possible to make some deductions about the evolution of the major metabolic pathways by comparing present-day organisms.

Bioenergetic processes

The first organisms are presumed to have been heterotrophic fermenters, utilising organic compounds already present in the primeval soup. Later, fermentation as a mode of life was overtaken by anaerobic photosynthesis and eventually by oxygen-producing photosynthesis and aerobic respiration. The cytochrome chain that operates in oxidative phosphorylation has many similarities with the electron-transport chain of photosynthesis and both may have a common origin. Broda (1975) suggests an order for the evolution of bioenergetic processes of:

$$\text{Fermentation---Anaerobic photosynthesis} \begin{cases} \text{Aerobic photosynthesis} \\ \\ \text{Oxidative respiration} \end{cases}$$

The evolution of new metabolic activities among the prokaryotes had a direct effect on their environment. In certain places the fermentative organisms may have depleted the supply of preformed organic chemicals and fixation of carbon dioxide by photosynthesis would have been an obviously advantageous development. The major event, or more accurately the major period of change, was that in which aerobic photosynthesis evolved in the blue-green algae (the cyanobacteria) leading thereby to an aerobic world. The present-day bacteria exhibit great metabolic variety but there may have been many other biochemical pathways that we shall never know. Competition between the populations of the early prokaryotic world may have resulted in the disappearance of some groups while others adapted to changing conditions in a way that allowed their descendants to survive. Some of these descendants are now to be found only in very specialised ecological niches. The strict anaerobes, obtaining all their energetic needs by fermentation, and the anaerobic photosynthetic bacteria may be the closest we shall get to representatives of the primitive world but even with these organisms many changes will have occurred during the long period of biological evolution.

Bacteria have continued to evolve ever since the time that they first appeared. At the present time we can observe them evolving still. Before describing experiments in adaptation and enzyme evolution we need to look at some of the main classes of bacteria and to consider how their genetic material is organised and transferred.

MAIN GROUPS OF BACTERIA

All bacteria, including the cyanobacteria, belong to the class of prokaryotes while plants, animals, fungi, protozoa and all other groups belong to the eukaryotes. The profound differences between these two classes of living organisms extend to the presence or absence of certain organic compounds, membrane and cytoplasmic structures and to the organisation of genetic material. The prokaryotes are haploid organisms with the major part of their DNA forming a single chromosome not contained within a nuclear membrane. Some characters are determined by extrachromosomal DNA and genetic exchange between bacteria may be carried out by a variety of processes some of which are described in more detail in the next section. The possibility of genetic transfer between bacteria belonging to different species, and to different genera, makes it difficult to assign an evolutionary progression by comparing the properties of present-day bacteria since the gene for a particular enzyme may have been acquired by a recent genetic event. The classical methods for identifying bacteria were based on morphology, staining reactions, response to oxygen and tests for biochemical activities. These have been joined by more detailed comparisons at the molecular level of relatedness of proteins and nucleic acids and, in certain groups of bacteria, by genetic analysis. This makes it possible on many occasions to take a new isolate and to assign it with confidence to a species already described. However, the pioneers of bacteriology were mainly concerned with organisms causing human and animal diseases and far less detailed analysis and taxonomic subdivision has been carried out on the free-living organisms. Bacteria can be divided into two major subgroups by their response to the Gram stain. Gram-positive bacteria contain peptidoglycan (murein) as the major polymer of the cell wall associated with teichoic acids. The Gram-negative bacteria also have peptidoglycan in their cell walls, but in smaller amounts, and the structure is more complex and includes layers containing lipoprotein and lipopolysaccharide. Both groups include rod-shaped and coccoid bacteria carrying out aerobic or anaerobic metabolism. The following examples indicate some of the groups that have been used for biochemical studies. Fuller descriptions of these and other groups of bacteria will be found in textbooks of microbiology such as Stanier, Adelberg & Ingraham (1977).

Woese suggests that certain groups of bacteria should be reclassified as Archaebacteria. The methanogenic bacteria are strict anaerobes whose main catabolic activity is the conversion of carbon dioxide to methane. The methanogens differ from the main groups of bacteria in ribosomal RNA homology; they lack peptidoglycan and possess unusual lipids. some extreme halophiles and thermoacidophiles have similar biochemical characteristics. Woese, Magrum & Fox (1978) conclude that these Archaebacteria differ from both prokaryotes and eukaryotes and represent a third major line of biological descent.

Gram-positive bacteria

Strict anaerobes (Fermenters). The anaerobic *Clostridia* are rod-shaped and form heat-resistant endospores. They obtain energy by fermentation and the growth of many species is inhibited by oxygen even at very low concentrations. Carbohydrates serve as fermentation substrates for some species and a variety of end-products, including butanol and acetone, may be produced. The proteolytic species ferment amino acids and produce ATP by associated substrate-level phosphorylations.

Facultative anaerobes (Fermenters). The lactic acid bacteria include the genera *Lactobacillus* (short rods) and *Streptococcus*. These are fermentative organisms but are oxygen-tolerant and can be grown on the surface of solid media in air. They do not synthesise cytochromes and are unable to carry out oxidative phosphorylation. The bacteria of this group have complex growth requirements and this is not a relic of primitive life but is the result of loss of biosynthetic abilities. Morishita, Fukada, Shirota & Yura (1974) found that *L. casei* could regain the ability to make at least seven amino acids by independent single site mutations. This indicated that the structural genes for the biosynthesis of these amino acids were present in *L. casei* although in an inactive form. The lactic acid bacteria are examples of organisms that have lost biochemical pathways during the course of evolution. The genes for the biosynthetic pathways had not been discarded but had become silent and non-expressed. Reactivation of silent genes to fulfil either their original function or a new role in metabolism may have been important in the evolution of microbial enzymes and can be observed experimentally.

The *Staphylococci*, as their name implies, are spherical in shape. Unlike the *Lactobacilli* they are capable of both fermentative and oxidative metabolism.

Strict aerobes. The *Bacilli* are rod-shaped, form endospores and carry out oxidative respiration. They are found in soil and have few, if any, requirements for organic growth factors. Included in this group are some extreme thermophiles with optimum growth temperatures of about 60°C. Many organic compounds can be used as growth substrates by members of this group.

Gram-negative bacteria

Anaerobic photosynthetic bacteria. Photosynthesis is carried out under anaerobic conditions and no oxygen is produced. Organisms with this mode of life may have been dominant on the earth in the early days of life but these bacteria are now restricted to a few specialised ecological niches. The *green sulphur bacteria* use hydrogen sulphide, other sulphur compounds or hydrogen gas, as the electron donor in photosynthesis and are strict autotrophs. The *purple sulphur bacteria* also carry out anaerobic photosynthesis with reduced sulphur compounds acting as electron donors but can assimilate a few organic compounds anaerobically in the light. The *purple non-sulphur bacteria* are capable of living auto-trophically in the light with organic compounds acting as the electron donors but can also grow aerobically in the dark obtaining energy by oxidation of organic compounds via oxidative phosphorylation. During aerobic growth the synthesis of photosynthetic pigments is repressed so that on transfer from an anaerobic to an aerobic environment the pigment content of the cells is progressively decreased. The purple non-sulphur bacteria have some of the characteristics to be expected of organisms in transition from an anaerobic to an aerobic world.

Aerobic photosynthetic bacteria. The *blue-green bacteria* carry out oxygen-producing photosynthesis closely resembling that of higher plants. They comprise a heterogenous group with respect to morphology which includes filamentous and gliding forms as well as unicellular rods and cocci. Many are nitrogen-fixers and these are of economic importance in biological nitrogen fixation particularly in the tropics. The blue-green bacteria occupy a very special position in evolution and are considered to have been the main agents in the establishment of an oxygen atmosphere on the earth.

Facultative anaerobes. Many bacteria, including *Escherichia coli* and *Klebsiella aerogenes* are able to grow both aerobically and anaerobically with a variety of organic compounds. Catabolic pathways are channelled into the tricarboxylic acid cycle and during aerobic growth most of the ATP is derived by oxidative phosphorylation. These species belong to the enteric group of bacteria which also grow anaerobically by fermenting various sugars. Some enteric bacteria, in particular *Salmonella* and *Shigella* species, are pathogenic to man. *E. coli* and *K. aerogenes* have been widely used for studies on metabolic pathways and also for experimental work on enzyme evolution. These are short rods and are motile by means of peritrichate flagella.

Strict aerobes. Among the aerobic Gram-negative bacteria the *Pseudomonas* species have been studied in most detail. They are able to metabolise very many organic compounds by oxidative reactions and obtain energy by oxidative phosphorylation. Some strains can use nitrate instead of oxygen as the terminal electron acceptor and may reduce nitrate to nitrogen thereby effecting denitrification. They are ubiquitous in soil and water and are frequent isolates from enrichment cultures with unusual chemicals provided as growth substrates. In shape they are short rods with polar flagella. *Pseudomonas aeruginosa* is an opportunist pathogen for man and other species are plant pathogens. The *Acinetobacter* are also metabolically versatile bacteria and like the *Pseudomonas* species can be readily isolated from soil. They vary in shape during growth from short rods to cocci.

THE BACTERIAL GENOME

During the last three decades many advances in molecular biology have been based on studies in microbial genetics but before that time the mechanism of inheritance in bacteria was obscure. In 1945 Dubos, who suggested that the simplest possible arrangement for bacteria would be 'a single gene string existing as a rod or granule', was not very far from the truth. The variations seen by bacteriologists in the morphology and

biochemistry of bacteria and the absence of an obvious sexual cycle had
made it difficult to explain how bacteria could vary so much and also
retain genetic continuity. The fundamental difficulties can be ascribed to
the confusion of *genotype* (the total genetic information carried by an
organism) with *phenotype* (the actual characteristics observed). There
was also the unavoidable situation that with bacteria the experimenter
was dealing with populations and not individuals and that the population
might contain a mixture of individual bacteria that differed among
themselves. The second problem could be partly overcome by selecting a
single bacterial cell and growing from it a colony or *clone* of bacteria with
identical genetic make-up. As we shall see later this technique is not
always successful in eliminating variation. The first problem was resolved
by analysis of variation with respect to a) changes in the genetic material
itself and b) effects due to phenotypic response to environmental changes.
Although the resolution of genetic and phenotypic factors may present
some difficulties this is no longer a conceptual problem. We are now
certain that bacterial genes are arranged on a chromosome although we
know that the total bacterial genome is rather more complex than the
structure envisaged by Dubos (1945). Much of our information about the
bacterial genome has come from studies on genetic exchange.

Genetic exchange in bacteria

Transformation.   The first demonstration that genetic material could be
transferred between bacteria came from the finding that a mouse injected
with a non-virulent strain of *Pneumococcus* would die if a heat-killed
virulent strain were to be injected at the same time. The significance of
this observation became clear when Avery, MacLeod & McCarty (1944)
showed that the substance that *transformed* a non-virulent *Pneumococcus*
into a virulent strain was the DNA extracted from the virulent strain.
Transfer of genetic material by transformation was later established for
Haemophilus species and it was shown that the ability to ferment
carbohydrates as well as resistance to drugs could be acquired by this
means. Transformation can also be carried out with *Bacillus*. Since the
bacteria of this group, unlike *Pneumococcus* and *Haemophilus*, can be
grown in minimal salt medium the technique of genetic analysis by
transformation is useful for mapping genes for metabolic enzymes. In
transformation the recipient bacterium takes up one of the strands of a
linear segment of double-stranded DNA of about 1–10 megadaltons
(Md) in molecular weight. Only certain strains of *B. subtilis* appeared to

be able to do this and then only when they were in a state of 'competence'. None of the Gram-negative bacteria, with the exception of *Haemophilus*, appeared to be capable of being transformed but this has now been achieved by treating the cells to make them more permeable. *Escherichia coli*, *Pseudomonas aeruginosa* and *Acinetobacter calcoaceticus* can be transformed with DNA preparations.

Transduction. Bacteriophages may act as agents for genetic transfer by carrying bacterial genes from a donor to a recipient. Temperate bacteriophages can remain dormant within a bacterial cell for many generations but when some of the individuals in the bacterial population are lysed fragments of bacterial DNA may be packaged into the phage particles instead of the phage DNA. In a transduction experiment the lysate is filtered to remove intact bacteria and mixed with a suspension of recipient bacteria. The phage particles carrying donor DNA inject this into the recipient bacterium and recombination may occur between the incoming DNA and the homologous region of the recipient chromosome. The main limitation is the size of the DNA fragment since it has to be about the same as the normal phage DNA.

In generalised transduction about $1/10^5$ phage particles carry bacterial DNA in place of phage DNA and this may be derived from any part of the bacterial chromosome. Specialised transduction occurs with bacteriophages whose phage DNA (prophage) becomes inserted at specific sites on the bacterial chromosome. When the prophage is released from the attachment site some hybrid DNA segments, containing both bacterial and bacteriophage DNA may be produced by inaccurate crossing over. Bacteriophage λ has an attachment site on the chromosome of *Escherichia coli* adjacent to the genes for the catabolism of galactose. From a strain of *E. coli* carrying the λ prophage at that site a few phage particles will be released carrying λ-*gal* DNA.

Bacteriophages are very common in nature and can be isolated from soil and sewage. Transducing phages are known for *Bacillus*, *Pseudomonas* and *Staphylococcus* species.

Conjugation and the bacterial chromosome. Genetic exchange between two bacterial mutants was first demonstrated with *Escherichia coli* K12. It was later found that transfer of chromosomal genes required the presence of a sex factor in the donor strain and that transfer was unidirectional. The extensive studies on genetic exchange in *E. coli* K12, using conjugation and transduction, have made it the organism for which chromosomal maps have been established in most detail. It has also been used for fundamental studies in the molecular biology of gene structure and expression and to examine the nature of mutation and genetic and phenotypic variation.

E. *coli* contains a single chromosome. It consists of a DNA duplex of about 2.7×10^3 Md forming a closed circle about 1 mm in total length. The circular nature of the *E. coli* chromosome was first established by genetic studies and later confirmed by physico-chemical observations. It is generally accepted that this is the pattern for all bacteria. The DNA of the chromosome of *E. coli* is sufficient to code for up to 4000 genes. Bachmann, Low & Taylor (1976) list 650 gene loci mapped by that date.

Circular genetic maps have been established for species of *Acineto-bacter* (Towner, 1978), *Bacillus* (Young & Wilson, 1975), *Pseudomonas* (Watson and Holloway, 1978), *Rhizobium* (Beringer, Hoggan & Johnson, 1978) and *Streptomyces* (Hopwood, Chater, Dowding & Vivian, 1973).

Plasmids

Plasmids are extrachromosomal genetic elements that can exist in an autonomous state within a bacterial cell and carry genes concerned with their own replication. An account of the biology of plasmids is given by Williams (1978).

Sex factors. Sex factors are plasmids that have the property of mobilising the bacterial chromosome. All plasmids consist of covalently closed circles of DNA and as a minimum requirement carry genes for replication. Such genetic entities are called *replicons*. Sex factors also carry genes for the transfer of themselves and part or all of the bacterial chromosome to a suitable recipient. These genes are called *tra* (transfer genes). Plasmids with sex factor activity may also carry genes for various other properties but if they have no other known functions they can be identified only by their sex factor activity. The first, and best known, sex factor is F which is present in male or donor strains of *E. coli* either in an autonomous state in the cell (F^+) or integrated into the bacterial chromosome (Hfr). There are many sites at which integration can occur and all Hfr strains are able to transfer chromosomal genes at a high frequency from the site of integration. In conjugal transfer from an Hfr strain the origin of transfer is within the sequence of the F factor and proceeds through the bacterial chromosome so that if, as is usual, the chromosomal transfer is incomplete only part of the F factor genes are transferred to the recipient. Recombination of the genes from the donor bacterium with the chromosome of the recipient requires activities specified by the recombination genes, in particular *recA*.

The *E. coli* sex factor F may be considered as an average sized plasmid of 63 Md. The sex factor FP2 of *P. aeruginosa* is about 60 Md.

Col factors.   Many plasmids were first recognised because they carried genes for properties that were readily lost or readily acquired by bacterial populations. The phenomenon of colicinogeny was first described by Fredericq (1957) who observed that certain strains of *E. coli* produced diffusable substances *colicins* that were lethal to other strains. This is now known to be very general among bacteria and many species produce *bacteriocins* that kill other strains of the same or related species (Konisky, 1978).

The genes for colicin production are carried on plasmids called Col factors. These are self-replicating autonomous entities but vary considerably in the other determinants carried. The Col factors may be divided into two main groups. One group includes the smaller plasmids such as ColEI with a molecular weight of 4.2 Md which is present in about 10–15 copies per cell. The other group includes the larger plasmids such as ColI with a molecular weight of 62 Md which resembles F in being present in the cell as only one or two copies per chromosome. The smaller Col factors carry very few genes and are not transmissible but the larger Col factors are capable of self-transmission and may also act as sex factors and mobilise the bacterial chromosome.

Drug resistance plasmids.   An important finding in the late nineteen fifties was that resistance to certain drugs and antibiotics appeared to be spreading very rapidly through bacterial populations and by a mechanism quite distinct from mutation and selection. This type of drug resistance appeared in enteric bacteria belonging to the genera *Shigella* (dysentery agents) and *Salmonella* (typhoid and food-poisoning organisms) and appeared to be related to the use of these drugs in treating infections. Many of the drug-resistant bacteria carried resistance to more than one drug and it was established that the genetic determinants for such drug resistance were not chromosomal but were carried on plasmids. Many classes of drug resistance plasmids or R factors are now known and although they have been studied in most detail in the enteric bacteria they are not confined to this group. They are classified according to the drug resistance and other genetic determinants present on the plasmid.

The drug resistance plasmids first isolated had a fairly limited host range but many could be transferred to other species of enteric bacteria. A surprising finding was that the plasmid RP1, isolated from *P. aeruginosa* (P group plasmid), could be transferred not only to the unrelated species *E. coli*, but also to a wide range of other Gram-negative bacteria (Olsen & Shipley, 1973). The frequency of transfer of all the drug resistance plasmids is host dependent even within different strains of the same species. Some of these plasmids can act as sex factors and the extent of this activity is also strain dependent. RP4 was used to map the chromosome of *Acinetobacter calcoaceticus* (Towner, 1978). The P group plasmid R68 has sex factor activity for the PAT strain of *P. aeruginosa* but gives very few recombinants with PAO strains. Haas & Holloway (1978) selected variants with enhanced sex factor activity for strain PAO and found that they could also promote chromosomal exchange between other strains of *P. aeruginosa*. Further, the R68.45 variant could be used as a sex factor for *Rhizobium leguminosarum* (Beringer, Hoggan & Johnston, 1978).

The classification of plasmids into incompatibility groups (Table 4.1) was first applied to drug resistance plasmids. It was found that some could not be maintained in an *E. coli* strain carrying F while others could not coexist (were incompatible) with the plasmid ColI. Some R plasmids were classified as F-type and others as I-type and it was found that bacteria

Table 4.1. *Examples of plasmids: extrachromosomal genetic elements*

Plasmid	Size megadaltons	Incompatibility group	Phenotype determined[a]	Isolated from
F	63	IncF1	sex factor *E. coli*	*E. coli*
ColI	65	IncI	colicin I	*E. coli*
R1	62	IncF11	Ap Cm Km Sm Su	*S. typhimurium*
R64	73	IncI	Sm Tc	*S. typhimurium*
RP4	36	IncP-1	Cb Km Nm Tc	*P. aeruginosa*
CAM	92[b]	IncP-2	camphor utilisation	*P. putida*
OCT	27[c]	IncP-2	octane utilisation	*P. putida*
SAL	51	IncP-9	salicylate utilisation	*P. putida*
TOL	78[d]	IncP-9	toluene and xylene utilisation	*P. putida*

[a] Drug resistances: Ap, ampicillin; Cm, chloramphenicol; Km, kanamycin; Sm, streptomycin; Su, sulphonamide; Tc, tetracycline; Cb, carbenicillin; Nm, neomycin.
[b] Most CAM+ strains carry CAM plasmid of 92 Md and factor K, 64 Md but some CAM+ strains carry plasmid of 160 Md in which CAM and K have recombined.
[c] OCT+ strains carry OCT, 27 Md and factor K.
[d] TOL plasmid from *P. putida* (*arvilla*) but other TOL+ strains carry plasmids ranging in size from 60 to 170 Md.

carrying these plasmids produced F-like sex pili or I-like sex pili respectively. The sex pili are thought to be concerned with DNA transfer in conjugation and their presence can be most readily detected by the sensitivity of bacteria on which they are present to F-type or I-type male-specific phages which attach to the sex pili and infect the bacteria. Many more incompatibility groups have now been recognised and examples are given in Table 4.1. Specific sex pili have been observed for many of these groups and group-specific phages have been isolated.

Catabolic plasmids. We have seen that the genes for a variety of characters may be carried on plasmids rather than on the bacterial chromosome. The genes for all essential biosynthetic activities are located on the chromosome but this is not necessarily the case for catabolic genes. We discussed earlier that some groups of bacteria are able to utilise a very large number of different compounds as growth substrates and this raises the question whether the genes for all these catabolic pathways can be accommodated on the chromosome. Gunsalus and colleagues observed that several *Pseudomonas putida* strains were unstable with respect to growth with certain compounds. The loss of ability to grow on a particular compound can be ascribed to a number of different causes but one possibility was that the genes for that pathway were located on an extrachromosomal element that could be eliminated without affecting growth on alternative substrates. Rheinwald, Chakrabarty & Gunsalus (1973) showed that the genes for camphor utilisation were carried by a plasmid (CAM) that could be transferred to other *Pseudomonas* species. Within a few years many more catabolic plasmids were identified by genetic and biochemical methods and later the plasmid DNA was isolated and examined. Catabolic plasmids are now known for the metabolism of naphthalene (Dunn & Gunsalus, 1973), octane (Chakrabarty, Chou & Gunsalus, 1973) toluene, *m*-xylene, *p*-xylene (Williams & Murray, 1974) and nicotine-nicotinate (Thacker, Rørvig, Kahlon & Gunsalus, 1978). Some of their properties are shown in Table 4.1. Although catabolic plasmids appear to be particularly abundant in the *Pseudomonas* group they are also to be found in the enteric bacteria. Sucrose and lactose fermentation by *Klebsiella* and *Salmonella* strains may be dependent on the presence of plasmids (Reeve & Braithwaite, 1974; Wohlhieter *et al.*, 1975).

The catabolic plasmids provide a reservoir of genetic information in a bacterial population. This is essentially a population gene pool in contrast to each individual bacterium carrying the complete set of genetic information. In nature this arrangement allows rapid adaptation to changes in the chemical environment by the transmission of plasmids throughout the population when needed. Plasmid and chromosomal genes can operate together and strains with novel growth properties can be constructed in the laboratory by plasmid transfer. Plasmid genes can be mutated or rearranged or hybrids constructed carrying genes from two or more plasmids. During the course of evolution of catabolic pathways plasmids may have been very important agents of change.

Other plasmids.   The pathogenicity of some *E. coli* strains is known to be due to the presence of plasmids and this may also be the case for other species. The virulence of *Agrobacterium tumifaciens*, which produces crown gall tumours, is related to the presence of a large plasmid (see Williams, 1978).

Resistance to heavy metals is frequently plasmid-borne and has been described in *Staphylococcus* and in enteric bacteria. The sex factor FP2 from *P. aeruginosa* confers mercury resistance.

Plasmids are usually first recognised by the growth phenotypes that they determine but when bacterial cells are disrupted it is frequently found that a number of small circular DNA molecules are present. If their role has not been established they are referred to as cryptic plasmids. Although in some cases these may be artefacts of isolation, in others their presence reflects the existence of a variety of independent replicons in bacterial cells.

Methods for comparing plasmids.   Plasmids are described and classified by the characters that they determine. Biological relationships between them can be established by compatibility studies. Isolation of plasmid DNA gives the opportunity to carry out more detailed comparisons. The very large catabolic plasmids may be up to 200 Md while the small non-transmissible plasmids like ColEI are only 4.2 Md. Cleavage by restriction endonucleases provides a powerful tool since the number and size of the restriction fragments will depend on the number and location of the sequences for the site-specific enzymes on the plasmid DNA.

Hybridisation of DNA can be used to detect homology between plasmids. A more detailed comparison can be made by examination of heteroduplexes formed by hybridising single strands of two different plasmids (see Williams, 1978). Regions of non-homology are revealed as lengths of single-stranded DNA. This is of particular value for comparing closely related DNA species (see Fig. 4.1).

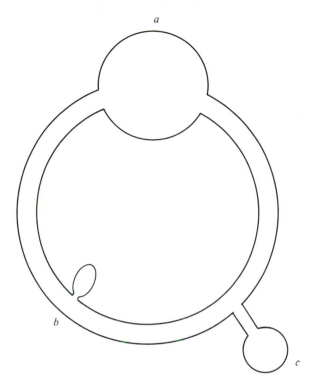

Fig. 4.1. Diagrammatic representation of heteroduplex formed between two related plasmids. Single-stranded loops indicate regions of non-homology. Region *a* shows the effect of different DNA sequences in the two plasmids; region *b* shows the effect of a short DNA insertion in one of the plasmids: region *c* is characteristic of the insertion of a transposon bounded by inverted repeat sequences of DNA (see Bukhari, Shapiro & Adhya, 1977; Williams, 1978 for examples).

GENE TRANSLOCATION

Transfer of genes by transformation, transduction or conjugation results in the production of recombinant bacteria by enzymic processes operating on homologous regions of DNA. It is now known that transfer of certain genes and of short sequences of DNA may occur by a different process, not involving the normal recombination enzymes.

Insertion elements (*IS sequences*)

A class of genetic elements with unusual properties was first identified as a result of the finding that certain mutations in the galactose operon (galactose utilisation) were not due to the usual types of single site or deletion mutations but were due to the insertion of unrelated sequences of DNA. They were also found to appear independently in different genetic locations in the bacterial chromosome and in plasmids and in the genome of bacteriophages. These insertion elements varied in size from 800 to 1400 nucleotides and could be inserted in either orientation. IS2 in one orientation causes chain termination while in the other orientation it can act as a promoter. Identification of IS sequences can be most readily carried out by heteroduplex mapping. The IS sequence of a known genetic element can be used to identify any IS elements in a DNA preparation from another source. In this way it was found that IS sequences are normally present at a number of locations on the chromosome and at key sites on plasmids. Of particular interest was the finding that IS sequences are present on the sex plasmid F and at the sites at which it integrates into the bacterial chromosome. Further, sequences resembling IS elements were found in plasmids at the boundaries of antibiotic genes and at the boundaries between transfer genes and resistance genes.

Transposons

Transposons were identified because it was observed that some drug resistance determinants could transpose, or jump, from one plasmid to another or from plasmid to chromosome. The gene coding for β-lactamase (penicillinase) in *E. coli* transposed intact in this way (Hedges & Jacob, 1974; Bennett & Richmond, 1976). Since the early observations transposons have been identified for resistance to ampicillin, kanamycin, tetracycline, streptomycin and chloramphenicol (Table 4.2). Analysis by heteroduplex mapping showed that some of the transposons have terminal repeated sequences at each end. These vary in length from 100 to 1500 base pairs and some of the terminal sequences have been identified as IS sequences.

The genetic rearrangements of the drug-resistance transposons does not require the enzymes of the normal recombination systems. No extensive sequence homology is required and the process is independent of the functions coded for the *recA* gene. Analysis of transposons structure by hereroduplex mapping can indicate the extent of the terminal repeat sequences and if they are present in opposing orientations the pairing of the homologous regions gives the balloon and stem appearance shown in Fig. 4.1.

Table 4.2. *Transposons*

Transposable element	Size megadaltons	Plasmid[a] origin	Resistance[b] markers
Tn1	3.2	RP4	Ap
Tn3	3.2	R1	Ap
Tn4	13	R1	Ap, Sm, Su
Tn5	3.5	JR67	Km
Tn6	2.7	JR72	Km
Tn7	9	R483	Tp, Sm
Tn9	1.7	pSM14	Cm
Tn10	5.5	R100	Tc

[a] Plasmid RP4 was first isolated from *P. aeruginosa*. All other plasmids were isolated from enteric bacteria.
[b] Resistance markers: Ap, ampicillin; Sm, streptomycin; Su, sulphonamide; Km, kanamycin; Tp, trimethoprim; Cm, chloramphenicol; Tc, tetracycline. Data from Cohen (1976) and Bennett & Richmond (1978).

Evolution of chromosomes and plasmids

Observations of drug resistance phenotypes indicated that many plasmids were genetically unstable and the discovery of transposons and IS elements suggested a way in which plasmids could lose or acquire genes for drug resistance. The essential genes for a plasmid are those for replication and if the plasmid also carries transfer genes it has the basic genetic information to be a transmissible replicon or transfer factor. It was known that some drug resistance plasmids dissociate into transfer factors and non-transmissible resistance determinants and Cohen (1976) outlined a mechanism for segregation or association of resistance genes and transfer factors. This is outlined in Fig. 4.2.

The discovery that the *tol* genes of the *P. putida* TOL plasmid could be transposed to the drug resistance plasmid RP4 suggests that a similar mechanism might account for the evolution of catabolic plasmids. Jacoby, Rogers, Jacob & Hedges (1978) obtained a hybrid RP4-TOL plasmid in a conjugal transfer from a *P. aeruginosa* strain carrying both TOL and RP4 to another strain of *P. aeruginosa*. Among transconjugants selected for growth on *m*-toluate was one that had acquired carbenicillin and kanamycin resistance stably linked to the Tol+ character. Further analysis showed that the *tol* genes had been inserted into RP4 with the loss of the tetracycline resistance. The *tol* genes appeared to behave as a transposon and similar results were subsequently obtained by transfer into other plasmids. Gene transposition provides an additional

132   *P. H. Clarke*

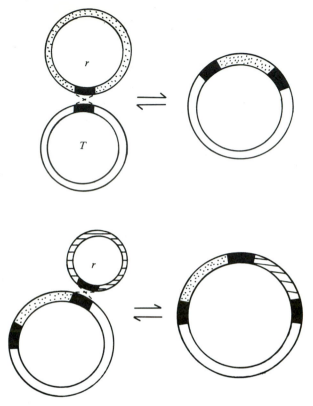

Fig. 4.2.   Diagrammatic representation of association and dissociation of drug resistance plasmids. The resistance transfer factor *T* associates with a replicon containing drug resistance genes *r* to form a transmissible drug-resistance plasmid. (Addition of transposons can produce a plasmid carrying multiple resistances.)

evolutionary mechanism by means of which new functions can be acquired and the content of the chromosomal or plasmid genome expanded. A transposon carrying a lactose operon was detected on a plasmid originally isolated from *Yersinia enterolytica*. Cornelis, Ghosal & Saedler (1978) showed that it possessed inverted repeat sequences and carried genes that appeared to be homologous with the *lac, I, Z, Y* genes of *E. coli*.

An outline of the possible role of insertion elements and transposons in the evolution of chromosomes and plasmids is given by Shapiro (1977) and details of research in this field will be found in the book edited by Bukhari, Shapiro & Adhya (1977).

GENE EXPRESSION

The genome of a bacterium determines its potential properties but the genes are not all expressed at the same time. Although primitive bacteria may have translated all the nucleotide sequences coding for polypeptides equally well, the modern bacteria have complex regulatory systems to ensure economy of protein synthesis and efficient use of nutrients. There are good reasons to suppose that the metabolic pathways themselves evolved before their regulatory systems. For the evolution of new metabolic activities in present-day bacteria it is frequently necessary to change both the metabolic enzymes and their regulation.

Biosynthetic pathways

The biosynthetic pathways for amino acids are remarkably similar throughout living organisms. In biochemical textbooks it will be found that some of the biosynthetic pathways were established by experiments with mutants of *Neurospora crassa*. At a later date other pathways were established by experiments with *E. coli* and the reason for the change over was that bacterial genetics had then become possible and it is faster and easier to work with bacteria rather than moulds. The high degree of similarity of biosynthetic pathways reflects the basic unity of biochemistry. Although detailed comparative studies have shown minor differences in the enzyme reactions of amino-acid biosynthetic pathways in different groups of bacteria (Jensen, 1976) even a superficial investigation reveals differences in the regulation of synthesis of these enzymes. In *E. coli* the expression of many biosynthetic genes is repressed by end-products of the pathway. For example the biosynthesis of tryptophan from chorismate requires five enzymes and if tryptophan is present in the medium the synthesis of all five enzymes is repressed. The genes for the *trp* enzymes (tryptophan synthesis) are arranged close together on the chromosome and are transcribed as a single unit. Genes that are subject to unified control form an *operon* (regulatory unit). A regulatory gene, *trpR*, codes for a protein, the *trp repressor* which combines with tryptophan and binds at a site on the chromosome adjacent to the structural genes for the enzymes and prevents them being expressed (negative control of gene expression). When tryptophan is absent the repressor is unable to bind at the operator site. RNA polymerase binds to the *promoter* site and transcribes the DNA of the structural genes into *trp* messenger RNA. *E. coli* has evolved very sophisticated systems for the regulation of gene expression and an additional fine control operates at very low tryptophan concentrations involving the *attenuator* region preceding the first of the structural genes. In other species of bacteria the arrangement of the *trp* genes and their regulation takes different forms (Crawford, 1975).

The clustering of genes for the biosynthesis of amino acids was at one time thought to be evidence for the evolution of biosynthetic pathways by stepwise retrograde evolution. Horowitz (1945, 1965) suggested that as the primeval soup became depleted of amino acids an enzyme appeared which catalysed the final step. Then, as the immediate precursor became depleted, gene duplication followed by divergence gave rise to a gene for the penultimate enzyme and so on until a complete pathway had evolved. It is now clear that the tight clustering of biosynthetic genes found in the enteric bacteria is concerned with the regulatory system of a group of bacteria that intermittently encounter relatively high concentrations of amino acids in their natural environment. Comparative studies have shown that a variety of other patterns have evolved for regulating biosynthetic pathways. In those species normally found in soil and water, feedback inhibition may be more significant than repression of synthesis (Clarke, 1979).

Catabolic enzymes

The classical example of the regulation of expression of genes for catabolic enzymes is the *lac* operon of *E. coli*. Three structural genes are under *negative* control by the *lac* repressor. The presence of lactose, or certain β-galactosides, in the growth medium results in enzyme *induction*. β-galactosides combine with the *lac* repressor, coded by the *lacI* gene, resulting in its removal from the *lac* operator. In addition, a general regulatory system exerts *positive* control over the *lac* operon and certain other operons of *E. coli*. The *crp* gene codes for a protein which binds cAMP (cyclic 3′,5′-adenosine monophosphate) and acts at the promoter site to facilitate the initiation of transcription by RNA polymerase. Other catabolic genes may be under positive control by an operon-specific activator protein. These include the *ara* (arabinose) operon and the *dsd* (serine deaminase) operon of *E. coli* and the *ami* (amidase) gene of *P. aeruginosa*. The inducer specificity is not usually exactly the same as the substrate specificity. In many instances an enzyme may be induced by the product of the reaction rather than the substrate (Clarke, 1979).

EVOLUTION OF NEW METABOLIC ACTIVITIES

Regulation and evolution

The normal regulatory controls can be altered by mutations. A biosynthetic pathway that is normally under repression control then becomes *derepressed*. This may provide a temporary advantage to the organism under particular cultural conditions. For example amino-acid analogues are frequently lethal agents for *E. coli* since they mimic the action of the amino acid by repressing enzyme synthesis. A derepressed mutant may be obtained by selecting for bacteria that can grow in the presence of amino-acid analogues. However, in other media the derepressed mutant may be at a growth disadvantage since it cannot reduce the level of the biosynthetic enzymes in response to an increase in the exogenous or endogenous levels of the amino acid. Similar considerations apply to the loss of response to feedback inhibition. These factors are most important in organisms like *E. coli* in which significant changes may occur in enzyme levels and activities in response to environmental changes. Catabolic enzymes that are inducible in wild type strains are produced in the absence of inducer by *constitutive* mutants. Constitutive mutants may therefore be able to utilise a compound that is a substrate but not an enzyme inducer. However, the synthesis of catabolic enzymes when they are not required may result in a lower growth rate.

The catabolism of natural and unnatural carbohydrates

Some sugars and polyhydric alcohols are commonly found in nature while others are found rarely, if at all. An examination of the catabolism of natural and unnatural pentoses and pentitols was started by Mortlock and colleagues who found that organisms that could grow on the natural isomers might adapt to grow on compounds not usually encountered in the natural environment. This topic was the subject of an excellent review by Mortlock (1976) and only a few examples will be described in this chapter. A consistent finding was that while enzymes were able to accept unnatural isomers as substrates these compounds were not usually inducers. Frequently the first step towards acquiring a novel catabolic activity was a mutation in a regulator gene to *constitutive* synthesis of an enzyme that was normally *inducible* (Lin, Hacking & Aguilar, 1976).

Klebsiella aerogenes utilises L-fucose for growth and the initial reaction is carried out by L-fucose isomerase which converts it to L-fuculose.

Mutants of *K. aerogenes* that were able to utilise D-arabinose (an unnatural pentose) were found to have become constitutive for L-fucose isomerase which converts D-arabinose to D-ribulose. In this case the further metabolism of D-ribulose was assured since it is itself the inducer of D-ribulokinase and D-ribulokinase-5-phosphate 3-epimerase which convert it to D-xylulose-5-phosphate. Ribitol and D-ribulose are normal metabolites of *K. aerogenes*. A regulatory mutation leading to the *constitutive* synthesis of an enzyme that is normally produced by this organism in response to a different substrate is sufficient to attain this new catabolic activity (Fig 4.3).

E. coli strain K12 can also give rise to mutants that are able to utilise D-arabinose. This strain does not utilise ribitol and D-ribulose and the regulatory mutation giving the D-arabinose-utilising phenotype was slightly different. The mutation results in the induction of three of the enzymes of L-fucose catabolism by D-arabinose. These enzymes have some activity towards the intermediates produced from D-arabinose and enable this sugar to be utilised for growth.

Of the four pentitols, ribitol and D-arabitol are metabolised by many species of bacteria but few are able to attack xylitol or L-arabitol. *Klebsiella* (*Aerobacter*) *aerogenes* utilises ribitol by the pathway shown in Fig. 4.3. D-Ribulose is the product-inducer of ribitol dehydrogenase and also induces the two subsequent enzymes of the pathway. A xylitol-utilising mutant was found to be constitutive for ribitol dehydrogenase for which xylitol is also a substrate. This strain grew very slowly on xylitol as sole carbon source but faster growth was achieved by a mutation in the structural gene giving increased affinity for xylitol and by a further mutation giving an increased rate of xylitol uptake (Wu, Lin & Tanaka, 1968). This system exhibits several facets of the process of evolving new catabolic activities. The specificity of metabolic enzymes is seldom absolute and this flexibility allows them to accept substrates other than those for which their activity is optimal. The specific regulatory systems place severe constraints on metabolic capacities but, if the regulatory controls can be altered by mutation, the residual activities of enzymes towards poor substrates can be brought into play. In the case of ribitol dehydrogenase the activity towards xylitol at physiological concentrations is very low and even with the high levels of enzyme produced by the constitutive strain the activity of this enzyme was rate-limiting for growth. Growth in continuous culture provides a very powerful method for selecting mutants with increased ability to utilise poor growth substrates. Hartley and colleagues used the *K. aerogenes* ribitol

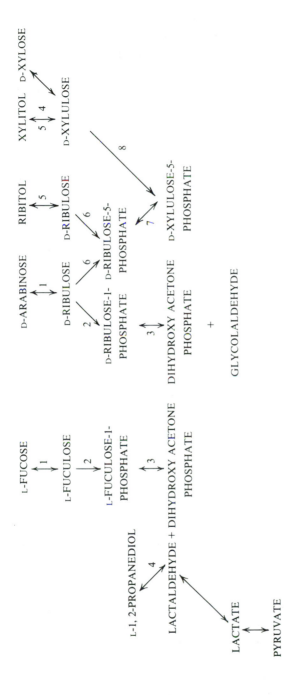

Fig. 4.3. Pathways of pentose and pentitol catabolism in *K. aerogenes* and *E. coli*. A mutant of *K. aerogenes* utilises D-arabinose via 1) L-fucose isomerase, 6) D-ribulokinase, 7) D-ribulose-5-phosphate-3-epimerase. A mutant of *E. coli* K12 utilises D-arabinose via the L-fucose pathway enzymes, 1) isomerase, 2) L-fuculokinase, 3) L-fuculose-1-phosphate aldolase. A mutant of *E. coli* K12 utilises L-1, 2-propanediol via enzyme 4) propanediol-lactaldehyde oxidoreductase, produced constitutively. Another mutant oxidises xylitol via enzymes 4), propanediol-lactaldehyde oxidoreductase, and 8) D-xylulokinase. A mutant of *K. aerogenes* utilises xylitol via enzyme 5), ribitol dehydrogenase, and 8), D-xylulokinase.

dehydrogenase constitutive mutant of Wu *et al.*, (1968) to select strains
with faster growth rates on xylitol in continuous culture. In one series of
experiments they isolated mutants that produced very high levels of
enzyme but showed no improvement in the specific activity of the
enzyme. The mutants in the chemostat had achieved increased enzyme
activities by producing more of a poor enzyme rather than the same
amount of a fitter enzyme (Hartley, 1974). This appeared to have been
due to duplication of the ribitol dehydrogenase gene. Later, other ribitol
dehydrogenase mutants were isolated which produced enzymes with
higher affinities for xylitol. These had one or two mutations in the
structural gene for ribitol dehydrogenase making it easier for xylitol to be
accepted at the active site of the enzyme (Hartley, Altosaar, Dothie &
Neuberger, 1976).

   Reiner (1975) found that although *E. coli* K12 was unable to grow on
ribitol other strains of *E. coli*, including strain C, were able to do so. The
difference between the two strains could have been due to the loss of
ribitol-utilising ability by strain K12, or the recent acquisition of this
ability by strain C. Reiner (1975) found that the genes for ribitol
catabolism mapped on the chromosome of strain C between genes *metG*
and *his*. Using transduction, Rigby, Gething & Hartley (1976) transferred the ribitol-utilisation genes from *K. aerogenes* to *E. coli* K12 and
found that they were inserted into its chromosome at a site near the *his*
genes. This suggests that there is some homology between the ribitol
operon of *K. aerogenes* and the *his* region of the *E. coli* K12 chromosome
and that the inability of *E. coli* K12 to grow on ribitol is due to loss of that
character during evolution of this strain.

A new version of a well-known enzyme

A surprising finding was that of Campbell, Lengyel & Langridge (1973),
who observed that a strain of *E. coli* K12 in which the *lac* operon had
been deleted had regained the ability to ferment lactose. Whereas the *lac*
genes map at 8 min on the chromosome the genes for the new enzyme *ebg*
(evolved β-galactosidase) mapped at 58 min. This system has been
studied in detail by Hall who has shown that the wild type is induced by
lactose to make an enzyme which has weak β-galactosidase activity. For a
strain to acquire the capacity to utilise lactose for growth it is essential to
have two mutations; one in the regulatory gene *ebgR* to give either a
constitutive phenotype, or a higher rate of induction by lactose, and a
second mutation in the *ebgA* gene to produce an enzyme with a higher
specific activity (Hall & Hartl, 1974) (Fig. 4.4).

$$\text{Lactose} + H_2O \xrightarrow[\text{ebgA ebgR}]{\text{lacIZYA}} \text{Glucose} + \text{Galactose}$$

Fig. 4.4. Genes for β-galactosidases of *Escherichia coli* K12. The wild type strain hydrolyses lactose via the β-galactosidase coded by *lacZ* and the activity of the *ebgA*^{0} enzyme is insignificant. Mutations in the *ebg* genes can result in lactose utilisation by strains in which the *lacZ* gene has been deleted (Hall & Hartl, 1974).

The 'improved' enzymes with higher specific activities for lactose are the result of single site mutations in the *ebgA* gene. When *E. coli* K12 carries a normal *lac* operon the *ebg* system is quite irrelevant as far as lactose utilisation is concerned. In the experiments that have been described the *lac* operon has been deliberately removed by genetic manipulation and this has given freedom to this hitherto quiescent system to become activated and to take over a role as the essential enzyme for growth on lactose. The *ebg* enzyme is also capable of taking on other metabolic roles. New *ebg* enzymes have been evolved that are able to hydrolyse methyl-β-D-galactoside, lactulose and lactobionate (Hall, 1978). The so-called silent regions of bacterial chromosomes may carry other such genes. These genes may be remnants of past metabolic activities but may have been retained in an almost inert form to provide a reservoir of genetic material to be drawn on if required.

Evolution of a family of enzymes

Pseudomonas aeruginosa grows on the 2- and 3-carbon amides, acetamide and propionamide, which are hydrolysed by an inducible aliphatic amidase to ammonia and the corresponding acids (Fig. 4.5). The amidase inducer specificity differs from the substrate specificity and the interaction of these two factors imposes severe limitations on the amides that can be utilised for growth. For example, although the wild type enzyme can hydrolyse butyramide the activity is very weak and the affinity for this

$$CH_3CONH_2 + H_2O \xrightarrow[\text{amiR amiE}]{} CH_3COOH + NH_3$$

Fig. 4.5. Hydrolysis of acetamide by the inducible amidase of *Pseudomonas aeruginosa*. Mutations in the amidase genes *amiR* and *amiE* produce strains that are able to utilise other amides including butyramide, valeramide and phenylacetamide.

amide is low (K_m about 500 mM). Growth of the wild type on butyramide is ruled out even at high concentrations since butyramide is not only unable to act as an inducer but competes with inducing amides and represses amidase synthesis. Other amides exhibit different properties; for example, lactamide is a good inducer but a poor substrate while formamide is a weak inducer and a poor substrate. By exploiting the properties of various amides as substrates, inducers or amide analogue repressors it has been possible to evolve a family of mutants with novel amide growth phenotypes (Clarke, 1974). The enzyme is regulated by positive control by the *amiR* gene product and is also subject to catabolite repression by other carbon compounds such as succinate. The mutants with novel amide growth phenotypes may carry mutations in the amidase structural gene, *amiE*, the amidase regulator gene, *amiR*, or in both (Fig. 4.6).

Growth on the 4-carbon amide butyramide can be attained by a mutation in the regulator gene (*amiR*) that results in constitutive amidase synthesis. This is an example of a new growth phenotype being achieved by producing a high level of an enzyme with poor activity for the novel substrate (Strain CB4 in Fig. 4.6). An alternative solution is a mutation in the amidase structural gene (*amiE*) to give an enzyme with a higher affinity for butyramide. This is the case for mutant B6 which was derived from the constitutive strain C11 which was unable to grow on butyramide because it retained sensitivity to repression by butyramide.

Successive mutations in the amidase structural gene gave rise to mutants able to utilise valeramide (5-carbon), hexanoamide (6-carbon) and phenylacetamide. This series of mutants showed that it was possible to introduce progressive changes in the substrate specificity of an enzyme by introducing single site mutations in a series of steps. This mechanism for changes in enzyme specificity is of obvious advantage for bacterial evolution. Mutants PhV1 and PhV2 each had three mutations in the *amiE* gene but the amidases were not identical in physicochemical properties or in affinities for their amide substrates. Other phenyl-acetamide-utilising mutants were derived from different parents and carried one or two mutations in *amiE*. It was clear that there was no unique route for the evolution of a phenylacetamidase from the acetamidase of *P. aeruginosa*. The wild type strain utilises a very limited range of amides as growth substrates but the system is flexible enough for it to acquire a variety of new amide growth phenotypes as a result of a very small number of mutations in the amidase genes.

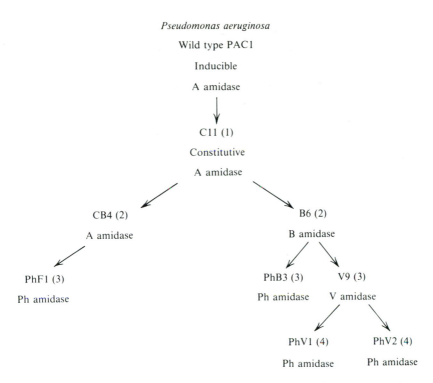

Fig. 4.6. Family of amidase mutants derived from *Pseudomonas aeruginosa* PAC1. The numbers of mutations are given in brackets. C11, *amiR11 amiE⁺*; CB4, *amiR11, 37 amiE⁺*; B6, *amiR11 amiE16*; PhB3, *amiR11 amiE16, 67*; V9, *amiR11 amiE16, 30*; PhV1, *amiR11 amiE16, 30, 78*; PhV2, *amiR11 amiE16, 30, 79*. All strains carrying *amiR11* are constitutive.

Figure 4.6 shows the family tree of some of the amidase mutants and indicates the numbers of mutations that have occurred at each stage.

EVOLUTION OF NEW METABOLIC PATHWAYS

Vertical and horizontal evolution

We have seen that mutations in the structural genes of enzymes can allow them to take on a new role. This may allow an organism to utilise a novel growth substrate if suitable enzymes are available for the further metabolism of the intermediates produced. If the organism retains a copy of the original gene as well as the mutated gene its metabolic potential will have been increased. The importance of gene duplication in enzyme evolution had been suggested by the similarity of sequence and structure of enzymes for related functions such as the NADH-dependent dehydrogenases (see also Hartley, 1974). In such a family of enzymes it is possible to envisage an archetypal enzyme of broad specificity coded by a gene which became duplicated several times and by divergent mutation gave rise to a group of enzymes with high specificities and high rates of reaction. This is horizontal evolution.

If we look at the enzyme of a catabolic pathway it is also possible to envisage evolution by duplication of genes of consecutive steps of the pathway. This vertical evolution is essentially similar to the stepwise retrograde evolution suggested by Horowitz (1945, 1965) for the amino-acid biosynthetic pathways. There are however difficulties in envisaging the evolution of catabolic pathways in this way. Catabolic pathways provide energy at a limited number of steps of the pathway and the first reactions of the pathway frequently require energy input. Only the entire pathway is of value to the organism. If on the other hand, it is envisaged that evolution proceeded from the last steps of the pathway then other difficulties arise. Many intermediates are unstable and are not readily taken up by bacterial cells.

In the case of vertical evolution we would expect some resemblance between two successive enzymes of a catabolic pathway even though the catalytic reactions were different. Jeffcoat & Dagley (1973) pointed out that one of the objections to the difficulty of the different chemical transformations for successive steps of a catabolic pathway could be overcome if the catalytic mechanisms had a similar basis. They showed that in the catabolism of D-glucarate by *E. coli* and *K. aerogenes* the action of a hydrolase is followed by an aldolase and that both these reactions require similar electron shifts. They are arguing for a common origin for a hydrolase and an aldolase and were able to show that the purified D-glucarate hydrolase of *K. aerogenes* had weak aldolase activity. These results suggest that at certain steps of a catabolic pathway the duplication of a gene could lead to an enzyme with a new catalytic role.

An alternative to vertical evolution is horizontal evolution. It could be imagined that an entire pathway could arise by the duplication of a pre-existing pathway. This would require the duplication of each of the genes for enzymes responsible for the successive steps of the pathway. The duplication of the gene for a hydrolase would give rise to a new hydrolase and the duplication of a gene for a dehydrogenase would give a new dehydrogenase. We have already seen that mutation can give rise to an enzyme with new substrate specificity. Divergence in substrate specificity is also in accord with comparative studies on enzyme proteins.

Patchwork pathways

An example of a novel catabolic pathway formed from enzymes of pre-existing pathways was given earlier in discussing the adaptation of *K. aerogenes* to growth on D-arabinose. The first step of the pathway in the D-arabinose-utilising mutant is carried out by L-fucose isomerase (L-fucose pathway) and the next two steps by D-ribulokinase and D-ribulose-5-phosphate 3-epimerase (ribitol pathway). In this example a new catabolic pathway has been assembled, without duplication and mutation of enzyme structural genes, by altering the regulation of L-fucose isomerase to make it constitutive. An independent and inducible catabolic pathway for D-arabinose would be a possibility if duplication and mutational divergence occurred in the structural and regulator genes for L-fucose isomerase. Constitutive mutants usually have a growth disadvantage with respect to inducible strains and the gain of a new metabolic activity by mutation to a constitutive or derepressed state might be an intermediate stage in evolving a novel inducible pathway.

The L-fucose catabolic enzymes have provided other novel pathways. An end-product of L-fucose catabolism in *E. coli* under anaerobic conditions is L-1, 2-propanediol (Fig. 4.3). The final step of this pathway is the reduction of lactaldehyde produced from L-fuculose-1-phosphate by an aldolase reaction. A mutant that produced lactaldehyde dehydrogenase (oxidoreductase) constitutively was able to utilise propanediol as a substrate for aerobic growth. Other propanediol-utilising mutants with faster growth rates were also altered in the regulation of the earlier enzymes of the L-fucose pathway. This family of mutants provides another example of a novel pathway using pre-existing enzymes (Lin *et al.*, 1976).

The propanediol oxidoreductase of *E. coli* can also oxidise xylitol to D-xylose. Xylitol can be taken into the bacteria by the uptake system for D-xylose but does not induce its formation. A xylitol-utilising mutant of *E. coli* obtained by Wu (1976) was constitutive for xylitol uptake (D-xylose transport) and constitutive for oxidation to D-xylose (propanediol oxidoreductase; L-fucose pathway). The further metabolism of xylitol was carried out by the D-xylose pathway enzymes induced by D-xylose (Fig. 4.3). In this example the employment of an enzyme that had been first thought of as a terminal dehydrogenase for lactaldehyde in a fermentative pathway is being utilised for the first step in the catabolism of xylitol under aerobic conditions. These and other examples of the use of pre-existing enzymes for new functions illustrate the way in which novel catabolic pathways may be assembled by a succession of single mutations in unrelated pathways (see Clarke, 1978).

Plasmids and novel catabolic pathways

The transmissible catabolic plasmids enable genes for catabolic enzymes to be spread rapidly through bacterial populations. If the bacteria can support more than one type of catabolic plasmid the genetic information for several different catabolic pathways can be assembled together in a single organism. Chakrabarty (1976) has developed this approach with plasmids carrying genes for the catabolism of various petroleum-related compounds. The genes for naphthalene, salicylate, camphor and octane catabolism are carried on plasmids found in strains of *Pseudomonas putida*. Transfer of NAH, SAL and the CAM–OCT hybrid plasmid to *P. aeruginosa* produced a strain that utilised crude oil more effectively than strains carrying these plasmids singly.

The genes carried by the plasmid and the genes carried on the chromosome may code for two different pathways for the catabolism of the same organic compounds. This is known to be the case for certain enzymes concerned with the metabolism of certain aromatic compounds. For naphthalene and toluene the NAH and TOL plasmids code for enzymes that convert the metabolites to catechols which are metabolised further by the *meta* cleavage enzymes, also coded by plasmid genes. Catechols may be produced from compounds such as benzoate by enzymes coded by chromosomal genes and other chromosomal enzymes may break down catechol by the *ortho* cleavage pathway. The possibility of alternative convergent and divergent pathways specified by plasmid and chromosomal genes gives a very great flexibility in the evolution of new catabolic pathways. An interesting development in strain constructions was the transfer of the TOL plasmid from *P. putida* to a *Pseudomonas* isolate to obtain a strain that catabolised 4-chlorocatechol by the initial activity of enzymes coded by the TOL plasmid and completed the breakdown by enzymes coded by genes of the recipient strain (Reinecke & Knackmuss, 1979).

Evolution in mixed populations of bacteria

In nature bacteria are found in pure culture only in extreme environmental conditions such as high temperature or high salinity. In most situations bacterial strains occur in competition, and in association, with a mixture of other microbial species. In a study on the isolation of microorganisms able to degrade the herbicide Dalapon (2,2'-dichloropropionate) Senior, Bull & Slater (1976) obtained a stable mixed population containing three primary utilisers and four secondary utilisers. The population remained stable in chemostat culture for some time but later a new primary utiliser was isolated that appeared to have been derived from one of the strains in the original microbial association.

The evolution of bacterial enzymes in nature

Biological recycling of organic compounds produced by plants and animals has always depended on the evolution of new catabolic activities by microorganisms but, during the last fifty years, the demands on the evolutionary capacity of the microbial world have increased in a dramatic way. Many new chemicals have been introduced as pesticides or herbicides or casually released as by-products of industrial processes. What is remarkable is not that there are a few recalcitrant substances, such as DDT and the chlorinated biphenyls, but that so many xenobiotic compounds are totally degraded by soil and water microorganisms. We have indicated the importance of the interaction of different species in microbial communities. It is also necessary to remember that organic compounds are unlikely to be present in isolation so that some synthetic chemicals may be co-metabolised with related biologically produced compounds.

We can be certain that new metabolic activities are evolving in bacterial populations today. At one time the process of microbial enzyme evolution was considered in terms of enzyme mutation and selection, with gene duplication accounting for the expansion of the gene pool. We can now see that mutations in regulatory systems are as important as mutations in the structural genes for the enzymes themselves. The catabolic and drug-resistance plasmids play a very important role in microbial evolution and with the discovery of the IS elements and transposons we can begin to account for the rapidity with which microbial evolution takes place.

146 *P. H. Clarke*

REFERENCES

Avery, O. T., MacLeod, C. M. & McCarty, M. (1944). Studies on the chemical nature of the substance inducing transformation of pneumococcal types. Induction of transformation by a desoxyribonucleic acid fraction isolated from Pneumococcus Type III. *Journal of Experimental Medicine*, **79**, 137–58.

Bachmann, B. J., Low, K. B. & Taylor, A. L. (1976). Recalibrated linkage map of *Escherichia coli* K-12. *Bacteriological Reviews*, **40**, 116–67.

Bennett, P. M. & Richmond, M. H. (1976). Translocation of a discrete piece of deoxyribonucleic acid carrying an *amp* gene between replicons in *Escherichia coli*. *Journal of Bacteriology*, **126**, 1–6.

Bennett, P. M. & Richmond, M. H. (1978). Plasmids and their possible influence on evolution. In *The Bacteria*, vol. VI, ed. L. N. Ornston & J. R. Sokatch, pp. 1–69. New York: Academic Press.

Beringer, J. E., Hoggan, S. A. & Johnston, A. W. B. (1978). Linkage mapping in *Rhizobium leguminosarum* by means of R plasmid-mediated recombination. *Journal of General Microbiology*, **104**, 201–7.

Broda, E. (1975). In *The Evolution of the Bioenergetic Processes*. Oxford: Pergamon.

Bukhari, A. J., Shapiro, J. A. & Adhya, S. L. (eds.) (1977). *DNA, Insertion Elements, Plasmids and Episomes*. Cold Spring Harbor Laboratory.

Campbell, J. H. Lengyel, J. A. & Langridge, J. (1973). Evolution of a second gene for β-galactosidase in *Escherichia coli*. *Proceedings of the National Academy of Sciences USA*, **70**, 1841–5.

Chakrabarty, A. M. (1972). Genetic basis of the biodegradation of salicylate in *Pseudomonas*. Journal of Bacteriology, **112**, 815–23.

Chakrabarty, A. M. (1976). Plasmids in *Pseudomonas*. *Annual Review of Genetics*, **10**, 7–30.

Chakrabarty, A. M., Chou, G. & Gunsalus, I. C. (1973). Genetic regulation of octane dissimilation plasmid in *Pseudomonas*. *Proceedings of the National Academy of Sciences USA*, **70**, 1137–40.

Clarke, P. H. (1974). The evolution of enzymes for the utilization of novel substrates. In *Evolution in the Microbial World. Symposia of the Society for General Microbiology*, **24**, 183–217, ed. M. J. Carlile & J. J. Skehel. London: Cambridge University Press.

Clarke, P. H. (1978). Experiments in Microbial evolution. In *The Bacteria*, vol. VI, ed. L. N. Ornston & J. R. Sokatch, pp. 137–218. New York: Academic Press.

Clarke, P. H. (1979). Regulation of enzyme synthesis in the bacteria: a comparative and evolutionary study. In *Biological Regulation and Development*, ed. R. F. Goldberger, vol. 1, pp. 109–70. New York: Plenum.

Cohen, S. N. (1976). Transposable genetic elements and plasmid evolution. *Nature*, **263**, 731–8.

Cornelis, G., Ghosal, D. & Saedler, H. (1978). Tn 951: A new transposon carrying a lactose operon. *Molecular and General Genetics*, **160**, 215–24.

Crawford, I. P. (1975). Gene rearrangements in the evolution of the tryptophan synthetic pathway. *Bacteriological Reviews*, **39**, 87–120.

Dubos, R. (1945). In *The Bacterial Cell*, p. 28. Cambridge, Mass: Harvard Press.

Dunn, N. W. & Gunsalus, I. C. (1973). Transmissible plasmid coding early enzymes of naphthalene oxidation in *Pseudomonas putida*. *Journal of Bacteriology*, **114**, 974–9.

Fredericq, P. (1957). Colicins. *Annual Review of Microbiology*, **11**, 7–22.

Haas, D. & Holloway, B. W. (1978). Chromosome mobilization by the R plasmid R68.45: a tool in *Pseudomonas* genetics. *Molecular and General Genetics*, **158**, 229–37.

Hall, B. G. (1978). Experimental evolution of a new enzymatic function. II, Evolution of multiple functions for *ebg* enzyme in *E. coli*. *Genetics*, **89**, 453–65.

Hall, B. G. & Hartl, D. L. (1974). Regulation of newly evolved enzymes. I, Selection of a novel lactase regulated by lactose in *Escherichia coli*. *Genetics*, **76**, 391–400.

Hartley, B. S. (1974). Enzyme families. In *Evolution in the Microbial World*, *Symposia of the Society for General Microbiology*, **19**, 151–82, ed. M. J. Carlile & J. J. Skehel. London: Cambridge University Press.

Hartley, B. S., Altosaar, I., Dothie, J. M. & Neuberger, M. S. (1976). Experimental evolution of a xylitol dehydrogenase. In *Structure – Function Relationships of Proteins*, ed. R. Markham & R. W. Horne, pp. 191–200. Amsterdam: North Holland.

Hartman, J., Reinecke, W. & Knackmuss, H. J. (1979). Metabolism of 3-chloro, 4-chloro and 3,5-dichlorobenzoate by a pseudomonad. *Applied and Environmental Microbiology*, **37**, 421–8.

Hedges, R. W. & Jacob, A. E. (1974). Transposition of ampicillin resistance from RP4 to other replicons. *Molecular and General Genetics*, **132**, 31–40.

Hopwood, D. A., Chater, K. F., Dowding, J. E. & Vivian, A. (1973). Advances in *Streptomyces coelicolor* genetics. *Bacteriological Reviews*, **37**, 371–405.

Horowitz, N. H. (1945). On the evolution of biochemical synthesis. *Proceedings of the National Academy of Sciences, USA*, **31**, 153–7.

Horowitz, N. H. (1965). The evolution of biochemical synthesis – retrospect and prospect. In *Evolving genes and proteins*, ed. V. Bryson & H. J. Vogel, New York: Academic Press.

Jacoby, G. A., Rogers, J. E., Jacob, A. E. & Hedges, R. W. (1978). Transposition of *Pseudomonas* toluene-degrading genes and expression in *Escherichia coli*. *Nature*, **274**, 179–80.

Jeffcoat, R. & Dagley, S. (1973). Bacterial hydrolases and aldolases in evolution. *Nature, New Biology*, **241**, 186–7.

Jensen, R. A. (1976). Enzyme recruitment in evolution of new function. *Annual Review of Microbiology*, **30**, 409–25.

Konisky, J. (1978). The bacteriocins. In *The Bacteria*, ed. J. R. Sokatch & L. N. Ornston, vol VI, pp. 71–136. New York: Academic Press.

Lin, E. C. C., Hacking, A. J. & Aguilar, J. (1976). Experimental models of acquisitive evolution. *Bio Science*, **26**, 548–55.

Morishita, T., Fukada, T., Shirota, M. & Yura, T. (1974). Genetic basis of nutritional requirements in *Lactobacillus casei*. *Journal of Bacteriology*, **120**, 1078–84.

148   *P. H. Clarke*

Mortlock, R. P. (1976). Catabolism of unnatural carbohydrates by micro-organisms. *Advances in Microbial Physiology*, **13**, 1–53.

Olsen, R. H. & Shipley, P. (1973). Host range and properties of the *Pseudomonas aeruginosa* R factor R1822. *Journal of Bacteriology*, **113**, 771–80.

Reeve, E. C. R. & Braithwaite, J. A. (1974). The lactose system in *Klebsiella aerogenes* V9A. 4. A comparison of the *lac* operons of *Klebsiella* and *Escherichia coli*. *Genetical Research, Cambridge*, **24**, 323–31.

Reinecke, W. & Knackmuss, H. J. (1979). Construction of haloaromatics utilising bacteria. *Nature*, **277**, 385–6.

Reiner, A. M. (1975). Genes for ribitol and D-arabitol catabolism in *Escherichia coli*: their loci in C strains and absence in K12 and B strains. *Journal of Bacteriology*, **123**, 530–6.

Rheinwald, J. G., Chakrabarty, A. M. & Gunsalus, I. C. (1973). Transmissible plasmid controlling camphor oxidation in *Pseudomonas putida*. *Proceedings National Academy of Sciences, USA*, **70**, 885–7.

Rigby, P. W. J., Gething, M. J. & Hartley, B. S. (1976). Construction of intergeneric hybrids using bacteriophage P1 CM: Transfer of the *Klebsiella aerogenes* ribitol dehydrogenase gene to *Escherichia coli*. *Journal of Bacteriology*, **125**, 728–38.

Schopf, J. W. & Barghoorn, E. S. (1967). Alga-like fossils from the early precambrian of South Africa. *Science*, **156**, 508–12.

Senior, E., Bull, A. T. & Slater, J. H. (1976). Enzyme evolution in a microbial community growing on the herbicide Dalapon. *Nature*, **263**, 476–9.

Shapiro, J. A. (1977). DNA insertion elements and the evolution of chromosome primary structure. *Trends in Biochemical Research*, **2**, 176–80.

Siegel, B. Z. & Siegel, S. M. (1970). Biology of the precambrian genes *Kakabekia*: New observations on living *Kakabekia barghoorniana*. *Proceedings of the National Academy of Sciences, USA*, **67**, 1005–10.

Stanier, R. Y., Adelberg, E. A. & Ingraham, J. L. (1977). In *General Microbiology*, 4th edn. London: Macmillan.

Thacker, R., Rørvig, O., Kahlon, P. & Gunsalus, I. C. (1978). NIC, a conjugative nicotine–nicotinate degradative plasmid in *Pseudomonas convexa*. *Journal of Bacteriology*, **135**, 289–90.

Towner, K. J. (1978). Chromosome mapping in *Acinetobacter calcoaceticus*. *Journal of General Microbiology*, **104**, 175–80.

Watson, J. M. & Holloway, B. W. (1978). Chromosome mapping in *Pseudomonas aeruginosa* PAT. *Journal of Bacteriology*, **133**, 1113–325.

Williams, P. A. (1978). The biology of plasmids. In *Companion to Microbiology*, ed. A. T. Bull & P. M. Meadow, pp. 77–108. London: Longman.

Williams, P. A. & Murray, K. (1974). Metabolism of benzoate and the methyl benzoates by *Pseudomonas putida* (*arvilla*) mt-2: Evidence for the existence of a TOL plasmid. *Journal of Bacteriology*, **120**, 416–23.

Woese, C. R. & Magrum, L. J. & Fox, G. E. (1978). Archaebacteria. *Journal of Molecular Evolution*, **11**, 245–52.

Wohlhieter, J. A., Lazere, J. R., Snellings, N. J., Johnson, E. M., Synenki, R. M. & Baron, L. S. (1975). Characterization of transmissible genetic elements from sucrose-fermenting *Salmonella* strains. *Journal of Bacteriology*, **122**, 401–6.

Wu, T. T. (1976). Growth of a mutant of *Escherichia coli* K–12 on xylitol by recruiting enzymes for D-xylose and L-1, 2-propanediol metabolism. *Biochemica et Biophysica Acta*, **428**, 656–63.

Wu, T. T., Lin, E. C. C. & Tanaka, S. (1968). Mutants of Aerobacter aerogenes capable of utilizing xylitol as a novel carbon source. *Journal of Bacteriology*, **96**, 447–56.

Young, F. E. & Wilson, G. A. (1975). Chromosomal map of *Bacillus subtilis*. In *Spores*, ed. P. Gerhardt, R. N. Costilov & H. L. Sadoff, pp. 596–614. Washington: American Society of Microbiologists.

5

The photosynthetic apparatus

K. K. RAO, D. O. HALL & R. CAMMACK

PLANT SCIENCES DEPT, KING'S COLLEGE,
68 HALF MOON LANE, LONDON SE 24 9JF, UK

Photosynthesis is the process by which the sun's energy is converted to chemical energy. The ultimate source of all metabolic energy and most of the 'technological energy' on our planet is the sun, and photosynthesis is essential for the maintenance and development of all forms of life on the earth. Though, recently, there have been attempts to use inorganic materials for solar energy conversion e.g., carbon dioxide and nitrogen fixation, this review will devote attention only to photosynthesis carried out by living organisms. In this context photosynthesis may be defined as the process by which biological organelles capture photon energy from the sun and utilise this energy for the synthesis of ATP and reductants such as $NADH_2$ or $NADPH_2$. The ATP and reductant are subsequently used for the fixation of carbon dioxide (and also nitrogen and sulphur) into biological components essential for the growth and metabolism of the organism. The essential components of the photosynthetic apparatus are (1) a pigment (or pigments) to harvest the light, (2) a reaction centre where the light energy is transformed into chemical energy (as redox compounds) and (3) a membranous system wherein ATP and reductant are synthesised as the result of electron (and proton) migration through an electric potential gradient. The evolutionary synthesis of the photosynthetic apparatus was a major event in the course of biological evolution and we will attempt to bring together current theories, some of them controversial, regarding this event in this brief review.

ORIGIN AND EVOLUTION OF LIFE ON EARTH

Geological studies on ancient rocks using radioactive isotope dating techniques suggest that the earth probably originated 4.6×10^9 years ago (see Broda, 1975, for a discussion). The oldest microfossils found in the Onverwacht series are about 3.4×10^9 years old. During this period of chemical evolution abiogenic synthesis of organic compounds and the appearance of primitive forms of life, which used these simple organic molecules as carbon and energy sources, would have taken place. It is generally accepted that the earth's early atmosphere was nonoxygenic and reducing (see Berkner & Marshall, 1965). The prebiotic atmosphere would have consisted mainly of hydrogen and nitrogen with lesser amounts of ammonia, hydrogen sulphide, methane, hydrogen cyanide, water etc. Electric discharges, ultraviolet and cosmic radiation and volcanic heat were some of the forms of energy available for the synthesis of simple organic acids, sugars and amino acids from the gaseous constituents of the atmosphere. Whether ancient forms of life originated in volcanic areas or in primeval ocean 'soups' these primitive organisms would have used simple organic compounds as carbon and energy sources and the earliest mode of energy metabolism would have been anaerobic fermentations and ATP production by substrate level phosphorylation. This type of energy metabolism is found in a wide range of extant bacteria such as the *Clostridium* species which are obligate anaerobes capable of fermenting sugars and organic acids and fixing atmospheric nitrogen. The anaerobic fermenters, probably, were the only forms of life prevailing on the earth one billion years after its formation.

The second stage in the development of living organisms would be the evolution of photosynthetic bacteria, probably in an aquatic environment, which could directly use solar radiation as an energy source and metabolise both organic and inorganic compounds. The photosynthetic bacteria, thus, need not compete with the anaerobic fermenters for the depleting supply of organic molecules. The development of photosynthesis is associated with the formation of chlorophyll, the light-capturing molecule. Porphyrins which form the basis of chlorophyll pigments can be synthesised abiotically. By subjecting a mixture of methane, ammonia and water vapour to high frequency electric discharges Simionescu, Mora & Simionescu (1978) were able to obtain porphyrins along with amino acids, purines and sugars. These authors also achieved synthesis of porphyrins by thermal condensation of formaldehyde and pyrroles in the presence of mineral salts. It seems likely that tetrapyrrole compounds were formed at quite an early stage of chemical evolution. Green and purple sulphur bacteria can grow photosynthetically using sulphides and carbon dioxide. Sulphides must have been abundant in the prebiotic atmosphere as well as in aquatic environments. These photosynthetic bacteria oxidise sulphides to sulphates. Probably soon after this the anaerobic sulphate reducers represented by *Desulfovibrio* species would have evolved. These sulphate reducers could metabolise sulphate back to sulphide thus producing ecosystems where the two types exist together feeding on each others' wastes. Schidlowski (1979) has obtained isotopic evidence for dissimilatory sulphate reduction in the upper Archean rocks of the Aldan Shield, Siberia ($\sim 3 \times 10^9$ yr old) and in the Woman River banded iron formations of Canada (2.75×10^9 yr). He narrows the possible time for the appearance of sulphate respirers to the interval of $2.8–3.1 \times 10^9$ yr and suggests that photosynthesis is older than sulphate respiration. The origin of methanogens, a unique group of bacteria, would have occurred when an anaerobic atmosphere rich in carbon dioxide and hydrogen developed on earth (Fox *et al.*, 1977). These bacteria may have coevolved with the photosynthetic bacteria.

Once the photosynthetic sulphur bacteria had become established, the next stage in evolution would have been the origin of the blue-green algae or cyanobacteria. The important feature of photosynthesis by the cyanobacteria is that they can use the energy-poor and abundantly available substrate, water, as electron donor and evolve oxygen as a by-product. In order to accomplish this the cyanobacteria developed an additional pigment system (called photosystem II) which catalysed the splitting of water and evolution of oxygen. This involved the use of a second quantum of light energy, but since light and water were universally available the cyanobacteria had a great evolutionary advantage over the photosynthetic bacteria. The origin of cyanobacteria was a giant step in evolution since it marked the appearance and subsequent gradual accumulation of oxygen in the biosphere and thus the development of aerobic life. Some of the present-day cyanobacterial species are quite similar to those found in micro fossils 3.1×10^9 yr old (Schopf, 1974) and it is also possible that these algae may have developed extensively only 2×10^9 yr ago. The notion that blue-green algae evolved from photosynthetic bacteria by the incorporation of an additional photosystem is given important support by the recent discovery by Cohen, Padan & Shilo (1975) of the cyanobacterium *Oscillatoria limnetica*, which is capable of a facultative bacterial-type anoxygenic carbon dioxide assimilation (using sulphide as the source of electrons) and plant-type oxygenic photosynthesis (using water).

Simultaneous with the development of blue-green algae from the purple bacteria there occurred the development of a new group of photosynthetic bacteria, the purple non-sulphur bacteria. These purple non-sulphur bacteria can grow anaerobically in light, assimilating carbon dioxide, but they can also grow aerobically using organic molecules as substrates. It is possible that these non-sulphur phototrophs are the forerunners of present-day aerobic bacteria and nitrate respirers (see John & Whatley, 1977), and also possibly of mitochondria. The higher plants and red and green algae probably evolved from the blue-green algae. A scheme depicting the possible evolutionary development of organisms is shown in Fig. 5.1.

PROKARYOTES AND EUKARYOTES – THE ENDOSYMBIOTIC THEORY

On the basis of the internal architecture of the cell, its metabolism and reproduction, living organisms can be classified into two types, the eukaryotes and the prokaryotes. The eukaryotes are characterised by the presence of a nucleoplasm bounded by a membrane, of a nucleolus and of more than one chromosome. Eukaryotic cells reproduce by mitosis. Their cells contain organelles such as chloroplasts, mitochondria, endoplasmic reticulum, Golgi apparatus, lysosomes and microtubular systems. All the above characteristics are absent in the prokaryote. Most prokaryotes have a cell wall built up of peptidoglycan. The prokaryotic ribosomes are of the 70S type whereas the eukaryotes possess 80S-type ribosomes in the cytoplasm and 70S ribosomes in certain organelles, such as mitochondria and/or chloroplasts. All bacteria and blue-green algae have prokaryotic cell structure: hence the suggestion that blue-green algae should be named cyanobacteria (see Stanier, Adelberg & Ingraham, 1977). All other forms of organisms are eukaryotes e.g. green algae, higher plants, fungi, metazoa.

The marked differences in organisation and function between eukaryotic and prokaryotic cells have raised important questions regarding the origin of the two types and their interrelationships. It is generally accepted that the primordial organisms were anaerobic, fermentative, heterotrophic prokaryotes. According to Margulis (1970) selection pressures in the early Pre-Cambrian era led to an extensive adaptive divergence among the prokaryotes, mainly at the metabolic level. This led to the evolution of cell types like the coccoid blue-green algae and Gram-negative eubacteria that oxidised small organic acids via the Krebs' cycle. The first step in the origin of eukaryotes occurred when a fermentative anaerobe was invaded by Krebs'-cycle-containing eubac·teria (protomitochondria); the stabilisation of this association led to the formation of amoeboids, which contained mitochondria and from which all other eukaryotes descend. The nuclear and all other endoplasmic

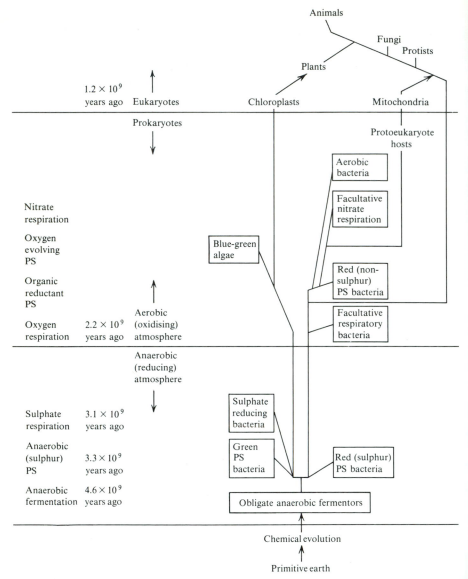

Fig. 5.1.   Simplified evolutionary tree adapted from protein sequence and other
data (see Hall, 1979 for details).

membranes evolved auto-genously after the steroid biosynthetic pathway
was made available by the presence of the protomitochondria. Margulis
termed this association between eubacteria and prokaryotes as *endosymbiosis*. Meanwhile ingestion, without digestion, of blue-green algae

by various eukaryotic heterotrophs under nutrient-poor conditions, led to the establishment of stable heritable populations containing the oxygen-evolving photosynthetic apparatus. The prokaryotic algae eventually became the obligatory symbiotic photosynthetic plastids in the origin of various lines of nucleated algae e.g. red and green algae and eventually all chlorophytes and higher plants. Thus the theory of serial endosymbiosis proposes that mitochondria and chloroplasts are evolved from fermentative prokaryotic bacteria and blue-green algae respectively. The endosymbiotic theory is not free from discrepancies, especially regarding the origin of mitochondria, and a number of different hypotheses have been put forward to explain the origin of the eukaryotic cell (Allsopp, 1969; Raff & Mahler, 1972; Taylor, 1974; Uzzell & Spolsky, 1974). We will only discuss this hypothesis with reference to the photosynthetic organelles (see Ragan & Chapman, 1978, for details).

CLASSIFICATION OF PHOTOSYNTHETIC ORGANISMS

The overall reaction of photosynthesis is the same in all photosynthesisers, an oxidation–reduction sequence initiated by light reactions, in which carbon dioxide (low energy) is reduced to carbohydrates (high energy) at the expense of hydrogen donors. It can be represented by the general equation proposed by van Niel (1941):

$$2H_2A + CO_2 \rightarrow (CH_2O) + 2A + H_2O$$

Depending on the nature of H_2A, photosynthetic organisms are divided into two broad groups.

(1) The photosynthetic bacteria, which are obligate anaerobes when grown in the light using H_2A a substrate more reducing than water, e.g. sulphide, thiosulphate, organic molecules such as acetate, malate etc. There are three major classes of photosynthetic bacteria: *a*, the green or brown bacteria (represented by the group Chlorobiaceae); *b*, the purple sulphur bacteria (Chromataceae) and *c*, the purple non-sulphur bacteria (Rhodospirillaceae).

(2) Algae and plants which can grow aerobically in light and in which H_2A can be water with the result that oxygen is evolved as a by product. Two light reactions are involved in oxygenic photosynthesis.

The major characteristics of photosynthetic organisms are given in Table 5.1.

Table 5.1. *Characteristics of photosynthetic organisms*

Group	Electron donor	Photosynthetic pigments	Growth conditions and other properties
Green photo-synthetic bacteria (Chlorobiaceae)	H_2S $Na_2S_2O_3$ H_2	Bchl *a*, Bchl *c*, Bchl *d* or Bchl *e* (brown bacteria) carotenoids	Light, autotrophic. Strict anaerobes, non motile (except *Chloroflexus*) e.g.: *Chlorobium limicola*
Purple sulphur bacteria (Chromataceae)	H_2S $Na_2S_2O_3$ H_2 organic substrates	Bchl *a* or Bchl *b* carotenoids	Light, autotrophic or dark, heterotrophic strict anaerobes, some are motile, e.g.: *Chromatium*
Purple non-sulphur bacteria (Rhodospirillaceae)	H_2 organic substrates	Bchl *a* or Bchl *b* carotenoids	Light, autotrophic or heterotrophic. Dark, aerobic respirer. e.g.: *Rhodospirillum rubrum*
Blue-green algae (Cyanobacteria)	H_2O	Chlorophyll *a* Phycobilins	Light, autotrophic, some species will grow heterotrophically in dark. e.g.: *Anacystis nidulans*
Eukaryotic algae	H_2O	Chlorophyll *a*, chlorophyll *b* or chlorophyll *c*, Phycobilins (red algae)	Light, autotrophic, some will grow heterotrophically in dark. e.g.: *Chlorella*
Higher plants	H_2O	Chlorophyll *a* and chlorophyll *b*	Light, autotrophic

PHOTOSYNTHETIC BACTERIA

The ability of purple bacteria to grow and move in light was first observed by Englemann about 1880 and he suggested that these bacteria were photosynthetic organisms. Since then many photosynthetic bacteria have been discovered and characterised. Their natural habitat is aquatic. They are strict anaerobes when growing photosynthetically; as already pointed out they do not evolve oxygen by photosynthesis. The mechanism of light absorption, energy transfer and electron transport are similar in all photosynthetic bacteria. All of them possess bacteriochlorophylls (Bchl; the structure of chlorophyll *a* is shown in Fig. 5.2 and the differences between chlorophyll *a* and other forms of chlorophyll are given in Table 5.2) and carotenoids which harvest light and then transfer this excitation energy to a reaction centre containing a Bchl *a*–protein complex. The Bchl *a*–protein complex in turn transfers the energy to a few specialised

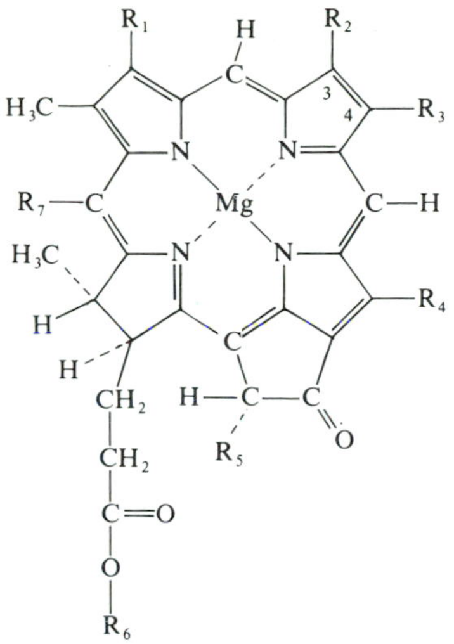

R_1 H R_2

H_3C—

R_7—C Mg C—H

H_3C

H

H

CH_2 H—C—C

CH_2 R_5 O

C=O

O

R_6

Fig. 5.2.   Structure of chlorophyll *a*. The nature of 'R' is given in Table 5.2.

Bchl molecules in the reaction centre, which are called P840, P870 etc. depending on the wavelength in nanometers of their maximum absorption. Charge separation occurs at these special Bchl molecules followed by electron transport resulting in the production of ATP, $NADH_2$ or reduced ferredoxin. The components of the electron-transport chain include quinones, cytochromes and iron-sulphur proteins.

The green bacteria

In all probability these are the most ancient group of photosynthetic organisms. They are strict anaerobes growing by the oxidation of sulphides or thiosulphates in light and fixing carbon dioxide. *Chlorobium limicola* and *Prosthecochloris aestuarii* are members of this group which have been studied in detail. The photosynthetic pigments of the green bacteria are located predominantly in special structures called the *chlorobium vesicles* (Cohen-Bazire, 1963) which underlie and are firmly attached to the cytoplasmic membrane. The major pigments of these bacteria, Bchl *c*, Bchl *d* or Bchl *e*, are light-harvesting molecules and are located in the chlorobium vesicles. The exact location of the other photosynthetic components such as reaction-centre chlorophyll, quinones, cytochromes, iron-sulphur proteins etc. is still uncertain; they may be located in the cytoplasmic membrane attached to the chlorobium

Table 5.2. *Chemical differences between chlorophyll* a *and other chlorophylls (Adapted from Stainer* et al. *1977)*

Chlorophyll	R_1	R_2	R_3	R_4	R_5	R_6	R_7
a	$-CH=CH_2$	$-CH_3$	$-C_2H_5$	$-CH_3$	$-C-OCH_3$ (C=O)	Phytyl	$-H$
b	$-CH=CH_2$	$-C=O$ ($\overset{\mid}{H}$)	$-C_2H_5$	$-CH_3$	$-C-OCH$ (C=O)	Phytyl	$-H$
Bchl a	$-\overset{\parallel}{\underset{O}{C}}-CH_3$	$-CH_3{}^a$	$-C_2H_5{}^a$	$-CH_3$	$-\overset{\parallel}{\underset{O}{C}}-OCH_3$	Phytyl or geranyl-geranyl	$-H$
Bchl b	$-\overset{\parallel}{\underset{O}{C}}-CH_3$	$-CH_3{}^b$	$=CH-CH_3{}^b$	$-CH_3$	$-\overset{\parallel}{\underset{O}{C}}-OCH_3$	Phytyl	$-H$
Bchl c	$-\overset{\overset{H}{\mid}}{\underset{OH}{C}}-CH_3$	$-CH_3$	$-C_2H_5$	$-C_2H_5$	$-H$	Farnesyl	$-CH_3$
Bchl d	$-\overset{\overset{H}{\mid}}{\underset{OH}{C}}-CH_3$	$-CH_3$	$-C_2H_5$	$-C_2H_5$	$-H$	Farnesyl	$-H$
Bchl e	$-\overset{\overset{H}{\mid}}{\underset{OH}{C}}-CH_3$	$-C=O$ ($\overset{\mid}{H}$)	$-C_2H_5$	$-C_2H_5$	$-H$	Farnesyl	$-CH_3$

[a] No double bond between C-3 and C-4; additional $-H$ atoms at C-3 and C-4.
[b] No double bond between C-3 and C-4; additional $-H$ atom at C-3.

vesicles. The Bchl *a*–protein complex isolated from *P. aestuarii* is a trimer of molecular weight approximately 140 000 (Olson, 1978*a*). The function of the Bchl *a*–protein complex *in vivo* is to accept excitation energy from the light-harvesting pigment and transfer this excitation energy to the reaction centre Bchl P840.

The intact and functional photosynthetic apparatus of green bacteria disintegrates very readily during isolation (Pierson & Castenholz, 1978) and so the exact mechanism of photosynthesis in green bacteria is still not clearly understood. The site of photophosphorylation is also still not known. Isolated membrane fractions show little or no ability to synthesise ATP. Cell-free particles prepared from *Chlorobium limicola* f. *thiosulfatophilum* are able to photoreduce ferredoxin using sodium sulphide, cysteine, glutathione or ascorbate–DPIP (Evans & Buchanan, 1965). The reduced ferredoxin in turn can photoreduce NAD (Buchanan & Evans,

1969). A probable scheme of photosynthetic electron transport in green
bacteria is shown in Fig. 5.3.

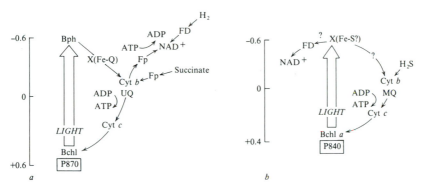

Fig. 5.3.   Photosynthetic electron transport in bacteria, adapted from Olson
(1978*b*). *a*, Purple bacteria (Bchl *a* or *b*; Bph *a* or *b*; P870, P890 or P960).
b, Green bacteria.

Carbon dioxide fixation in green bacteria.   The range of primary products
of $^{14}CO_2$ fixation observed in short exposure experiments in photosynthetic bacteria are variable. In many experiments performed with
Chlorobium, $^{14}CO_2$ is primarily incorporated into amino acids, particularly glutamate. Although low activities of the key enzymes of the
Calvin cycle (Fig. 5.12) were obtained in cell-free extracts of *Chlorobium*
strains the contribution of this cycle to total carbon dioxide assimilation in
green bacteria appears negligible. Evans, Buchanan & Arnon (1966)
proposed, on the basis of enzymatic data, a new type of carbon dioxide
assimilation cycle, dependent on photoreduced ferredoxin. This pathway, called the reductive tricarboxylic acid cycle (Fig. 5.4) is essentially a
reversal of the oxidative tricarboxylic cycle wherein the ferredoxin-
dependent enzymes, pyruvate synthase and α-ketoglutarate synthase
catalyse the reductive carboxylation of acetyl- and succinyl coenzyme A
respectively.

$$\text{Acetyl CoA} + CO_2 + Fd_{red} \xrightarrow[\text{synthase}]{\text{Pyruvate}} \text{Pyruvate} + CoA + Fd_{ox}$$

$$\text{Succinyl CoA} + CO_2 + Fd_{red} \xrightarrow[\text{synthase}]{\alpha\text{-ketoglutarate}} \alpha\text{-ketoglutarate} + CoA + Fd_{ox}$$

The overall cycle involves the fixation of four molecules of carbon
dioxide. However, no activity of citrate lyase, the key enzyme of this

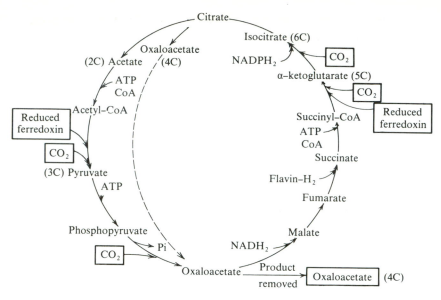

Fig. 5.4. The reductive carboxylic acid cycle for fixation of carbon dioxide in photosynthetic bacteria (Evans, Buchanan & Arnon, 1966).

cycle could be demonstrated in *Chlorobium*. The pathway of carbon dioxide fixation in the green bacteria still remains to be elucidated (see Fuller, 1978).

Chloroflexus aurantiacus (*filamentous, gliding green bacteria*).   Pierson & Castenholz (1971) first described filamentous phototrophic bacteria, found in hot springs, containing Bchl *a* and Bchl *c* in a ratio similar to *Chlorobium*. Further work on 16 isolates from hot springs in Japan, USA, Iceland and New Zealand (Pierson & Castenholz, 1974) showed that these thermophilic flexibacteria possess chlorobium vesicles and carry out an anoxygenic photosynthesis. This family has been named *Chloroflexus*. Although *Chloroflexus* may be capable of photo-autotrophic growth with sulphide as the electron donor, it is primarily a photo-organotrophic bacterium. The natural habitats of *Chloroflexus* are hot springs where they grow along with cyanobacteria. When *Ch. aurantiacus* is grown under fully aerobic conditions, no chlorobium vesicles are seen, and high oxygen tension completely suppresses synthesis of bacterio-chlorophylls (Pierson & Castenholz, 1974). With its regulatory capacity to grow either by anoxygenic photosynthesis or by respiration when oxygen is present *Chloroflexus* represents, among the Chlorobiaceae, a counterpart to the purple non-sulphur bacteria. Some of the general characteristics regarding the morphology and physiology of green bacteria had to be revised after the discovery and inclusion of *Chloroflexus* in the group.

The purple photosynthetic bacteria (see Fig. 5.5)

Most purple bacteria are flagellated and exhibit photo and chemotaxis. There are two types: *a*, the sulphur bacteria (Chromataceae) which can use sulphide as photosynthetic electron donor e.g. *Chromatium* and *Thiocapsa* and *b*, the non-sulphur bacteria (Rhodospirillaceae) which are unable to use inorganic electron donors and depend on organic substrates e.g. *Rhodospirillum* and *Rhodopseudomonas*. These bacteria are more versatile in their metabolic behaviour. The photosynthetic pigments of purple bacteria are located on intracytoplasmic unit membrane systems that are continuous with the cytoplasmic membrane and may fill almost the entire volume of the cell. A particulate fraction containing the photosynthetic pigments of *Rhodospirillum rubrum* was prepared by differential centrifugation of alumina-ground cells by Pardee, Schachman & Stanier (1952) and designated *chromatophores*. The chromatophores appeared in electron micrographs as collapsed disc-shaped structures with a diameter of about 100 nm. The electronic absorption bands of the chromatophores were identical to those of whole cells. Isolated chromatophores were shown by Frenkel (1956) to participate in many photochemical reactions including photophosphorylation. Isolated chromatophores thus represented the photosynthetic apparatus of purple bacteria. In the last 25 years there have been significant improvements in the isolation and purification of photosynthetic units from a variety of purple bacteria. Due to these improved techniques of isolation, coupled with the development of electron paramagnetic resonance spectroscopy and picosecond flash spectrophotometry, we know quite a lot regarding the composition of, and electron transport pathway in, the photosynthetic apparatus of the purple bacteria.

Photosynthetic electron transport in Rhodospirillaceae

The primary electron-transfer reaction in bacterial photosynthesis occurs within a membrane-bound complex of pigments and proteins called a 'reaction centre'. The photons absorbed by antenna bacteriochlorophylls and carotenoids are funnelled to the reaction centre where

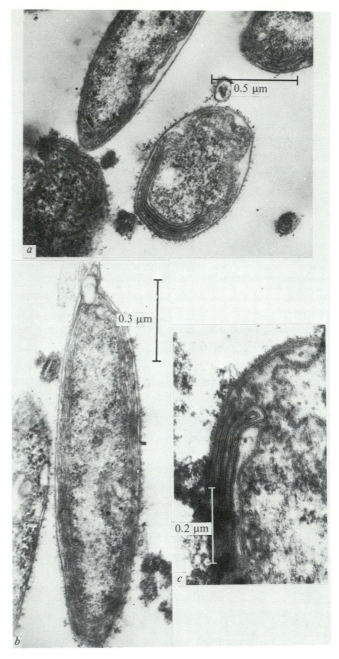

Fig. 5.5.   Electron micrographs of the purple non-sulphur bacterium, *Rhodo-microbium vannielli*. Note the looping of lamellar membrane. (Courtesy Dr C. Dow.)

photochemistry 'occurs'. The migration of energy to the reaction centre and its subsequent capture in an electron transfer reaction are very efficient; in many species at least 90% of the photons that are absorbed result in electron transfer (Parson & Cogdell, 1975). The reaction centre protein is located within the chromatophore membrane and can be isolated with detergents. Usually the reaction centre is a three subunit protein of total molecular weight 70 000. The protein contains four Bchl *a* (a few species of bacteria have Bchl *b* at the reaction centre which absorbs light at a different wavelength) and two bacteriophaeophytins (BPh; Mg-free Bchl), one iron atom and two quinones. Two of the Bchl *a*s absorb light of 800 nm and their role remains unclear. The other two Bchl *a*s absorb in the 865–885 nm region and together form a dimer or 'special pair' designated (Bchl)$_2$ (Dutton, Prince & Tiede, 1978). The (Bchl)$_2$ pair undergoes oxidation following light absorption and instantly reduces BPh. The electron is then transferred from reduced BPh to the 'primary acceptor' (X) which is one of the quinones associated with iron (QFe). From the primary acceptor the electrons are then transferred through cytochromes, iron-sulphur proteins and flavoproteins to NAD and ferredoxin. The cyclic electron transport pathway in purple bacteria is shown in Fig. 5.3. The redox potentials of the various components of the chain do not vary greatly from species to species except for the primary acceptor (Olson, 1978*b*). In the electron transfer from Bchl *a* to BPh *a* excitation energy is converted to chemical free energy which may be as high as 1.0 eV, but for the complete electron transfer from Bchl *a* to X the range is between 0.5 and 0.6 eV, a value not sufficient to oxidise water to oxygen. During the cyclic electron flow ADP is phosphorylated to ATP. In some bacteria phosphorylation produces pyrophosphate as well as ATP (Baltscheffsky, Baltscheffsky & Thore, 1971).

Carbon dioxide fixation in purple bacteria. The main pathway of CO_2 assimilation in purple bacteria is via the Calvin cycle (Fig. 5.12) though the reductive carboxylic cycle is also operative in these species.

BLUE-GREEN ALGAE (CYANOBACTERIA)

These are photosynthetic prokaryotes which have the capacity to split water. They have been recorded from the early Pre-Cambrian period (about 3×10^9 yr ago) and may be responsible for the initial production of oxygen in the atmosphere. They are of world-wide distribution, although they are seldom dominant except in areas such as thermal hot-springs which resemble a type of environment which may have occurred in Pre-Cambrian times (Stewart, 1978). Morphologically, cyanobacteria may be unicellular, colonial or filamentous. They are bounded by a cell wall composed of a protein and liposaccharide outer layer, and a peptidoglycan inner layer similar to that in prokaryotic bacteria. The photosynthetic apparatus of the cyanobacteria is located in flattened vesicles called *thylakoids* or photosynthetic lamellae which traverse most of the cytoplasm. Structurally the thylakoids appear tripartite, having a typical 'unit membrane' structure. The number and arrangement of the thylakoids vary according to the species and growth conditions of the algae. The photosynthetic pigments (chlorophyll *a*, phycobilins and carotenoids) and the electron-transfer components are located in these lamellae. The phycobiliproteins (phycocyanin, phycoerythrin and allophycocyanin) are water-soluble chromoproteins which function as accessory light-harvesting pigments. They are found associated with the photosynthetic apparatus of only two other groups, both eukaryotic: the *Rhodophyta* and *Cryptophyta*. (There is no allophycocyanin in *Cryptophyta*.) The mechanism of photosynthesis is similar in blue-green algae, eukaryotic algae and higher plants and will be described later.

Anoxygenic photosynthesis in cyanobacteria

Cohen *et al.* (1975) showed that the cyanobacterium, *Oscillatoria limnetica* isolated from the sulphide-rich layers of the Solar Lake, Israel, can carry out oxygenic photosynthesis as well as a facultative bacterial-type anoxygenic carbon-dioxide photo-assimilation with sulphide as electron donor. This was the first illustration of a prokaryotic bacterium performing bacterial- and plant-type photosynthesis. This anoxygenic reaction was later found in many other cyanobacteria (see Padan, 1979). In an aerobic atmosphere *O. limnetica* carries out PSII-mediated oxygenic photosynthesis inhibited by DCMU. However in the presence of high sulphide concentrations (3 mM) the bacterium assimilates carbon dioxide by PSI-mediated electron transport, insensitive to DCMU, with sulphide as the electron donor. If sulphide is removed, *O. limnetica* immediately shifts to oxygenic photosynthesis. Thus, the photosynthetic apparatus of *O. limnetica* operates facultatively, both oxygenically and anoxygenically. In addition to photo-assimilation of carbon dioxide, sulphide donates electrons for photosynthetic hydrogen evolution in *O. limnetica*. The pattern of sulphide oxidation of these cyanobacterial strains is different from photosynthetic bacteria; photosynthetic bacteria oxidise sulphide to sulphate gaining eight electrons for each hydrogen sulphide molecule, while *O. limnetica* gains only two electrons for each hydrogen sulphide. Also, other sulphur-containing electron donors which are used by photosynthetic bacteria do not seem to serve cyano-bacterial photosynthesis (Padan, 1979). *O. limnetica* which can readily shift between aerobic photo-autotrophic and anaerobic photo-autotrophic growth patterns may represent an intermediate physiology between that of eukaryotic algae and photosynthetic bacteria.

Evolutionary interrelationships between photosynthetic and cyanobacteria

The photosynthetic bacteria and cyanobacteria are morphologically similar (Table 5.3). The cell walls of both types contain muramic acid and diamino-pimelic acid or lysine (Echlin & Morris, 1965). The photosynthetic apparatus of both groups possesses light-harvesting pigments, reaction-centre chlorophylls and electron-transport factors such as quinones, cytochromes, ferredoxins (and other iron-sulphur proteins) and flavoproteins. Members of all species are able to metabolise hydrogen, i.e., they contain hydrogenase. Nitrogen fixation is carried out by purple photosynthetic bacteria and many species of cyanobacteria. The enzymes for nitrogen fixation, under both aerobic and anaerobic conditions, in many species of cyanobacteria, is associated with the formation of specialised cells called *heterocysts* which are apparently free from the accessory pigments, phycobiliproteins, which normally are present in the non-specialised cells (Fig. 5.6). The heterocysts provide an environment which protects the nitrogenase (which catalyses nitrogen fixation) from oxygen inactivation. The photosynthetic membranes of the heterocysts contain only Photosystem I (PSI) and operate a bacterial type photosynthesis (see Stewart, 1978). Calvin-cycle type carbon dioxide fixation is either absent or catalytically inactive in the heterocysts. All species of photosynthetic and cyanobacteria so far examined contain Fe and/or Mn superoxide dismutase (see Hall, 1977), the enzyme which can remove any toxic superoxide radical formed as a result of photochemical reactions. Thus the photosynthetic and cyanobacteria are structurally and functionally related.

The question is whether these different species of photosynthetic prokaryotes evolved separately from a common ancestral bacterium or whether they evolved sequentially. In attempting to answer this question let us examine some of the differences between the members. The green sulphur bacteria (except *Chloroflexus*) are non-motile and part of their

Table 5.3. *Some comparative properties of purple photosynthetic bacteria and blue-green algae. (From Hall* et al. *1975)*

Photosynthetic bacteria	Photosynthetic apparatus	Growth conditions	Characteristics
(1) *Rhodospirillum rubrum*	Vesicles	Anaerobic, light	No difference in photosynthetic apparatus
Rhodopseudomonas viridis	Lamellae	Anaerobic, dark	Bchl *a* Bchl *b* in *Rh. viridis*
(2) *R. rubrum*	Vesicles	Anaerobic, light	Respiratory system present in both but
		Aerobic, dark	terminal oxidase different; vesicles and bacteriochlorophyll only under anaerobic conditions
(3) *R. rubrum*	Vesicles	Anaerobic, light	8Fe ferredoxin + 4Fe ferredoxin
		Aerobic, dark	4Fe ferredoxin only
(4) *Rh. spheroides*	Vesicles	Anaerobic, light	Bacteriochlorophyll + cytochrome *o*
		Aerobic, dark	No bacteriochlorophyll or vesicles + cytochrome *a*
(5) *Rh. capsulata*		Anaerobic, dark	Bacteriochlorophyll, but fewer vesicles
		Anaerobic, light	
		Aerobic, dark	
(6) *Ectothiorhodospira mobilis*	Stacked lamellae	Anaerobic, light (high concn. of salt and temperature)	Phosphorylation coupling factors interchangeable

Blue-green algae

(1) Facultative photoheterotrophs			Grow in light with CO_2 or glucose as C source
Facultative chemoheterotrophs			Grow in light or dark – with glucose as C source
(2) Fatty-acid biosynthesis			Desaturating enzyme systems
			(a) Only bacterial type
			(b) Both bacterial and algal types
			(c) Only algal type

(3) Polyunsaturated fatty acid content

Species diversity
(a) Low content (similar to other prokaryotes)
(b) Intermediate
(c) High content (chloroplast-like)

(4) Glycolipid content (mono- and di-galactosyl diglyceride)

Species diversity
(a) Blue-green algae and chloroplasts contain both
(b) Green photosynthetic bacteria contain mono- form
(c) Red photosynthetic bacteria contain neither

(5) Dependence on NADP as electron carrier

Species diversity
(a) Pentose phosphate pathway for substrate oxidation
(b) Tricarboxylic acid cycle for biosynthesis
(c) NADPH oxidation yields ATP

(6) pH requirements

Species diversity
(a) Blue-green bacteria do not grow below pH 4–5
(b) Eukaryotic algae grow below pH 4 (below 56°C)

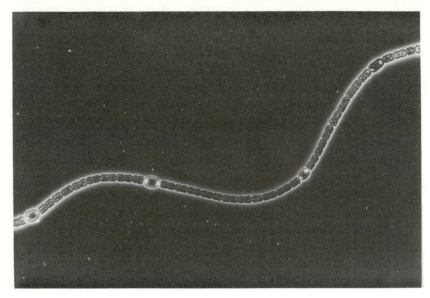

Fig. 5.6.   Electron micrograph of the blue-green alga, *Anabena* showing heterocysts (larger oval cells) and vegetative cells. Mag. × 700. (Courtesy I. J. Foulds.)

photosynthetic apparatus is located in non-membraneous chlorobium
vesicles. Many species of purple bacteria and blue-green algae are
flagellated and motile. The purple bacteria have a cellular organisation of
greater flexibility than the green bacteria; their photometabolism, like
that of the blue-green algae, is not confined to the cell periphery but
extends into the unit membranes found in the inner part of the cell. Both
green and purple sulphur bacteria are strict anaerobes; they do not
respire. They carry out only photosynthetic cyclic and non-cyclic (using
hydrogen sulphide) electron transport for producing ATP and $NADH_2$.
The Krebs' cycle is used only to synthesise glutamate and not ATP; they
do not employ the Krebs' cycle as a source of $NADH_2$ for respiration. On
the other hand, as shown in Table 5.3, the purple non-sulphur bacteria
are more versatile in their growth requirements and metabolism than the
green and purple sulphur bacteria. The purple non-sulphur bacteria can
grow in the light anaerobically and in the dark aerobically. They possess a
complete Krebs' cycle to generate $NADH_2$ when grown aerobically;
metabolic control is achieved by the fact that production of Bchl is
suppressed in the presence of oxygen. In many respects the purple
non-sulphur bacteria are more closely related to the cyanobacteria than
the rest of the group.

From the above discussion and from the fact that certain cyanobacteria such as *Oscillatoria* are able to carry out oxygenic photosynthesis with water as electron donor and anoxygenic (PSI-mediated) photosynthesis using sulphide as electron donor it seems logical to assume the evolutionary sequence:

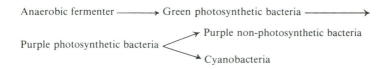

The closeness of anaerobic fermenters and green sulphur bacteria is borne out, as we shall see later, by a comparison of the ferredoxin sequences of *Clostridium pasteurianum* (anaerobic fermenter) and *Chlorobium* (green bacteria). It is noteworthy that the redox potential of the primary acceptor (X) of the green bacterial photosystem is low enough to reduce ferredoxin directly. However, there are differences of opinion. The discovery of the gliding, filamentous, thermophilic bacteria *Chloroflexus* has shown that there are representatives of the non-sulphur bacterial type among the green bacteria too! *Chloroflexus* is included among green bacteria because of the nature of its chlorophyll composition (Bchl *c* and Bchl *a*) and the presence of chlorobium vesicles. However, unlike other green bacteria, *Chloroflexus* can grow chemoorganotrophically by respiration. As in the case of the purple non-sulphur bacteria, the synthesis of Bchl is repressed in oxygen. It has the capacity for syntrophic growth e.g., in the presence of *Synechococcus lividus*, *Chloroflexus* grew well in inorganic media, deriving the needed organic nutrients from the cyanobacterium (see Pfennig, 1978). *Chloroflexus*, thus, could be an evolutionary forerunner of the blue-green algae.

A different scheme of evolution is proposed by Olson (1978*b*). According to Olson a common ancestor for aerobic non-photosynthetic bacteria, photosynthetic bacteria, blue-green algae and chloroplasts existed more than 3×10^9 yr ago below the surface of water. This ancestral prokaryote contained chlorophyll *a* in its photochemical reaction centre which drove a cyclic electron flow through cytochromes *c*, *b* and quinones. The entire apparatus was built into the cytoplasmic membrane and the electron flow was coupled to phosphorylation. Mutations enabled the development of non-cyclic electron chains which permitted the reduction of substances with redox potentials lower than that of the quinone pool, e.g. NAD and NADP. Photosynthetic carbon dioxide fixation could then be driven by ATP and NAD(P)H$_2$. Concurrently with the evolution of the electron transport system, there evolved light-harvesting chlorophyll proteins that fed excitation energy to the reaction centres. The shading of the underlying organisms in the available blue light made the mutational switch from chlorophyll *a* to Bchl *a*, this switch involved a loss of about 0.4 eV in the free energy stored in the primary photochemical conversion. The protoalgae that retained chlorophyll *a* were able to reduce ferredoxin and produce NADPH$_2$ (by photochemical reaction) using exogenous electron donors such as hydrazine and hydroxylamine. As the supply of exogenous electron donors for the reduction of ferredoxin diminished, modifications were made to some reaction centres in the linear system so as to form a stronger oxidant and a weaker reductant in the primary photochemical act – this was the evolution of Photosystem II (PSII). PSII was finally refined to enable the oxidation of water. The respiratory electron-transport chain in purple non-sulphur bacteria has evolved from the cyclic photosynthetic electron transport pathway of primitive purple bacteria.

Olson's proposed evolutionary scheme can be summarised as

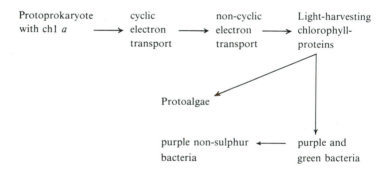

Biochemical evolution studies are still in their early stages and more comparative investigations of various 'biomolecules' are required before we can put forward, if at all, a definitive theory of prokaryotic evolution.

Krasnovsky (1976) has used inorganic components of the earth's crust and coenzymes and porphyrins of abiogenic and biogenic origin as photoreceptors of solar radiation. By the aid of inorganic photosensitisers (TiO$_2$, ZnO) he has constructed models of PSI and PSII and has put forward a hypothesis outlining plausible steps for the chemical evolution of photosynthesis.

PHOTOSYNTHESIS IN EUKARYOTES

Eukaryotes which perform photosynthesis include the four major groups of algae (red, brown and green algae and the diatoms) and the higher plants. [For a description of the classification of algae and the structure of their photosynthetic apparatus see Coombs & Greenwood (1976).] The ability of green plants to evolve oxygen in sunlight was discovered by Ingenhousz, a Dutch physician, about 200 years ago but most of our knowledge regarding the ultrastructure of the photosynthetic organelles and the mechanism of photosynthesis has only been acquired in the last few decades.

The chloroplasts

In green algae and in higher plants photosynthesis occurs within specific cytoplasmic organelles, called the chloroplasts. With the possible exception of C_4-plants the chloroplasts of all higher plants are similar in structure, composition and function. Pictures taken with the electron microscope show that plant chloroplasts are saucer-shaped bodies 4 to 10 nm in diameter and 1 nm in thickness with an outer envelope separating them from the rest of the cytoplasm (Figs. 5.7, 5.8). The outer envelope consists of two unit membranes separated by a space of about 10 nm. The chloroplast envelope represents a selectively permeable barrier which may regulate the movement in and out of the chloroplast of fixed carbon, reducing power, substrates and proteins synthesised in the cytoplasm.

Internally the chloroplast is composed of a system of lamellae or flattened *thylakoids* which are arranged in stacks in certain regions known as *grana* (Fig. 5.9). Each lamella in the chloroplast may contain two double-layer membranes. The grana are embedded in an amorphous, colourless matrix, the *stroma*. The stroma is largely composed of protein, the principal component being ribulose bisphosphate carboxylase (Fraction I protein), the enzyme which catalyses carbon dioxide fixation. Other enzymes of the reductive pentose phosphate pathway (Calvin cycle) are also located in the stroma. Within a chloroplast the grana are interconnected by a system of loosely arranged membranes called the stroma lamellae or *frets* (see Hall & Rao, 1977). The number of chloroplasts per cell, in higher plants, varies from one to more than a hundred depending upon the particular plant and growth conditions. The stacking of thylakoids into grana is not observed in many algal species; the types of photosynthetic organelles in eukaryotic algae are given in Table 5.4.

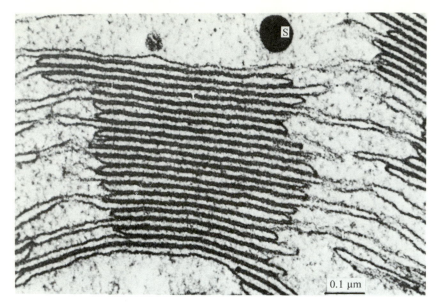

Fig. 5.7. A single granum within a chloroplast showing thylakoid stacks (t) and interconnecting stroma lamellae (l). S, lipid droplet in the stroma. (Courtesy A. D. Greenwood.)

By careful manipulation, it is possible to isolate intact chloroplasts from leaves and then to fractionate them into thylakoids (lamellae) and stroma. The lamellar membranes in which the chlorophyll pigments (chl *a* and chl *b*) are embedded are approximately half lipid and half protein in chemical composition. The proteins catalyse the enzyme and electron-transfer reactions and provide mechanical strength to the membranes while the presence of lipids facilitates energy migration and offers selective permeability to ions, sugars, substrates, etc. The photoreactions of light absorption and energy conversion and subsequent electron transport occur in the lamellae. The ATP and $NADPH_2$ produced by photochemical reactions are then utilised by the stroma enzymes to fix carbon dioxide.

Fractionation of chloroplast membrane components

As we will see later, photosynthesis in plants and algae involves the co-operation of two photochemical acts mediated by PSII and PSI

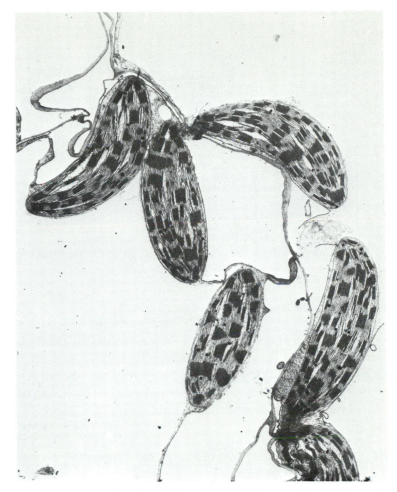

Fig. 5.8.   Maize chloroplasts. Mag. × 30 000. (Courtesy Dr G. Montes.)

present in the thylakoid membranes. Each photosystem contains a certain number of chlorophyll *a* and chlorophyll *b* molecules complexed with proteins. Polyacrylamide gel electrophoretic studies on detergent-treated maize thylakoids suggest that no significant amount of free chlorophyll exists in the chloroplast membranes *in vivo* (Markwell, Thornber & Boggs, 1979). The function of most of the chlorophyll *a* molecules and all the chlorophyll *b* molecules is to absorb light quanta (light-harvesting chlorophyll) and transfer the excitation energy to a small fraction of special chlorophyll *a* molecules, existing as dimers, at the photochemical reaction centre. In this context a photosynthetic unit is

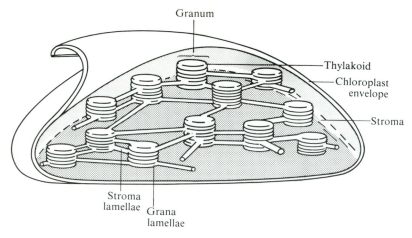

Fig. 5.9. Cut-away representation of chloroplast to show three dimensional structure (from Hall & Rao, 1977).

defined as the minimum number of chlorophyll molecules associated with PSI and PSII; usually this is about 400 chlorophylls per reaction centre.

By the disruption of thylakoid membranes with non-ionic detergents, sonication or a French pressure cell, and subsequent fractionation of the fragments, particles enriched in either PSI or PSII activity can be prepared (see Boardman, Anderson & Goodchild, 1978). However, unlike the case of the purple photosynthetic bacteria, PSI or PSII reaction centre

Table 5.4. *Organisation of thylakoids in the chloroplasts of different eukaryotic algal classes. (From Hall, 1973)*

	Usual number of thylakoids per band	Grana present	Degree of apposition of thylakoids	Interchange of thylakoids between bands
Rhodophyceae (reds)	1	No	–	–
Cryptophyceae (cryptomonads)	2	No	Loose	Common
Dinophyceae (dinoflagellate)	3	No	Variable	Uncommon
Crysophyceae (diatoms)	3	No	Variable	Common
Bacillariophyceae (diatoms)	3	No	Variable	Common
Xanthophyceae (yellow-green)	3	No	Variable	Common
Phaeophyceae (browns)	3	No	Loose	Common
Euglenophyceae (euglenoids)	3	No	Tight	Common
Prasinophyceae	2–4	Yes	Tight	Common
Chlorophyceae (greens)	2–6	Yes	Tight	Common

particles completely free of contaminants have not yet been obtained from plant chloroplasts. Chloroplast membranes may be solubilised by anionic detergents and then the components separated by poly-acrylamide gel electrophoresis. The availability of various types of detergents and the improvements in the technique of gel electrophoresis has enabled several investigators to separate and identify various components of the chloroplast membrane (see Arntzen, 1978; Machold, Simpson & Møller, 1979 & references therein). The membrane components thus resolved and identified include various iron-sulphur proteins, cytochrome *f*, reaction centre chlorophyll–protein complex, light-harvesting chlorophyll–protein complex, coupling factors (asso-ciated with ATP synthesising enzyme) etc. By comparing the poly-acrylamide gel electrophoresis pattern obtained from chloroplasts of selected mutants (defective in one or more photosynthetic reactions) with that of the corresponding wild-type it is possible to assign definite functions to various bands in the electrophoretogram (Simpson, Hoyer-Hansen, Chua & von Wettstein, 1977). This type of genetic approach combined with ultrastructural technique is becoming more and more useful in photosynthesis.

Photosynthetic electron transport in chloroplasts

The overall photosynthetic process in chloroplasts (and blue-green algae) is summarised by the equation:

$$2NADP + 2H_2O + 4ADP + 4Pi \xrightarrow[\text{membranes}]{\text{light}} 2NADPH_2 + 4ATP + O_2$$

(According to some investigators only two ATP molecules are synthesised per molecule of oxygen.)

A diagrammatic representation of the chloroplast electron transport chain is shown in Fig. 5.10. This pathway, generally known as the 'Z scheme' was first proposed by Hill & Bendall (1960). Hill & Bendall ordered the chloroplast electron-transport components (which had then been identified) on the basis of their oxidation–reduction potentials; the scheme has undergone some modifications after the discovery of addi-tional components. The idea that two light reactions are involved came from the work of Emerson, Chalmers & Cederstrand (1957) who showed that there was an enhancement effect in photosynthesis when two mono-chromatic light sources (red light, 680 nm and far-red light, up to 720 nm)

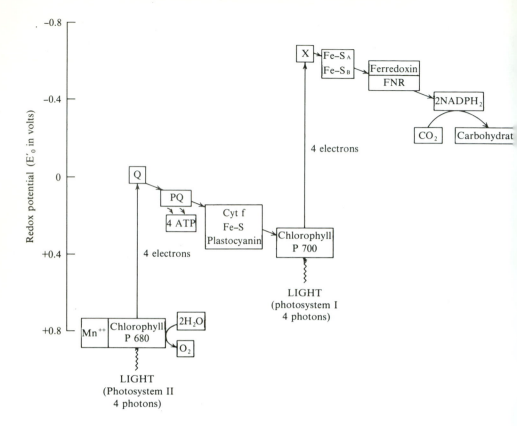

Fig. 5.10.　The Z scheme of photosynthesis (modified from Hill & Bendall, 1960). Fe–S, bound iron-sulphur protein; Q, quinone; PQ, plastoquinone; FNR, ferredoxin-NADP reductase.

of limiting light intensity were used in combination instead of red or far-red light alone.

Each photosystem in the Z-scheme has its own light-harvesting chlorophyll–protein complexes and reaction centre (chl a)$_2$ molecules; P680 at PSII and P700 at PSI. The overall potential change involved in the transfer of electrons from water to ferredoxin is about 1.2 eV. It is likely that within the photosynthetic membranes an even larger potential difference of about 1.6 eV exists between the primary oxidant of PSII ($E_m' \geq +0.9$ V) and the primary acceptor of PSI ($E_m' \leq -0.7$ V). Oxidation of water to oxygen occurs at PSII; the precise mechanism of this reaction is still not known though a manganese-containing complex, and chloride and bicarbonate ions are implicated. The initial photo-act at

PSII is the extraction of an electron from P680 to generate P680$^+$. The electron is then transferred within a millisecond to the primary electron acceptors of PSII (phaeophytin, Q_1, Q_2) and then to a plastoquinone pool. Plastoquinone, a lipophilic quinone with an isoprenoid side chain, acts as an electron buffer between the two light reactions and there is evidence that it translocates protons across the membrane to aid phosphorylation. A proton gradient, often as high as 2.5 to 3.5 pH units from the outside to the inside of the membrane, is built up, which is subsequently used for ATP synthesis via a chemiosmotic mechanism (Mitchell, 1966). The reduced plastoquinone passes its electron through an iron-sulphur protein, cytochrome f and plastocyanin to oxidised P700$^+$ chlorophyll dimer, the reaction centre trap of PSI. In some algae either reduced cytochrome f or plastocyanin can function as electron donor to P700$^+$.

In contrast to Q of PSII, the precise chemical nature of X, the primary acceptor of PSI is still not known, although it can be observed by low-temperature e.p.r. spectroscopy. It may be an iron-sulphur cluster with a redox potential of about -0.73 V. This primary reaction occurs in about 20 ns and then the electrons are passed within 100 ns to two membrane-bound iron-sulphur proteins with redox potentials of -0.59 and -0.55 V, then to ferredoxin (-0.42 V), to flavoprotein (-0.37 V) and finally to NADP to form NADPH$_2$ (see Bolton & Hall, 1979).

A different aspect of energy transduction in chloroplasts has been examined by Trebst (1974). Immunological studies using antibodies against specific chloroplast proteins have made it possible to localise these components on different sides of the thylakoid membrane. From these studies, together with the kinetic studies of Witt and his colleagues (see Witt, 1979) on the sequence of redox changes of chloroplast membrane components accompanying flash photolysis, Trebst has proposed a spatial model for photosynthetic electron transport. This model shows the reduction of NADP coupled to the pumping of protons from one side of the membrane to the other and indicates a strong sidedness of the membrane. A more recent version of this spatial model is shown in Fig. 5.11.

Energy conservation in chloroplast electron transport

The energy absorbed by the chloroplasts is stored and made available to the metabolic processes of the plant in the form of NADPH$_2$ and ATP. The synthesis of ATP as a result of light–induced electron flow from water

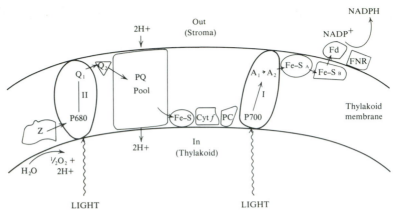

Fig. 5.11. Scheme showing probable arrangement of components in the thylakoid membrane. A_1, A_2, possible primary electron acceptors of PSI; Fd, ferredoxin; Fe–S, iron-sulphur protein; FNR, ferredoxin-NADP reductase; PC, plastocyanin; PQ, plastoquinone; Q, Quinone; Z, primary electron donor to PSII. A_2 may be the same as X in Fig. 5.10.

to $NADPH_2$ is known as *non-cyclic photophosphorylation*. Using specific inhibitors it is possible to deactivate PSII and then, by supplying artificial electron donors (such as redox dyes, cysteine etc), the PSII reaction can be bypassed. A ferredoxin-catalysed, PSI-mediated ATP synthesis known as *cyclic photophosphorylation* can also occur when the electrons cycle only around PSI with no net oxidation and reduction (see Gimmler, 1977). In many respects cyclic photophosphorylation by chloroplasts is similar to photophosphorylation in bacteria.

As in mitochondria, electron transport in chloroplasts is coupled to phosphorylation by a chemiosmotic mechanism (Mitchell, 1966) involving the generation of a membrane potential and proton gradient across the chloroplast membrane. Coupling factors and ATPase enzyme required for ATP synthesis are located in the membrane (see Evans, 1975; Hall, 1976; Jagendorf, 1977).

Carbon-dioxide fixation by chloroplasts: The Calvin cycle

The dark reactions of chloroplasts take place in the stroma. The pathway by which the $NADPH_2$ and ATP produced in the light reaction are used to convert carbon dioxide to carbohydrate was elucidated by Calvin and associates (see Bassham & Calvin, 1962). They exposed photosynthesising algae, *Chlorella* or *Scenedesmus*, to $^{14}CO_2$ for different lengths of time, separated the products by chromatography and identified them by autoradiography. For details of these fascinating experiments refer to Bassham (1962). The fixation of carbon dioxide to the level of sugar can be considered to occur in four distinct phases, as shown in Fig. 5.12.

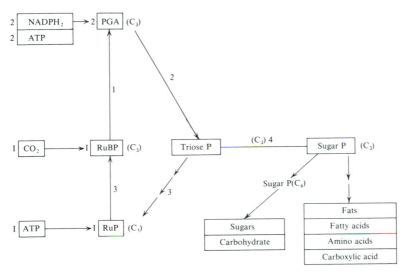

Fig. 5.12. Simplified scheme of carbon dioxide fixation by the Calvin cycle (from Hall & Rao, 1977). RuBP, ribulose bisphosphate; PGA, phosphoglyceric acid; RuP, ribulose-5-phosphate; Triose P, phosphoglyceraldehyde.

(1) Carboxylation Phase. Here carbon dioxide is added to the 5-carbon sugar, ribulose bisphosphate to form two molecules of phosphoglyceric acid. This reaction is catalysed by the enzyme ribulose bisphosphate carboxylase.

(2) Reduction phase. The phosphoglyceric acid formed is converted to a high energy 3-carbon sugar, triosephosphate, with the consumption of $NADPH_2$ and ATP (supplied by the membranes).

(3) Regeneration phase. The ribulose bisphosphate is regenerated for further carbon dioxide fixations by a complex series of reactions involving 3-, 4-, 5-, 6- and 7-carbon sugar phosphates (see Bassham, 1962; Hall & Rao, 1977).

(4) Product synthesis phase. The end-products of photosynthesis are mainly sugars but fats, fatty acids, amino acids and other organic acids have also been shown to be formed during carbon dioxide fixation. The enzymes for all these are found in the stroma. Since the Calvin cycle involves the utilisation and regeneration of a pentose phosphate

it is also called the pentose phosphate cycle. As the primary product of carbon dioxide fixation identified by radioactive labelling experiments is a 3-carbon sugar the reaction sequence is sometimes also referred to as the C_3 pathway. The net reaction can be represented by the equation

$$6CO_2 + 12NADPH_2 + 18ATP \rightarrow C_6H_{12}O_6$$
$$+ 6H_2O + 12NADP + 18ADP + 18Pi$$

The overall efficiency of the process in terms of the light quanta absorbed is about 30%.

The C_4 pathway of carbon dioxide fixation

Many tropical grasses and plants such as sugar-cane (Hatch & Slack, 1966) and maize are able to fix carbon dioxide into 4-carbon compounds e.g., malate and aspartate, in addition to their use of the Calvin fixation pathway. The leaves of these plants contain two distinct types of chloroplasts, those located in the mesophyll cells and those in the 'bundle sheath' cells of the leaf. Localisation studies of photosynthetic enzymes in C_4 plants indicate that enzymes for carbon dioxide fixation through the C_4 pathway are concentrated in the mesophyll cells while enzymes for carbon dioxide fixation through the C_3 cycle are abundant in the bundle sheath cells. The co-operative action of the two different cycles in the efficient utilisation of carbon dioxide is illustrated in Fig. 5.13. Generally plants possessing the C_4 pathway are more efficient in the utilisation of solar energy since the loss of carbon by photorespiration is very low in these species. Photorespiration is the light-stimulated release of carbon dioxide which had been previously fixed by leaves; in some temperate plant species photorespiration loss may be as high as 50% of gross photosynthesis. Since C_4 plants also have C_3 photosynthesis and since no angiosperms considered to be 'primitive' are C_4 plants, it appears that C_4 plants are both polyphyletic and recent in their origin (Laetsch, 1974).

Autonomy of chloroplasts

As early as 1885 Schimper proposed that chloroplasts do not arise *de novo* but are formed by the division of pre-existing plastids. Studies with the electron microscope have shown that, in vegetative reproducing cells of algae and plants, plastids multiply by division and are passed from parent cell to daughter cells. In detached leaves of the moss *Funaria* replication of nuclear DNA was shown to be followed by chloroplast

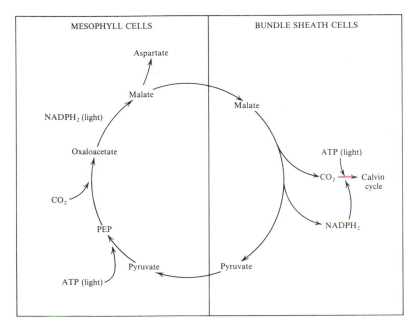

Fig. 5.13. Pathways of carbon dioxide fixation in C_4 plants (from Hall & Rao, 1977).

DNA replication. In addition to DNA and ribosomes, the chloroplasts possess tRNAs, amino acyl tRNA synthetases, initiation and elongation factors, mRNA etc. Chloroplast preparations capable of incorporating labelled amino acids into proteins, *in light*, have been isolated from algae and higher plants. All these factors point towards the genetic autonomy of chloroplasts. However, the continued growth and division of chloroplasts *in vitro* has not been established; this suggests that chloroplast development and function are dependent on nucleo-cytoplasmic activity.

The site of synthesis of a number of proteins participating in the light and dark reactions of photosynthesis has been elucidated recently. Two lines of approach are employed for this purpose.

(1) Antibiotics such as D-threo-chloramphenicol, lincomycin and streptomycin which inhibit protein synthesis by prokaryotic ribosomes (70S type) also inhibit protein synthesis by plastid ribosomes. Cycloheximide, on the other hand, inhibits protein synthesis in the 80S cytoplasmic ribosomes but has no effect on chloroplast ribosomes. By examining the effect of these inhibitors on the synthesis of chloroplast proteins during the development of etioplasts and chloroplasts it is possible to localise the site of synthesis of some of these proteins.

Alternatively, isolated chloroplasts can be supplied with specific inhibitors and the nature of protein synthesis followed. This technique has been applied by Ellis and Bradbeer and their associates to determine the site of synthesis of chloroplast and cytoplasmic proteins (see Ellis, 1976; Kirk & Tilney-Bassett, 1978).

(2) Many mutants are known in green algae (e.g. *Chlamydomonas*) and higher plants (e.g. tobacco and barley) which possess variegated chloroplast deficiencies; these have been termed *plastome* mutants. The classic work of Wildman and associates on tobacco mutants has shown that many of the characteristics of plastome mutants are maternally inherited from the chloroplast genes. By isolating selective chloroplast mutants which lack specific structure and/or function and comparing them with the wild type it is possible to define the role of individual genes in the biogenesis of the organelle (for details see Gillham, Boynton & Chua, 1978 and references therein).

The involvement of chloroplast-cytoplasm interactions in protein synthesis is exemplified in the synthesis and assembly of the key enzyme, ribulose bisphosphate carboxylase. This protein (identified in leaves by Wildman & Bonner in 1947) is an octamer of molecular weight 560 000 with eight large subunits each of molecular weight 55 000 and eight small subunits of molecular weight 12 000 each. By the use of mutants and inhibitor studies it is now established that the large subunit is coded by the chloroplast gene and the small subunit by a nuclear gene. These studies have revealed that most of the enzymes of the carbon dioxide fixation cycle and the electron transfer proteins, ferredoxin, ferredoxin-NADP reductase and plastocyanin are synthesised in the cytoplasm. In addition to the large subunit of ribulose bisphosphate carboxylase, cytochrome f and cytochrome b_{559} and b_{563} are synthesised in the chloroplasts. There are thirty or more polypeptides associated with the thylakoid membrane; we still do not know which of these polypeptides are coded by the nuclear genes and which by chloroplast genes. Also, we are still in the early stages of understanding the mechanism by which some of the cytoplasmically synthesised proteins are transported into and assembled in the chloroplasts.

Origin of chloroplasts

Earlier in this chapter we mentioned that the serial endosymbiosis theory of eukaryotic origin assumes that a primitive eubacteria took up blue-green algae to form the precursors of present-day chloroplasts. Let us examine how far the structure, composition and function of chloroplasts lend support to this hypothesis. We have already discussed the close similarities between the chloroplast and prokaryote ribosomes in terms of molecular mass and antibiotic sensitivity. The chloroplast DNA is different from the nuclear DNA in base composition, ease of renaturation and the lack of histones. Further, chloroplast DNA appears to be made up of closed circles of 40–45 nm in diameter; their average mass is about 1×10^8 daltons. Hybridisation studies have shown that rRNA and tRNA isolated from chloroplasts form hybrids only with chloroplast DNA. A protein synthesising system from *Escherichia coli* would make the large subunit of ribulose bisphosphate carboxylase when supplied with spinach chloroplast RNA but a wheat germ ribosome system would not, indicating that the prokaryotic 70S ribosome system only has the ability to translate chloroplast mRNA. The DNA-containing regions of chloroplasts are similar in appearance to nucleic acid-bearing regions of the blue-green algal cell. As in the prokaryotes, fatty acid synthesis in chloroplasts makes use of an acyl carrier protein.

Symbiotic associations between blue-green algae and eukaryotic hosts, ranging from simple fungi to the large angiosperm *Gunnera*, are widespread. The water fern, *Azolla*, harbours colonies of nitrogen fixing *Anabena*. The blue-green alga *Nostoc* forms symbionts with various fungi (lichens) and cycads (see Stewart, 1978). *Cyanophora paradoxa* is a small, unicellular, obligate photoautotrophic flagellate unable to degrade exogenous carbohydrates by respiration. Each cell of *Cyanophora* contains one to six endosymbiotic, blue-greenish structures called *cyanelles*. The cyanelles are similar in photosynthetic pigmentation and thylakoid structure to chloroplasts of red algae. The cyanelles reproduce in an autonomous way and the genome of the cyanelles resembles that of the chloroplasts rather than the blue-green alga. However, the wall structure characteristics of cyanelles show similarity with the prokaryotic algae e.g., the presence of cell-wall remnants as seen in the electron microscope and the sensitivity to lysozyme indicates the presence of peptidoglycan. *Cyanophora* can grow in hydrogen using a Knall gas-type reaction; it fixes carbon dioxide by the Calvin pathway. The cyanelles provide the eukaryotic cell with an organic carbon source and a supply of nitrogenous compounds. The cyanelles may be considered modern representatives of intermediates between chloroplasts of blue-green and red algae (Whatley, John & Whatley, 1979).

Moving on to the red algae (*Rhodophyta*), the photosynthetic apparatus is located in an extensive, but not contiguous, system of thylakoids which runs through the cytoplasm and is enclosed only by the cytoplasmic membrane. The interthylakoid space in red algae, e.g. *Porphyridium*, contains a regular array of granules each 30 to 40 nm in diameter which are attached in rows to the outer surface of each thylakoid. These aggregates, called *phycobilisomes*, contain the light-harvesting biliproteins. The amino-acid sequences of many biliproteins from blue-green and red algae are now known (MacColl & Berns, 1979). The data available show that the biliproteins from the two groups are quite similar and also suggest that the changes in biliproteins are more conservative than changes in cellular structure. From the above discussion it is reasonable to infer that at least the photosynthetic apparatus of the eukaryotic red algae and of *Cyanophora* originated from the blue-green algae.

The evidence to suggest that the chloroplasts of green algae and higher plants originated from blue-green algae is slender at present. Possibly the chloroplasts are polyphyletic in their origin. Chloroplasts of green algae and plants differ from those of the blue-green and red algae in two important respects, (a) in the presence of the lipid-soluble chlorophyll *b* as light harvesting pigment instead of water-soluble phycobiliproteins and (b) in the arrangement of thylakoids in stacks (grana). Is there a prokaryote with pigment composition similar to the chlorophyta? The answer may be 'yes'. Recently a group of prokaryotic algae which contains both chlorophyll *a* and chlorophyll *b* has been discovered (Lewin, 1976); these are called *Prochlorophyta*. The cell wall of *Prochloron*, a member of this group, resembles that of a prokaryote. The thylakoids of this alga, however, are present not as single lamellae, as in blue-green algae, but in pairs or sometimes, in thicker stacks. No prochlorophyte is known to occur as an endosymbiont; *Prochloron didemni* occurs as an extra-cellular symbiont of *Didemnum spp.*, or related ascidians. *Prochloron* contains chlorophyll *a* and chlorophyll *b* but lacks phycobilins. Possibly this could be the precursor of higher plant chloroplasts.

The organism *Cyanidium caldarium* is a special case of adaptation to thermophilic and acidophilic conditions and involves a phyletically unique eukaryote fitting into no known class of algae (Castenholz, 1979). *Cyanidium* is a chlorella-like unicell, dividing by cleavage after two mitotic divisions, but has a chloroplast with cyanobacterial pigments. Neither wall composition nor manner of division resemble any red alga. It seems to be a likely case of endosymbiosis which occurred after the evolution of *Chlorophyta*, may be within the last one billion years. Cyanobacteria do not tolerate high acidity but *Cyanidium* is an obligate acidophile growing at pH 2 to 3. Thus evolution and symbiosis seem to be necessary traits for survival.

PHOTOSYNTHESIS BY HALOBACTERIA

A unique type of photosynthetic process is exhibited by certain halobacteria typified by *Halobacterium halobium* (Stoeckenius *et al.*, 1979). These organisms live in extremely saline environments (3.5–5 M NaCl). They are aerobes and use the oxidation of amino acids as energy sources. When grown aerobically they synthesise lots of red carotenoids which are incorporated into the cell membrane and which protect the cell from photochemical damage. However, when grown at low oxygen tension *H. halobium* cells form patches of purple membrane on their surface which contain a single protein, *bacteriorhodopsin*. This protein is so called because, like rhodopsin of the eye it contains retinal which acts as a receptor of visible light. However, unlike rhodopsin, which dissociates when irradiated and only serves as a trigger for the visual process, bacteriorhodopsin can utilise light energy to generate a proton gradient across the cell membrane. This gradient can then be used to generate ATP, thus providing energy for a type of photosynthesis. Another important function is in the maintenance of other gradients of ions and amino acids across the membrane which are coupled with the hydrogen ion gradient. (Lanyi, 1978). *Halobacterium* maintains its osmotic balance by means of a high intracellular potassium ion concentration (about 3 M). The cell envelope is different from most other prokaryotes in having a wall consisting of single glycoprotein, and a membrane composed of isoprene derivatives. The intracellular metabolic pathways of *Halobacterium* also present unusual features. Although it does not carry out photosynthetic electron transport, *H. halobium* contains high concentrations of a soluble plant-type ferredoxin (see later) which can serve as an electron acceptor for α-keto-acid dehydrogenases (Kerscher & Oesterhelt, 1977). Some of the differences from other prokaryotes may partly reflect adaptations to high chloride concentrations; others may be due to the evolutionary history of the halophile. Just as oxygen deprivation derepresses synthesis of bacteriochlorophyll in purple non-sulphur bacteria, limiting concentration of oxygen derepresses purple membrane synthesis in the halobacteria.

PROTEIN STRUCTURE AND EVOLUTION

Many proteins are living 'fossils' in the sense that their structures have been conserved during the course of evolution spread over billions of years (see Schwartz & Dayhoff, 1978). Evolution is caused by favourable mutations, deletions, insertions and duplications in the genes, and the primary structures of proteins reflect these gene changes. When the sequences of homologous proteins from different animal species are compared with dates from their fossil record it is found that some proteins, such as fibrinopeptides, are evolving rapidly while others, such as cytochrome c, are evolving slowly. The rate of change of amino-acid sequences in a given protein appears to be fairly constant. Thus the relatedness between two species can be estimated by noting the time which has elapsed since they evolved from a common ancestor. As the construction of evolutionary trees from sequence studies forms chapter 3, we will not go into these details. For elucidating phylogenetic relationships from protein-sequence data the protein selected should be widely distributed and should have the same function in all species compared. It would be an advantage if the tertiary structure of the protein were known in addition to its primary sequence. The proteins which meet these requirements are ferredoxins, cytochrome c, plastocyanin and flavodoxin.

Ferredoxins

The ferredoxins are a group of iron-sulphur proteins containing non-haem iron and an equal number of sulphur ligands bonded to cysteinyl residues of the protein chain (see Hall, Rao & Cammack, 1975). They have been isolated and purified from diverse groups of bacteria, algae, plants and mammals. The active centres of ferredoxins consist of iron-sulphur clusters which participate in electron transfer during biological oxido-reductions. Spectroscopic and X-ray crystallographic studies have established two types of iron-sulphur clusters in ferredoxins, the [2Fe–2S] cluster found in plant and algal ferredoxins and the [4Fe–4S] cluster occurring in many bacterial ferredoxins (Fig. 5.14). Some ferredoxins may contain both types of cluster. Some of the properties of ferredoxins are listed in Table 5.5 and biological reactions catalysed by these proteins in Table 5.6. (See note added in proof.)

Since ferredoxins are relatively small molecules and easy to purify, the sequences of ferredoxins from a wide spectrum of organisms are known (see Yasunobu & Tanaka, 1973; Matsubara, Hase, Wakabayashi &

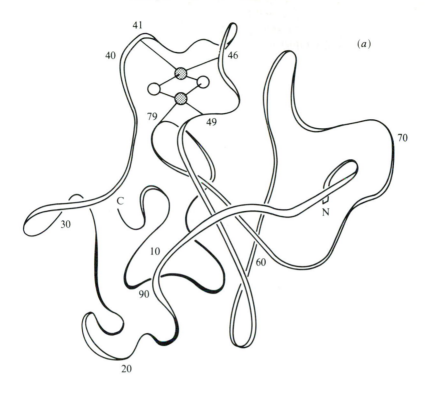

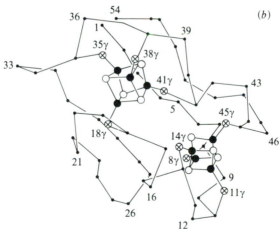

Fig. 5.14. Structure of ferredoxins from X-ray analysis. *a*, [2Fe–2S] ferredoxin from *Spirulina platensis* (from Tsukihara *et al.*, 1978). *b*, [4Fe–4S] ferredoxin from *Peptococcus aerogenes* (after Adman, Sieker & Jensen, 1973). Filled circles, iron; open circles, sulphur; crosses, Cys–S.

Table 5.5. *Physical properties of ferredoxins. (From Cammack, 1979.)* E_m *is the midpoint reduction potential, relative to the standard hydrogen electrode*

Protein	Typical source	Molecular weight $\times 10^{-3}$	Fe-S	E_m mV
2-Fe ferredoxin (Plant)	spinach	10.5	[2Fe-2S]	−420
2-Fe ferredoxin (blue-green algae)	*Spirulina platensis*	10.5	[2Fe-2S]	−390
2-Fe ferredoxin	*H. Halobium*	15	[2Fe-2S]	−345
2-Fe ferredoxin (monoxygenase)	Adrenal mitochondria	12	[2Fe-2S]	−270
2-Fe ferredoxin	*Ps. putida*	12.5	[2Fe-2S]	−240
2-Fe ferredoxin	*C. pasteurianum*	25	[2Fe-2S]	−300
Fe-S protein I	*Azotobacter vinelandii*	21	[2Fe-2S]	
Fe-S protein II	*A. vinelandii*	24	[2Fe-2S]	
HiPIP	*Chromatium vinosum*	9.5	[4Fe-4S]	+350
4-Fe ferredoxin	*Bacillus polymyxa*	8		
	Rhodospirillum rubrum	14.5	(4Fe-4S)	
6-Fe ferredoxin	*Thermus thermophilus*	9	[2Fe-2S] [4Fe-4S]	−530 −250
7-Fe ferredoxin I (Fe-S protein III)	*A. vinelandii*	14.5	7Fe, 7S	+340 −420
Ferredoxin IV	*R. rubrum*	14	8Fe, 8S	+355 −380
8-Fe ferredoxin	*C. pasteurianum* } *Peptococcus aerogenes* }	6	2[4Fe-4S]	−400
8-Fe ferredoxin	*Chr. vinosum*	10	2[4Fe-4S]	−490
Ferredoxin II (Fe-S protein IV)	*A. vinelandii*			−460
Ferredoxin I	*D. gigas*	18	3[4Fe-4S]	−455
Ferredoxin II	*D. gigas*	24	4[3Fe-3S]	−130
Fe-S protein	Mitochondrial outer membrane		[2Fe-2S]	+75

Wada, 1980). As early as 1966, when the first amino-acid sequence of ferredoxin from *Clostridium pasteurianum* became available, Eck & Dayoff proposed that this ferredoxin or its prototype may have participated in the metabolism of organisms at a very early stage of biochemical evolution. The arguments which support the hypothesis that a ferredoxin-type molecule was functioning in 'early cells' are as follows (see Rao & Hall, 1977). 1. Organisms which lived in this period would have been obligate anaerobic fermenters. Ferredoxins are essential catalysts in the present-day fermenters such as *Clostridia* where they catalyse electron transfer reactions at a potential near to that of the hydrogen electrode. 2. The clostridial-type ferredoxin molecule contains only 54 to 56 amino acids and the latter half of the molecule is homologous to the initial half suggesting gene duplication. In fact the size of the ferredoxin molecule is so small it could even be coded by tRNA. 3. *C. butyricum* ferredoxin is composed of only 11 different amino acids. These are the same amino acids which can be synthesised from methane, ammonia, hydrogen cyanide, formaldehyde, etc., under conditions which simulate the primitive environment. These are also the major amino acids detected in meteorites and lunar samples. 4. The active centres of these ferredoxins are two [4Fe–4S] clusters. The ease of formation and stability of these thiolate clusters (in the absence of oxygen) suggest that they may have been the first to be incorporated into any protein. (Under the reducing conditions of the primitive earth copper would have been bound up in insoluble sulphides.) A few synthetic clusters with the ferredoxin-type centre active in biological reactions are now known (see Adams *et al.*, 1980). The complete amino-acid sequences of 8Fe-ferredoxins from eight fermentative anaerobic bacteria have been determined. In addition the sequence of four ferredoxins from photosynthetic bacteria are known; three of these are from green sulphur bacteria and the other from the purple bacterium, *Chromatium*. Ferredoxins from the greens have 60 or 61 residues and that from *Chromatium* has 82 amino-acid residues, all four contain nine cysteines. It is striking that the positions of eight cysteine residues which bind the iron atoms are identical in these ferredoxins and in the *Clostridia* (Fig. 5.15). The sequences suggest a gradual evolution from fermenters to purple sulphur bacteria via the green bacteria. We still do not have a sequence of a purple non-sulphur bacterial ferredoxin. Ferredoxins containing single [4Fe–4S] and two [4Fe–4S] clusters have been purified and sequenced from aerobic bacteria also (see Matsubara *et al.*, 1980).

Ferredoxins of blue-green algae, green algae and higher plants all contain a single [2Fe–2S] centre and they are about 97 amino-acid residues long, so ferredoxins from all oxygen-evolving photosynthesisers are structurally and functionally identical. Halobacterial ferredoxins have an active site cluster similar to that of chloroplast ferredoxins though they do not function in the photoreduction of NADP. The sequence of halobacterium ferredoxin is homologous to that of a chloroplast ferredoxin (Fig. 5.16). The primary structures of 2Fe-ferredoxins from 28 species are known and the tertiary structure of *Spirulina platensis*

Table 5.6. *Biological reactions catalysed by ferredoxins (From Rao & Hall, 1977)*

Type of active centre and source of ferredoxin	Reaction	Other components required
I [2Fe-2S] (a) Algae and plants	1. Non cyclic photophosphorylation and NADP reduction $NADP^+ + H_2O + 2ADP + 2Pi \rightarrow NADPH_2 + \frac{1}{2}O_2 = 2ATP$	Ferredoxin-$NADP^+$ oxidoreductase Illuminated chloroplast membranes
	2. Cyclic photophosphorylation $ADP + Pi \rightarrow ATP$	Illuminated chloroplast electron transport chain
	3. Sulphite reduction $SO_3^{2-} \rightarrow S^{2-}$	Sulphite reductase } Photosynthetic membranes of algae and plants
	4. Nitrite reduction $NO_2^- \rightarrow NH_3$	Nitrite reductase }
	5. Fatty acid desaturation	Fatty acid desaturase NADPH oxidase
	6. Pyruvate decarboxylation $Pyruvate + CoA \rightleftharpoons Acetyl\ CoA + CO_2$	Pyruvate ferredoxin oxidoreductase ATP. In blue-green algae only
	7. Hydrogen metabolism $H_2 \rightleftharpoons 2H^+ + 2e^-$	Blue green algal membranes
(b) Animal mitochondria (adrenodoxin) (c) Bacteria (putidaredoxin)	8. Hydroxylation $R-H + O_2 + NAD(P)H \rightarrow ROH + H_2O + NAD(P)$	Pyridine nucleotide oxidoreductases
II [4Fe-4S] *Rhodospirillum rubrum* *Bacillus polymyxa*	9. Nitrogen fixation $N_2 + 3H_2 \rightleftharpoons 2NH_3$	Nitrogenase complex
Desulfovibrio gigas	10. Sulphite reduction Hydrogen metabolism	Sulphite reductase Hydrogenase, cytochrome c_3

III 2[4Fe-4S]
or 8Fe-8S
Fermentative and
photosynthetic bacteria

11. Phosphoroclastic reaction
   Pyruvate + Pi $\xrightarrow{\text{CoA}}$ Acetyl phosphate + CO_2 — Pyruvate dehydrogenase
12. CO_2 fixation (synthesis of α ketoacids) — Various α-ketoacid synthetases, TPP
   Acetyl CoA + CO_2 → Pyruvate + CoA
   Succinyl CoA + CO_2 → α-oxoglutarate + CoA
   Propionyl CoA + CO_2 → α-oxobutyrate + CoA
13. One-carbon metabolism $CO_2 \rightleftharpoons HCO_2^-$ — CO_2 reductase, formate dehydrogenase
14. Hydrogen metabolism — Hydrogenase
15. Nitrogen fixation — Nitrogenase
16. NAD reduction $NAD^+ + 2H^+ \rightarrow NADH_2$ — *Chlorobium* membranes + light

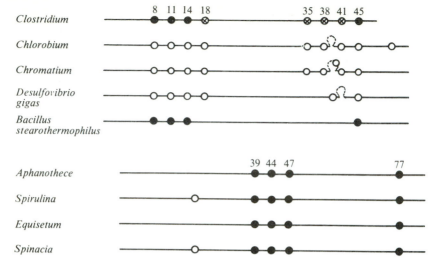

Fig. 5.15.   Cysteine residues in bacterial and plant ferredoxins. Circles denote the positions of cysteines in the amino-acid chain. In *Clostridium* ferredoxins containing two (4Fe–4S) centres, one centre is coordinated to the four cysteines shown as filled circles and the other centre to the four cysteines shown as crosses. In plant (and algal) ferredoxins the cysteines shown as filled circles are bound to the single (2Fe–2S) centre.

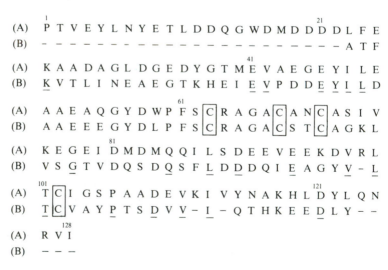

Fig. 5.16.   Comparison of amino-acid sequencies in *Halobacterium* ferredoxin (A: *Halobacterium halobium*) and chloroplast-type ferredoxin (B: *Nostoc muscorum*).

ferredoxin has recently been determined. A comparison of the primary structures of various 2Fe-ferredoxins with the X-ray structure of *S. platensis* ferredoxin reveals that the regions near the four iron-chelating cysteines are occupied always by invariant or semi-invariant residues (Matsubara *et al.*, 1980). Sequence comparisons show that the algal ferredoxins are an evolutionarily diverse group, the divergence occurring early in the evolutionary period. An evolutionary tree of all the ferredoxins derived from their sequences, assuming gene duplication, is shown in Fig. 5.17, from Matsubara *et al.* (1980). The relative positions of the chloroplast ferredoxins in this tree are in fair agreement with the taxonomic classification of algae and plants. If the symbiotic theory of chloroplast evolution is valid the evolutionary tree of plant ferredoxins would depict the radiation of blue-green algae followed by the development of symbiosis between one of these blue-greens and an ancestor of *Scendesmus* and the higher plants. Although the general features of the ferredoxin tree are more consistent with this theory than the alternative hypothesis (which assumes the origin of chloroplasts by compartmentalisation of the DNA within the cytoplasm of an evolving protoeukaryote) the tree by itself does not enable us to distinguish between the two theories because no ferredoxin sequences are available from the eukaryotic cytoplasm (Schwartz & Dayhoff, 1978). The construction of evolutionary trees from ferredoxin sequences is becoming more complicated due to the recent discoveries of more than one type of ferredoxin in some organisms.

Plastocyanins

Plastocyanins are electron transfer proteins widely distributed among plants (and algae) and associated with a chlorophyll-containing subchloroplast fragment (Boulter *et al.*, 1977). The presence of plastocyanin in any bacteria has not yet been reported. It is a protein of molecular weight about 10 500 and contains one atom of copper per molecule. It has a mid-point redox potential of about $+370\,mV$ and functions as an electron transfer agent to P700. Complete sequences of plastocyanins are known only from a few higher plants. From the data available on the complete and partial sequences of plastocyanins as well as cytochrome *f* (*c*-type) of cyanobacteria and of eukaryotes, Aitken (1976) suggests that the rate of evolution of the proteins in oxygenic photosynthetic prokaryotes is much less than the rates of evolution of corresponding proteins in eukaryotic algae and higher plants. The cyanobacteria have evolved very

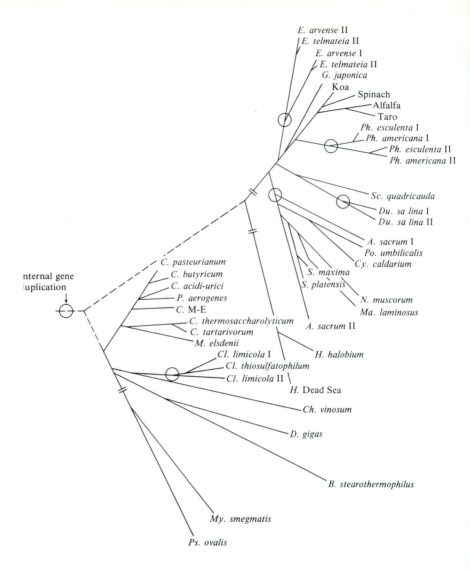

Fig. 5.17.   Phylogenetic tree based on ferredoxins (from Matsubara *et al.*, 1980). The most primitive are the anaerobic fermentative bacteria represented by the *Clostridium* spp. (*C. pasteurianum*, *C. butyricum*, *C. acidi urici*, *C. M-E.*, *C. thermosaccharolyticum* and *C. tartarivorum*) and *Peptococcus aerogenes* and *Megasphaera elsdenii*. These are followed by the green photosynthetic bacteria, *Chlorobium limicola* and *Chlorobium thiosulfatophilum* and then the purple photosynthetic bacterium, *Chromatium vinosum* and the anaerobic sulphate reducer, *Desulfovibrio gigas*. The aerobes, *Bacillus stearothermophilus*, *Myco-*

little in morphology and physiology since the oxygenic mode of photosynthesis was established in the Pre-Cambrian era.

Cytochromes c

c-Type cytochromes are present in all organisms which carry out respiration and/or photosynthesis. The amino-acid sequences of cytochromes c from a wide spectrum of prokaryotes and eukaryotes are known; it is one of the most extensively sequenced proteins. The chain length varies from 82 amino-acid residues in some photosynthetic and aerobic bacteria to 134 residues in the aerobic and nitrate respiring *Paracoccus denitrificans*. Further, the X-ray structures of cytochromes c from different classes of organisms such as horse, tuna, bonito, *Rhodospirillum rubrum*, *P. denitrificans*, *Pseudomonas aeruginosa* and *Chlorobium thiosulfatophilum* are known, mainly from the work of Dickerson and coworkers (see Dickerson, 1980a). The determination of the tertiary structure has enabled the alignment of the primary sequences of cytochromes, with the proper insertions, deletions and loops, so that the haem-containing active site occupies the same position in all the cytochrome c molecules compared. Evolutionary trees based on such alignments have been constructed. The primary and tertiary data suggest that the c-type cytochromes of photosynthetic bacteria, nitrate respirers and animal mitochondria are all members of an evolutionary family derived from a common ancestral gene. *Paracoccus* cytochrome c is related to both *R. rubrum* cytochrome and to eukaryotic mitochondrial cytochrome c indicating that the nitrate respirer may be a forerunner of the present-day mitochondria. Thus cytochrome c has played a vital role in tracing phylogenetic relationships among organisms.

Recently, from a comparison of c-type cytochrome sequences of 12 species of Rhodospirillaceae, Ambler *et al.* (1979a) found that in contrast to higher organisms there was no correlation between a species'

---

bacterium smegamatis and *Pseudomonas ovalis* evolved at a much later stage. Further gene duplication led to the evolution of *Halobacteria* such as *H. halobium* and *H.* Dead Sea. The ferredoxins from blue-green algae (*Aphanothece sacrum*, *Spirulina maxima* and *S. platensis*, *Nostoc muscorum*, *Mastigocladus laminosus*), red algae (e.g., *Porphyra umbilicalis*), green algae (*Scenedesmus quadricauda*, *Dunaliella salina*, *Cyanidium caldarum*) and higher plants (*Equisetum arvense* and *E. telmateia*, *Phytolacca americana* and *Ph. esculenta*, *Gleichenia japonica*, etc.) are functionally and structurally closed related.

cytochrome *c* sequence and its phylogenetic position in Bergey's Manual. The sequence differences of these cytochromes *c*-551 between similar species are as large as those between mitochondrial cytochromes *c* of mammals and insects; so, the deduction of phylogenetic information from the sequence of homologous proteins, according to Ambler *et al.*, is not straightforward. To explain these observations they have put forward the hypothesis that bacterial evolution proceeds by the assimilation of genes for single functions or for whole metabolic pathways from other organisms as well as by the gradual process of mutation and selection (see also Ambler, Meyer & Kamen, 1979*b*). Dickerson (1980*b*) has replied to this hypothesis by pointing out that Bergey's Manual gives a classification based on determinative characteristics of bacteria and has very little evolutionary or phylogenetic meaning. He has put forward strong

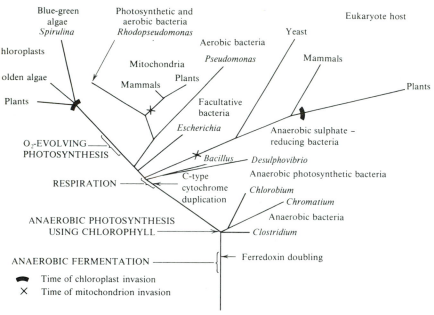

Fig. 5.18. Composite evolutionary tree proposed by Schwartz & Dayhoff (1978) based on ferredoxin, cytochrome *c* and 5S ribosomal RNA sequences. The tree shows the anaerobic fermenters as the primitive organisms followed by the anaerobic photosynthesisers, the green and purple sulphur bacteria. The time of 'chloroplast invasion' is depicted in two branches to comply with the computer calculations of the three different types of 'information molecules'. The mitochondria may have originated by the loss of chlorophyll from non-sulphur photosynthetic bacteria. Copyright 1978 by the American Association for the Advancement of Science.

arguments in favour of a revised classification of Rhodospirillaceae based on their molecular traits, such as cytochrome-c sequence similarities, rather than on gross morphology. Dickerson's revised classification of Rhodospirillaceae, based on a matrix of cytochrome-c sequence similarities, finds support in the 16S ribosomal RNA sequence comparisons of these bacteria determined by Woese, Gibson & Fox (1980). The lesson from these studies is that to build a proper evolutionary tree we should collect and assemble data from a diverse variety of 'information molecules' found in the organisms. A composite evolutionary tree based on ferredoxin, c-type cytochromes and 5S ribosomal RNA sequences is shown in Fig. 5.18.

CONCLUDING REMARKS

In this review we have attempted to give brief descriptions of the photosynthetic apparatus and the mechanisms by which various types of organisms use sunlight as an energy source. We have also discussed and compared the properties of various components of the photosynthetic system so that the reader can make his own conclusions regarding the evolutionary relationships between various groups of organisms. Both photosynthesis and biochemical evolution are rapidly expanding research areas and we have not been able to give due credit to all important publications. Most of the references are either to review papers or chapters in books and we apologise for any omissions.

REFERENCES

Adams, M. W. W., Rao, K. K., Hall, D. O., Christou, G. & Garner, C. D. (1980). Biological activity of synthetic molybdenum-iron-sulphur, iron-sulphur and iron-selenium analogues of ferredoxin-type centres. *Biochimica biophysica acta*, **589**, 1–9.

Adman, E. T., Sieker, L. C. & Jensen, L. H. (1973). The structure of a bacterial ferredoxin. *Journal of biological chemistry*, **248**, 3987–96.

Aitken, A. (1976). Protein evolution in cyanobacteria, *Nature*, **263**, 793–6.

Allsopp, A. (1969). Phylogenetic relationships of the prokaryotes and the origin of the eukaryotic cell. *New phytologist*, **68**, 591–612.

Ambler, R. P., Daniel, M., Hermoso, J., Meyer, T. E., Bartsch, R. G. & Kamen, M. D. (1979a). Cytochrome c_2 sequence variation among the recognised species of purple nonsulphur photosynthetic bacteria. *Nature*, **278**, 659–60.

Ambler, R. P., Meyer, T. E. & Kamen, M. D. (1979b). Anomalies in amino acid sequences of small cytochromes c and cytochrome c' from two species of purple photosynthetic bacteria. *Nature*, **278**, 661–2.

Arntzen, C. J. (1978). Dynamic structural features of chloroplast lamellae. In *Current topics in bioenergetics*, ed. R. D. Sanadi & L P. Vernon, **8**, pp. 112–160, London: Academic Press.

Baltscheffsky, H., Baltscheffsky, M. & Thore, A. (1971). Energy conversion reactions in bacterial photosynthesis. In *Current topics in bioenergetics*, ed. R. D. Sanadi, **4**, pp. 273–325. New York: Academic Press.

Bassham, J. A. (1962). The path of carbon in photosynthesis. *Scientific American*, June, 3–14.

Bassham, J. A. & Calvin, M. (1962). *The photosynthesis of carbon compounds*. New York: Benjamin.

Berkner, L. V. & Marshall, L. C. (1965). History of major atmospheric components. *Proceedings of the national academy of sciences of the USA*, **53**, 1215–25.

Boardman, N. K., Anderson, J. M. & Goodchild, D. J. (1978). Chlorophyll-protein complexes and structure of mature and developing chloroplasts. In *Current topics in bioenergetics*, ed. R. D. Sanadi & L. P. Vernon, **8**, pp. 35–109. London: Academic Press.

Bolton, J. R. & Hall, D. O. (1979). Photochemical conversion and storage of solar energy. *Annual review of energy*, **4**, 353–401.

Boulter, D., Haslett, B. G., Peacock, D., Ramshaw, J. A. M. & Scawen, M. D. (1977). Chemistry, function and evolution of plastocyanin. In *International review of biochemistry*, ed. D. H. Northcote, **13**, pp. 1–41. Baltimore: University Park Press.

Broda, E. (1975). *The evolution of the bioenergetic process*. Oxford: Pergamon Press.

Buchanan, B. B. & Evans, M. C. W. (1969). Photoreduction of ferredoxin and its use in $NAD(P^+)$ reduction by a subcellular preparation from the photosynthetic bacterium, *Chlorobium thiosulfatophilum*. *Biochimica biophysica acta*, **180**, 123–9.

Cammack, R. (1979). Functional aspects of iron-sulphur proteins. In *Metalloproteins: structure, function and clinical aspects*, ed. U. Weser, pp. 162–84. Stuttgart: Thieme Verlag.

Castenholz, R. W. (1979). Evolution and ecology of thermophilic microorganisms. In *Strategies of microbial life in extreme environments*, Dahlen Conference, ed. M. Shilo, pp. 373–92. Berlin: Dahlen Conference.

Cohen-Bazire, G. (1963). Some observations on the organisation of the photosynthetic apparatus in purple and green bacteria. In *Bacterial Photosynthesis*, ed. H. Gest, A. San Pietro & L. P. Vernon, pp. 89–109. Ohio, USA: Antioch Press.

Cohen, Y., Padan, E. & Shilo, M. (1975). Facultative anoxygenic photosynthesis in the cyanobacterium *Oscillatoria limnetica*. *Journal of bacteriology*, **123**, 855–61.

Coombs, J. & Greenwood, A. D. (1976). Compartmentation of the photosynthetic apparatus. In *Topics in photosynthesis*, ed. J. Barber, **1**, pp. 1–51. Amsterdam: Elsevier–North Holland Biomedical Press.

Dickerson, R. E. (1980*a*). The cytochromes *c*: an exercise in scientific seren-
dipity. In *Evolution of protein structure and function*, ed. D. S. Sigman &
M. A. B. Brazier, UCLA forum in medical science, **21**, pp. 00–00. New York:
Academic Press.

Dickerson, R. E. (1980*b*). Evolution and gene transfer in purple photosynthetic
bacteria. *Nature*, **283**, 210–2.

Dutton, P. L., Prince, R. C. & Tiede, D. M. (1978). The reaction centre of
photosynthetic bacteria. *Photochemistry and Photobiology*, **28**, 939–49.

Echlin, P. & Morris, I. (1965). The relationship between blue-green algae and
bacteria. *Biological reviews*, **40**, 173–87.

Ellis, R. J. (1976). Protein and nucleic acid synthesis by chloroplasts. In *Topics in
Photosynthesis*, ed. J. Barber, vol. 1, pp. 335–64. Amsterdam: Elsevier/North
Holland Biomedical Press.

Emerson, R., Chalmers, R. & Cederstrand, C. (1957). Some factors influencing
the long-wave limit of photosynthesis. *Proceedings of the national academy of
sciences of the USA*, **43**, 133–37.

Evans, M. C. W. & Buchanan, B. B. (1965). Photoreduction of ferredoxin and its
use in carbon dioxide fixation by a subcellular system from a photosynthetic
bacterium. *Proceedings of the national academy of sciences of the USA*, **53**,
1420–5.

Evans, M. C. W., Buchanan, B. B. & Arnon, D. I. (1966). A new ferredoxin-
dependent carbon reduction cycle in a photosynthetic bacterium. *Proceedings of
the national academy of sciences of the USA*, **55**, 928–34.

Evans, M. C. W. (1975). The mechanism of energy conversion in photosynthesis,
Science progress, **62**, 543–58.

Fox, G. E., Magrum, L. J., Balch, W. E., Wolfe, R. S. & Woese, C. R. (1977).
Classification of methanogenic bacteria by 16S ribosomal RNA characterization. *Proceedings of the national academy of sciences of the USA*, **74**,
4537–41.

Frenkel, A. W. (1956). Photophosphorylation of adenine nucleotides by cell-free
preparations of purple bacteria. *Journal of biological chemistry*, **222**, 823–34.

Fuller, R. C. (1978). Photosynthetic carbon metabolism in the green and purple
bacteria. In *The photosynthetic bacteria*, ed. R. K. Clayton & W. R. Sistrom, pp.
691–707. London: Plenum Press.

Gillham, N. W., Boynton, J. E. & Chua, N-H. (1978). Genetic control of
chloroplast proteins. In *Current topics in bioenergetics*, ed. R. D. Sanadi & L. P.
Vernon, **8**, pp. 211–60. New York: Academic Press.

Gimmler, H. (1977). Photophosphorylation *in vivo*. In *Encyclopedia of plant
physiology*, ed. A. Trebst & M. Avron, New Series, **5**, pp. 448–72. Berlin:
Springer-Verlag.

Hall, D. O. (1973). Origin and development of life in relation to solar energy.
Anais da academia Brasileria de ciencias, **45**, 71–83.

Hall, D. O., Cammack, R., Rao, K. K., Evans, M. C. W. & Mullinger, R. (1975).
Biochemical society transactions, **3**, 361–8.

Hall, D. O., Rao, K. K. & Cammack, R. (1975). The iron-sulphur proteins:
structure, function and evolution of a ubiquitous group of proteins. *Science
progress*, **62**, 285–317.

Hall, D. O. (1976). The coupling of photophosphorylation to electron transport in isolated chloroplasts. In *Topics in photosynthesis*, ed. J. Barber, **1**, pp. 135–70. Amsterdam: Elsevier/North-Holland Biomedical Press.

Hall, D. O. (1977). Iron-sulphur proteins and superoxide dismutases in the biology and evolution of electron transport. In *Advances in chemistry series, No. 162. Bioinorganic Chemistry II*, ed. K. N. Raymond, pp. 227–49. American Chemical Society.

Hall, D. O. & Rao, K. K. (1977). *Photosynthesis*. 2nd edn. London: Edward Arnold.

Hall, D. O. (1979). Solar energy use through biology – past, present and future. *Solar energy*, **22**, 307–28.

Hase, T., Wakabayashi, S., Matsubara, H., Kerscher, L., Oesterhelt, D., Rao, K. K. & Hall, D. O. (1977). *Halobacterium halobium* ferredoxin: A homologous protein to chloroplast-type ferredoxins. *FEBS Letters*, **77**, 308–10.

Hatch, M. D. & Slack, C. R. (1966). Photosynthesis by sugar cane leaves. A new carboxylation reaction and the pathway of sugar formation. *Biochemical Journal*, **101**, 103–11.

Hill, R. & Bendall, F. (1960). Function of the two cytochrome components in chloroplasts: a working hypothesis. *Nature*, **186**, 136–7.

Jagendorf, A. T. (1977). Photophosphorylation. In *Encyclopaedia of plant physiology*, ed. A. Trebst & M. Avron, New Series, **5**, pp. 307–37. Berlin: Springer-Verlag.

John, P. & Whatley, F. R. (1977). The bioenergetics of *Paracoccus denitrificans*. *Biochimica Biophysica Acta*, **463**, 129–53.

Kerscher, L. & Oesterhelt, D. (1977). Ferredoxin is the coenzyme of α-ketoacid oxidoreductases in *Halobacterium halobium*. *FEBS Letters*, **83**, 197–201.

Kirk, J. T. O. & Tilney-Bassett, R. A. E. (1978). *The plastids*. Amsterdam: Elsevier/North-Holland Biomedical Press.

Krasnovsky, A. A. (1976). Chemical evolution of photosynthesis. *Origins of life*, **7**, 133–43.

Laetsch, W. M. (1974). The C_4 syndrome: A structural analysis. *Annual Review of Plant Physiology*, **25**, 27–52.

Lanyi, J. K. (1978). Light-energy conversion in *Halobacterium halobium*. *Microbiol. Rev.* **42**, 682–706.

Lewin, R. A. (1976). Prochlorophyta as a proposed new division of algae. *Nature*, **261**, 697–8.

MacColl, R. & Berns, S. (1979). Evolution of the biliproteins. *Trends in biochemical science*, **4**, 44–7.

Machold, O., Simpson, D. J. & Møller, B. C. (1979). Chlorophyll-proteins of thylakoids from wild-type and mutants of barley (*Hordeum vulgare*, L.). *Carlsberg research communications*, **44**, 201–17.

Malkin, R. & Bearden, A. J. (1978). Membrane-bound iron-sulfur centres in photosynthetic systems. *Biochimica biophysica acta*, **505**, 147–81.

Margulis, L. (1970). *Origins of eukaryotic cells*. New Haven: Yale University Press.

Markwell, J. P., Thornber, J. P. & Boggs, R. T. (1979). Higher plant chloroplasts: evidence that all the chlorophyll exists as chlorophyll–protein complexes. *Proceedings of the national academy of sciences of the USA*, **76**, 1233–5.

Matsubara, H., Hase, T., Wakabayashi, S. & Wada, K. (1980). Structure and function of chloroplast- and bacterial-type of ferredoxins. In *Evolution of protein structure and function*, ed. D. S. Sigman & M. A. B. Brazier, UCLA forum in medical science, **21** (in press). New York: Academic Press.

Mitchell, P. (1966). *Chemiosmotic coupling in oxidative and photosynthetic phosphorylation.* Bodmin, Cornwall, England: Glynn Research.

Olson, J. M. (1978*a*). Bacteriochlorophyll a – proteins from green bacteria. In *The photosynthetic bacteria*, ed. R. K. Clayton & W. R. Sistrom, pp. 161–97. London: Plenum Press.

Olson, J. M. (1978*b*). Pre-Cambrian evolution of photosynthetic and respiratory organisms. In *Evolutionary biology*, ed. M. K. Hecht, W. C. Steere & B. Wallace, **11**, pp. 1–37. New York: Plenum Publishing Corporation.

Padan, E. (1979). Facultative Anoxygenic Photosynthesis in Cyanobacteria. *Annual reviews of plant physiology*, **30**, 27–40.

Pardee, A. B., Schachman, H. K. & Stanier, R. Y. (1952). Chromatophores of *Rhodospirillum rubrum*. *Nature*, **169**, 282–3.

Parson, W. W. & Cogdell, R. J. (1975). The primary photochemical reaction of bacterial photosynthesis, *Biochimica biophysica acta*, **416**, 105–49.

Pfennig, N. (1978). General physiology and ecology of photosynthetic bacteria. In *The photosynthetic bacteria*, ed. R. K. Clayton & W. R. Sistrom, pp. 3–18, London: Plenum Press.

Pierson, B. K. & Castenholtz, R. W. (1971). Bacteriochlorophylls in gliding filamentous prokaryotes from hot springs. *Nature, New Biology*, **233**, 25–7.

Pierson, B. K. & Castenholz, R. W. (1974). A phototropic gliding filamentous bacterium of hot springs, *Chloroflexus aurantiacus*, gen. and sp. Nov., *Archiv für Mikrobiologie*, **100**, 5–24.

Pierson, B. K. & Castenholz, R. W. (1978). Photosynthetic apparatus and cell membranes of the green bacteria. In *The photosynthetic bacteria*, ed. R. K. Clayton & W. R. Sistrom, pp. 179–98. London: Plenum Press.

Raff, R. A. & Mahler, H. R. (1972). The non-symbiotic origin of mitochondria. *Science*, **177**, 575–82.

Ragan, A. M. & Chapman, D. J. (1978). *A biochemical phylogeny of the protists.* New York: Academic Press.

Rao, K. K. & Hall, D. O. (1977). Chemistry and evolution of ferredoxins and hydrogenases. In *The evolution of metalloenzymes, metalloproteins and related materials*, ed. G. J. Leigh, pp. 39–64. London: Symposium Press.

Schidlowski, M. (1979). Antiquity and evolutionary status of bacterial sulfate reduction: sulfur isotope evidence. *Origins of life*, **9**, 299–311.

Schopf, J. W. (1974). Palaeobiology of the Pre-Cambrian: the age of blue-green algae. In *Evolutionary biology*, ed. T. Dobzhansky, M. K. Hecht & W. C. Steere. **7**, pp. 1–43, New York: Plenum Press.

Schwartz, M. & Dayhoff, M. O. (1978). Origins of prokaryotes, eukaryotes, mitochondria and chloroplasts. *Science*, **199**, 395–403.

Simionescu, C. I., Mora, R. & Simionescu, B. C. (1978). Porphyrins and the evolution of biosystems. *Bioelectrochemistry and bioenergetics*, **5**, 1–17.

Simpson, D., Hoyer-Hansen, G., Chua, N-H. & von Wettstein, D. (1977). The use of single gene mutants in barley to correlate thylakoid polypeptide composition with the structure of the photosynthetic membrane. *Proceedings of the 4th International Congress of Photosynthesis, Reading*, 537–48.

Stanier, R. Y., Adelberg, E. A. & Ingraham, J. L. (1977). *General Microbiology*, 4th edn. London: Macmillan Press.

Stewart, W. D. P. (1978). Nitrogen-fixing cyanobacteria and their association with eukaryotic plants. *Endeavour*, New Series, **2**, 170–9.

Stoeckenius, W., Lozier, R. H. & Bogomolni, R. A. (1979). Bacteriorhodopsin and the purple membrane of halobacteria. *Biochimica biophysica acta*, **505**, 215–78.

Taylor, F. J. R. (1974). Implications and extensions of the serial endosymbiosis theory of the origin of eukaryotes. *Taxon*, **23**, 229–58.

Trebst, A. (1974). Energy conservation in photosynthetic electron transport of chloroplasts. *Annual review of plant physiology*, **25**, 423–58.

Tsukihara, T., Fukuyama, K., Tahara, H., Katsube, Y., Matsuura, Y., Tanaka, N., Kakudo, M., Wada, K. & Matsubara, H. (1978). X-ray analysis of ferredoxin from *Spirulina platensis*. *Journal of biochemistry (Tokyo)*, **84**, 1645–7.

Uzzell, T. & Spolsky, C. (1974). Mitochondria and plastids as endosymbionts: a revival of special creation? *American scientist*, **62**, 334–43.

Van Niel, C. B. (1941). The bacterial photosyntheses and their importance for the general problem of photosynthesis. *Advances in Enzymology*, **1**, 263–328.

Whatley, J. M., John, P. & Whatley, F. R. (1979). From extracellular to intracellular: the establishment of mitochondria and chloroplasts. *Proceedings of the Royal society of London series B*, **204**, 165–87.

Witt, H. T. (1979). Energy conversion in the functional membrane of photosynthesis. Analysis by light pulse and electric pulse methods. *Biochimica biophysica acta*, **505**, 355–427.

Woese, C. R., Gibson, J. & Fox, G. E. (1980). Do genealogical patterns in purple photosynthetic bacteria reflect interspecific gene transfer? *Nature*, **283**, 212–4.

Yasunobu, K. T. & Tanaka, M. (1973). The types, distribution in nature, structure-function and evolutionary data of the iron-sulfur proteins. In *Iron-sulfur proteins*, ed. W. Lovenberg, **2**, pp. 27–130. New York: Academic Press.

Note added in proof. There is recent evidence that certain ferredoxins contain a [3Fe–3S] cluster (see Cammack, R., (1980) *Nature*, **286**, 442 and references therein), with a six-membered ring of iron and sulphide atoms. Thus *Azotobacter vinelandii* ferredoxin I contains [4Fe–4S] and [3Fe–3S] clusters. In *D. gigas* ferredoxin, the same polypeptide can accommodate either [4Fe–4S] or [3Fe–3S] clusters, with widely differing midpoint reduction potentials. This observation has significant implications for the evolution of ferredoxin function. See Table 5.5, p. 188.

6

Major histocompatibility antigens

J. R. L. PINK

DEPARTMENT OF PATHOLOGY, UNIVERSITY OF GENEVA,
GENEVA 1211, SWITZERLAND

How major histocompatibility antigens (MHAs) function is still a mystery, although their basic structure is established. The MHAs are simple glycoproteins of molecular weight about 45 000 which are present, in non-covalent association with a smaller polypeptide of molecular weight 12 000 (β_2-microglobulin), at the surface of most nucleated vertebrate cells. Their most unusual feature (from an evolutionary standpoint) is their very extensive genetic polymorphism. Mammals, as a rule, have two or three loci coding for MHAs, but the MHA coded by each locus exists in 10, 20 or more antigenically different forms. Each allele is inherited and expressed codominantly according to simple Mendelian rules. What is the evolutionary process maintaining this very large number of alleles?

This question cannot yet be answered in detail, but I would like to approach it by discussing, first, what is known of the genes coding for the synthesis of MHAs and related antigens, and their organisation in different species; next, the structural basis for the antigenic polymorphism of MHAs; and finally, the evidence that MHAs play a vital role in antiviral immunity, and that natural selection acts to maintain MHA polymorphism. Since a complete list of references to these topics would probably be longer than the text, the bibliography is limited to reviews and articles published rather recently.

THE MAJOR HISTOCOMPATIBILITY COMPLEX

MHAs were first identified, in mice, as antigens which led to exceptionally rapid skin graft rejection if they were present on a graft in an allelic form not carried by the recipient. The genes coding for these antigens were located in a small region of the mouse chromosome 17, which was called the *H*-2 complex (Klein, 1979). Homologous regions, called major histocompatibility complexes or MHCs, have since been identified in many mammals, and in two non-mammalian species (chickens and the clawed toad, *Xenopus laevis*) (Götze, 1977).

Mammalian MHCs typically contain two or more genes coding for MHAs, plus a larger number of genes coding for other proteins. Strains of inbred mice which differ only at particular loci in the MHC region have been developed, and used to prepare antisera against products of individual MHC genes. At the same time, interest in the practical problems of graft rejection has led to characterisation of products of the human MHC (the *HLA* complex) (Bodmer & Bodmer, 1978). In this way, four main classes of MHC product (including the MHAs) have been defined. All are mentioned here, because it seems quite possible (although there is room for doubt on this issue) that a common evolutionary origin will be demonstrated for the four classes.

The MHAs. As already mentioned, these consist of pairs of polypeptides, of which only the larger, MHC-coded chains are highly polymorphic. The smaller β_2-microglobulin chain is not coded in the MHC (at least in man, where chromosome 6 carries the MHC and chromosome 15 the β_2-microglobulin gene). MHAs are membrane-bound proteins which are present on cells of most tissues, in varying quantities: the highest amounts (about 500 000 molecules per cell) are found on lymphocytes in spleen, lymph nodes and other peripheral lymphoid organs, whereas the amounts on nerve cells and non-nucleated red blood cells are very low. Some thymic lymphocytes, as discussed in more detail on p. 217, also lack detectable MHAs.

In the human *HLA* complex, three loci coding for MHAs – *HLA-A, B* and *C* – have been defined. Different alleles of these loci are numbered (*A*1, *A*2, and so on). About 20 antigenically distinct allelic products of the *A* locus, and a similar number for the *B* locus, have been defined; fewer alleles of the more recently discovered *C* locus have been described. The mouse *H*-2 complex also includes three MHA loci – *H-2D, K* and *L* – whose alleles are denoted by small letter superscripts: D^b, D^d, D^k, and so on. It became obvious as soon as wild mouse populations were studied that the number of alleles at each locus would exceed 26, so that alleles of wild mouse *H*-2 loci are numbered D^{w1}, D^{w2}, etc. In other species, not surprisingly, nomenclature has tended to follow the HLA rather than the H-2 usage.

MHA-like antigens. Several MHA-like products of the MHC have been identified only recently, and are not well characterised biochemically. All have been recognised because of their genetic polymorphism. They are distinguishable from the MHAs by their tissue distributions, but they share some structural features with the MHAs.

Some of these products, e.g. the mouse TL and Qa-2 antigens, and what seem to be their homologues in guinea-pigs, resemble MHAs in consisting of a heavy chain, of molecular weight about 40 000, non-covalently joined to a β_2-microglobulin-like chain of molecular weight 12 000 (Vitetta & Capra, 1978; Schwartz, Cigen, Berggard & Shevach, 1978). The TL antigen, however, is present on most thymic lymphocytes, but not on peripheral cells: this distribution is exactly opposed to that of the H-2D and K antigens. A human TL-like protein, associated with a smaller chain which resembles β_2-microglobulin in size, but not in antigenic properties, has also been described (Ziegler & Milstein, 1979); whether this protein is MHC-coded is not known. The Qa-2 antigens are present on peripheral T lymphocytes and epithelial cells, but in lower amounts than MHAs. One protein which is probably not MHC-coded, the mouse F9 antigen, also has a similar two-chain structure; the smaller chain, however, is antigenically distinct from β_2-microglobulin (Vitetta & Capra, 1978). The gene coding for F9 is thought to be on the same chromosome as the MHC, about 15 centimorgans distant from it. Little more is known about any of these antigens, but it seems probable, from the structural and genetic data, that they are evolutionarily related to MHAs.

The Ia antigens. The Ia antigens of mice, and their homologues in other species, also consist of pairs of polypeptides of which one is genetically highly polymorphic (Schwartz & Cullen, 1978; McMillan *et al.*, 1978; Cook *et al.*, 1979). In mice, the smaller, polymorphic β-chains (mol. wt 25–30 000) are non-covalently bound to less polymorphic α-chains (mol. wt about 33 000). The α-β dimers are copurified with a third, non-polymorphic chain, Ii (mol. wt 31 000), but the stoichiometry of this complex is not known (Jones, Murphy, Hewgill & McDevitt, 1979). At least two, and probably more, loci in the mouse *H-2* complex code for polymorphic Ia antigens, each of which has a large number of different allelic forms. Ia antigens have a more limited tissue distribution than MHAs, being present (as far as is known) only on certain types of lymphocytes, macrophages or monocytic cells, and some specialised epithelial cells. The Ia antigens control, in an unknown manner, the interactions between these cell types during lymphocyte ontogeny and during antibody responses. Polymorphism at Ia loci is associated with high or low responses to a variety of antigens, a point which is important for our purposes, since it may lead to natural selection for Ia poly-morphism; and, because Ia and MHA loci are closely linked, the effects of this selection may be difficult to distinguish from selective effects acting only on MHA loci.

In their extensive genetic polymorphism (which can also give rise to rapid skin graft rejection), genetic control by the MHC, two-chain structure and role in immunity, Ia antigens show many resemblances to MHAs. However, no clear evolutionary homology between Ia antigens and MHAs has emerged from the limited amino-acid sequence data on Ia antigens, and not enough is yet known of their function and structure in different species to warrant a detailed description in this article.

Complement components. Some components of serum complement, which is another complex of proteins with an important role in immune reactions, are coded by genes in the MHC. For example, genetic variation in human C2, C4 and factor B of the alternative complement pathway is MHC-controlled (Lachmann & Hobart, 1978). A C3 gene also maps about 12 units away from the mouse MHC. No evolutionary relationships between complement components and any other MHC products have yet been established.

Other MHC products are present on mature red cells, but not on lymphocytes. These have been described in mice, humans and chickens, although no inter-species homology has been demonstrated (Pink *et al.*, 1977). The human antigens have turned out to be complement components (C4) absorbed from serum, and the same may perhaps be true for the mouse and chicken antigens.

Genetic organisation of the MHC

The order of genes coding for different MHC products has been established in various mammals (Fig. 6.1). The prevalence of duplicated genes coding for MHAs is striking, although, as will be discussed (p. 212), this picture may be due to independent duplications in different species. The recombination rates between different loci vary, with some gene pairs (e.g. HLA-A and B, or H-2K and D) being separated by 1–0.5 cross-over units, while recombination between other pairs (e.g. HLA-

Species	Name of MHC	Gene order
Mouse	H–2	
Man	HLA	
Rhesus monkey	RhLA	
Dog	DLA	

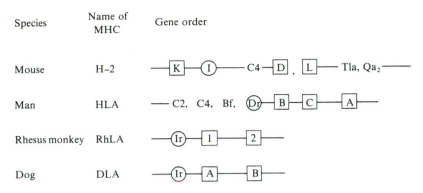

Fig. 6.1. Genetic organisation of the MHC of some mammals. Genes coding for MHAs are denoted by squares, with the name or number of the particular gene enclosed. Genes, or sets of genes, coding for Ia-like antigens (or uncharacterised products with similar functions) are denoted by circles. C2, C4 and Bf are components of complement. The Tla and Qa_2 genes code for the MHA-like TL and Qa_2 antigens discussed in the text. The order of genes separated by commas is not established. Gene order within the MHCs of other species has not been determined, but there is biochemical evidence for the existence of duplicated MHA genes in rats and guinea-pigs, and serological evidence for such a duplication in chimpanzees (see Götze, 1977).

B, C) is rare. The duplicated MHA genes thus differ from other gene families (for example, those coding for hemoglobins and immunoglobulin constant regions), where duplicated genes are very tightly linked.

Fig. 6.1 also shows that gene linkage and order within the mammalian MHC have been conserved during evolution. An important exception is the unusual gene order in the mouse *H*-2 complex. In this case, genes coding for Ia antigens lie between duplicated MHA genes. In other species (humans, monkeys, and dogs), on the contrary, genes coding for Ia-like products lie outside duplicated MHA genes. Perhaps a genetic inversion within the *H*-2 complex, or a translocation of one of the duplicated mouse MHA genes, led to this unusual arrangement.

MAJOR HISTOCOMPATIBILITY ANTIGEN STRUCTURE

Biochemistry of MHAs

The most detailed structural work on MHAs has been carried out on products of the mouse H-2D and K loci (Maizels, Frelinger & Hood, 1978; Coligan *et al.*, 1979), and of the human HLA-A and B loci

(Trägårdh, Wiman, Rask & Peterson, 1978; Orr *et al.*, 1979*a*, *b*). For the human antigens, it has been possible to purify sufficiently large amounts of material for classical biochemical analysis, but for other species it has been much more convenient to label MHAs biosynthetically with radioactive amino acids in short-term cell cultures, and to characterise the radioactive material precipitated by specific antisera from a lysate of the cells made in the presence of a solubilising agent such as a non-ionic detergent.

MHAs in all species studied have a similar structure. The following description (Fig. 6.2) is based mainly on analysis of HLA-A2 and B7

Fig. 6.2.   Structure of an MHA. Schematic representation of the structure of the HLA-B7 molecule, which includes a single carbohydrate moiety (CHO) of molecular weight about 3000, two disulphide bridges (–S–S–), a hydrophobic region (boxed) associated with the cell membrane, and two regions susceptible to cleavage by papain (marked by arrows).

antigens by Strominger and his colleagues. The MHA heavy chain is about 340 amino acids long; it contains two disulphide bridges and about 7% carbohydrate, located in a single group at residue 86 of the chain. The molecule is held at the cell surface by a hydrophobic carboxy-terminal portion embedded in the plasma membrane. A small hydrophilic carboxy-terminal 'tail', including a phosphoserine residue, is located on the cytoplasmic side of the membrane (Pober, Guild & Strominger, 1978). Each heavy chain is non-covalently joined to one β_2-micro-globulin molecule; this association is necessary for the maintenance of the heavy chains' conformation, and, presumably, as a consequence of this requirement, for insertion of the heavy chain into the plasma membrane (Ploegh, Cannon & Strominger, 1979). The β_2-microglobulin molecule itself is not a hydrophobic molecule; in fact, treatment of living cells with a proteolytic enzyme such as papain will release a large amino-terminal fragment of the heavy chain (mol. wt. 34 000) bound to the complete β_2-microglobulin chain.

Fig. 6.3 shows the locations of the amino-acid sequence differences between large regions of one mouse and two human MHAs, and Tables 6.1 and 6.2 compare more limited amino-acid sequence data from the

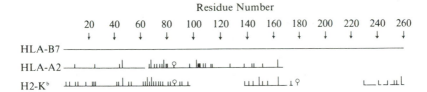

Fig. 6.3. Location of amino-acid sequence differences between MHAs from different loci and different species. Incomplete sequence data from HLA-A2 (Orr *et al.*, 1979*b*) and H-2K^b (Coligan *et al.*, 1979) are compared with the sequence of the HLA-B7 molecule (Orr *et al.*, 1979*a, b*). Vertical bars mark sequence differences from the HLA-B7 sequence. Where double-length vertical bars are used, two base changes are necessary to account for the observed differences. Locations of carbohydrate groups (circles) are also marked (the exact position of the second group in the H-2K^b molecule is unknown).

amino-terminal regions of MHAs from six different species. Three points can be made from these tables.

MHAs from different species are homologous. There is a 70–75% homology between the mouse and human MHA sequences (Fig. 6.3). The sequence differences occur throughout the regions compared, but seem to be more concentrated in certain areas, for example between residues 60 and 80 of the chains. The rate of accumulation of sequence differences since the mouse/human divergence is comparable to, or slower than, the evolutionary divergence rate of immunoglobulin constant regions; for example, homology between mouse and human μ chain constant region domains ranges from 48 to 78% (Kehry *et al.*, 1979).

A similar conclusion can be drawn from comparisons of amino-terminal sequences of MHAs from different species (Table 6.1). The data are still too limited to be useful in constructing an evolutionary tree; they are also biased, in the sense that some amino acids (e.g. aspartic acid, glycine) are difficult to detect using radioactive sequencing methodology, and are therefore excluded from most of the sequences shown. Nevertheless, the homology between, for example, chicken and mammalian MHAs is very striking.

MHC products from non-mammals other than chickens have not yet been characterised biochemically. However, in *Xenopus laevis*, a single chromosomal region controls the synthesis of polymorphic cell surface antigens with a wide tissue distribution, and polymorphism at this region

Table 6.1. *Partial amino-terminal amino-acid sequences of MHAs from different species. Gaps are unidentified residues. Where alternative residues at a position are given, they represent residues present in different allelic products (from humans and mice), or alternative residues found in a pool of MHAs from one or several strains (other species). Data from the compilation by Maizels, Frelinger & Hood (1978), and from Kimball, Coligan & Kindt (1979), Orr et al. (1979b), and Trägårdh et al. (1978)*

Species	Locus	1	2	3	4	5	6	7	8	9	10	11	12	13	14	15	16	17	18	19	20	21	22	23	24	25	26	27	28
Human	HLA-A	Gly	Ser	His	Ser	Met	Arg	Tyr	Phe	Phe	Thr	Ser	Val	Ser	Arg	Pro	Gly	Arg	Gly	Glu	Pro	Arg	Phe	Ile	Ala	Val	Gly	Tyr	Val
	HLA-B	Gly	Ser	His/Ser	Ser	Met	Val/Arg	Tyr	Phe	Phe	Thr	Ala/Ser	Val	Ser	Arg	Pro	Gly	Arg	Gly	Glu	Pro	Arg	Phe	Ile	Ser	Val	Gly	Tyr	Val
Mouse	H-2 K	Met/Gly	Pro	His	Ser	Leu	Arg	Tyr	Phe	His/Val	Thr	Ala	Val	Ser	Arg	Pro	Gly	Phe/Leu	Gly	Lys/Glu	Pro	Arg	Tyr/Phe	Ile/Met	Glu	Val	Gly	Tyr	Val
	H-2 D	Met	Arg/Pro	His		Leu	Arg	Tyr	Phe	Val	Thr	Ala	Val		Arg	Pro	Phe/Leu	Phe/Leu			Pro	Arg	Tyr	Ile		Val		Tyr	Val
Rat					Ile/Leu	Arg	Tyr	Phe	Tyr		Ala	Val											Phe	Ile	Ala	Val		Tyr	
Guinea-pig				His	Ile/Leu	Arg	Tyr	Phe	Tyr		Phe	Val				Pro							Phe			Val		Tyr	
Rabbit		Gly	Ser	His	Ser	Met	Arg	Tyr	Phe	Tyr	Ser	Val	Ser	Arg	Pro	Gly	Leu	Gly			Pro	Arg	Phe			Val	Gly	Tyr	Val
Chicken			Leu	His		Leu	Arg	Tyr	Phe/Ile	Arg/Phe	Ala	Met			Pro			Leu				Tyr/Phe	Val			Val		Tyr	Val

Table 6.2. *Partial amino-terminal amino-acid sequences of mouse MHAs. A solid line indicates sequence identity with H-2K^b. Gaps are unidentified residues. Bracketed residues are tentative identifications only. Data are from Coligan et al. (1978) and from the compilations by Maizels et al. (1979) and Vitetta & Capra (1978)*

Gene	Allele																	Residue number											
		1	2	3	4	5	6	7	8	9	10	11	12	13	14	15	16	17	18	19	20	21	22	23	24	25	26	27	28
H-2K	K^b	Gly	Pro	His	Ser	Leu	Arg	Tyr	Phe	Val	Thr	Ala	Val	Ser	Arg	Pro	Gly	Leu	Gly	Glu	Pro	Arg	Tyr	Met	Glu	Val	Gly	Tyr	Val
	K^d	Met				(—)			(—)		(—)					(—)		(—)			(—)	(—)							
	K^k	(—)							(—)	His													(Phe)						
	K^q									His													Phe	Ile					
	K^s																	Phe											
H-2D	D^b																												
	D^d	Met												(Thr)				Phe											
	D^q																						Ile						
	D^s			Arg																									

strongly affects skin-graft rejection time. This finding, together with the amino-acid homology between chicken and mammalian MHAs, strongly suggests that all vertebrate MHAs have a common evolutionary origin. An objection to this assumption is that many lower vertebrates (e.g. salamanders, sharks) do not exhibit rapid skin-graft rejection. This observation was originally taken as evidence for independent evolution of MHC-like systems in the precursors of contemporary amphibians, fish, birds and mammals (Cohen & Collins, 1977), but now seems much more likely to be due to inadequacies in our knowledge of graft rejection mechanisms in lower vertebrates.

Because some very primitive species (e.g. earthworms) are capable of rejecting skin grafts, some authors have considered that MHA recognition is, in terms of phylogeny, a very ancient process. However, there is no evidence for the existence of MHA-like molecules in invertebrates. In addition, invertebrates lack thymus epithelial tissues, which are necessary for the development of MHA function (see p. 216). Thus, the MHAs, like the immunoglobulins, probably developed their present form and function during early vertebrate evolution.

MHAs from different loci of the same species (e.g. HLA-A and B) are highly homologous. In fact, the homology between HLA-A and B (>80%) is greater than the homology of either antigen to MHAs of other species (75% or less). Similarly, H-2D and K are closely related (Table 6.2), whereas H-2 and HLA antigens are less so. These findings show that the several MHA loci present in many mammals arose by gene duplication, but raise the problem of exactly when the duplication(s) occurred: a single duplication at the outset of vertebrate evolution would be expected to produce sequences in which intraspecies homology (e.g. H-2D/K) was *less* than homology between corresponding loci of different species (Maizels, Frelinger & Hood, 1978). Thus, either several independent, recent duplications occurred during mammalian evolution, or – less likely – natural selection forced an unprecedented parallel evolution of different sequences in different species.

The inter-locus homology seen in Table 6.2 is surprising for a second reason: it is so strong that there is no evidence for locus-specific sequences, which would, for example, be typical of H-2D but not H-2K. This observation is in line with the fact that it has not been possible to raise antisera which distinguish H-2K molecules, as a class, from H-2D molecules. More sequence data are needed to discover which regions of the MHAs define them as products of one locus or another, or whether, in fact, there exist genetic mechanisms such that the products of two loci can share stretches of identical sequence.

Inter-allelic differences are correlated with multiple amino-acid substitutions. In the best-studied, amino-terminal, portion of MHAs, inter-allelic differences amount to 0–3 substitutions per 28 residues, which for a molecule of 350 residues would be equivalent to as much as 40 inter-allelic differences, if the substitutions were equally distributed along the chain (Table 6.2). Results from peptide mapping experiments on different H-2 antigens are also consistent with a figure of 20 or more inter-allelic substitutions, at least for some pairs of alleles. The sequence data from larger portions of H-2 and HLA molecules (Fig. 6.3) suggest that interallelic differences are probably not distributed randomly, but may be concentrated in regions (e.g. positions 60–80) where most inter-species and interlocus substitutions occur.

These substitutions presumably arose by the stepwise accumulation of point mutations. The mutation rate of *H*-2 genes, as measured by a skin-graft rejection assay for mutants with altered antigenic properties, is high, but not exceptionally so, and seems to vary from allele to allele: most *H*-2 mutants are derived from the H-$2K^b$ allele, whose mutation rate is estimated at about 2×10^{-4} per mouse generation, but other alleles (e.g. D^b) have given rise to detectable mutants at a much (5 to 10-fold) slower rate (Klein, 1978). These studies, of course, do not detect an unknown proportion of mutants which leave unaltered the molecule's antigenicity. It will be important to compare the present mutation rate figures with estimates made using other techniques, which can detect, for example, all charge-shift mutants in a given protein (Milstein *et al.*, 1977).

In at least one case (Nathenson *et al.*, 1979), a single point mutation (from arginine to an unknown amino acid at position 155 of the K^b chain) seems to be sufficient to cause a recognisable change in the antigenicity of an H-2 molecule; in another case, two substitutions (probably at positions 77 and 89) between mutant and parental H-2K sequences were found. These observations presumably reflect some unusual property of the three-dimensional structure of MHAs, which renders it particularly sensitive to alterations in certain areas of the molecule. It is interesting that one of these substitutions occurs in the region of the chain where inter-species and inter-locus differences are concentrated (Fig. 6.3). (Nathenson *et al.*, 1979).

It is clear from these and other results that the antigenic properties of MHAs depend on their amino-acid sequences rather than their

214   *J. R. L. Pink*

carbohydrate structure. However, it is still possible that different MHA
alleles may carry different amounts of carbohydrate. In fact, HLA-A2
has one carbohydrate group, while H-2K^b has two (Fig. 6.3). Also, there
are many observations of slight differences in the apparent molecular
weights of different MHA alleles (e.g. Ziegler & Pink, 1976); these might
be due to mutations involving asparagine residues to which carbohydrate
chains can be attached.

Homologies to immunoglobulins

β_2-microglobulin sequences from various mammals are homologous to
each other, and show that the molecule has been evolving at a rate
comparable to that of MHAs (Gates, Coligan & Kindt, 1979). More
surprisingly, there is a striking homology between β_2-microglobulin and
immunoglobulin (Ig) sequences (Tables 6.3, 6.4), so that the three-
dimensional structure of β_2-microglobulin probably resembles that of an
isolated immunoglobulin domain. The presence of a single interchain
disulphide loop, of size 50–70 residues, in each structure, adds weight to
this assumption.

The MHA chains also contain intrachain disulphide bridges, enclosing
similar-sized loops (Fig. 6.2). The sequences at the amino-termini of
various MHAs have given no clear evidence for Ig-MHA homology, but

Table 6.3. *Homology of human β_2-microglobulin ($\beta_2\mu$), and
the second disulphide-bridged loop of the HLA-B7 antigen
(B7.2), to various immunoglobulin light chain and γ chain
constant region domains. Figures represent percent identity
between pairs of sequences. In brackets are the numbers of gaps
or insertions which had to be introduced into each sequence in
order to maximise homology with other sequences. The fragment of the HLA-B7 antigen extends from residue 183 to 271 of
the chain. Human β_2-microglobulin has 99 residues. Con-
densed from Orr et al. (1979a)*

	Cλ	Cγ1	Cγ2	Cγ3	$\beta_2\mu$	B7.2
Cκ(1)	38	25	27	25	20	22
Cλ(2)		32	24	28	15	21
Cγ1(3)			23	27	13	18
Cγ2(3)				24	15	13
Cγ3(1)					24	21
$\beta_2\mu$(4)						23

Table 6.4. *Comparison of the sequences around a cysteine residue in HLA antigens (Trägårdh et al., 1978; Orr et al., 1979a) with sequences from β₂-microglobulin and immunoglobulin domains. The cysteine is residue 203 in HLA, and residue 25 in β₂-microglobulin. In all cases, it forms a disulphide bridge with a second cysteine about 60 residues away towards the C-terminus of the molecule. Homologies of HLA to other sequences are boxed. The arrow marks the position of a two-residue deletion which has been introduced into the Cγ2 sequence to maximise its homology with other sequences*

HLA	ALA	THR	LEU	ARG	CYS	TRP	ALA	LEU	GLY	PHE	TYR	PRO	ALA	GLU	ILE	THR	LEU	THR	TRP
B₂M	ASN	PHE	LEU	ASN	CYS	TYR	VAL	SER	GLY	PHE	HIS	PRO	SER	ASP	ILE	GLU	VAL	ASP	LEU
Cλ	ALA	THR	LEU	VAL	CYS	LEU	ILE	SER	ASP	PHE	TYR	PRO	GLY	ALA	VAL	THR	VAL	ALA	TRP
Cκ	ALA	SER	VAL	VAL	CYS	LEU	LEU	ASN	ASN	PHE	TYR	PRO	ARG	GLU	ALA	LYS	VAL	GLN	TRP
Cγ1	ALA	ALA	LEU	GLY	CYS	LEU	VAL	LYS	ASP	TYR	PHE	PRO	GLU	PRO	VAL	THR	VAL	SER	TRP
Cγ2	PRO	GLU	VAL	THR	CYS	VAL	VAL	VAL	ASP	VAL	SER	HIS	GLU	ASP	PRO	VAL	LYS	PHE	ASN	TRP
Cγ3	VAL	SER	LEU	THR	CYS	LEU	VAL	LYS	GLY	PHE	TYR	PRO	SER	ASP	ILE	ALA	VAL	GLU	TRP

sequences around the second disulphide-bridged loop of human MHAs do show a clear homology to Ig sequences (Tables 6.3, 6.4)(Trägårdh *et al.*, 1978; Orr *et al.*, 1979*b*). The significance of this homology is not yet clear, but it might suggest that the β_2-microglobulin molecule binds to the MHA chain in the region of this loop.

MAJOR HISTOCOMPATIBILITY ANTIGEN FUNCTION

The accidental discovery by Doherty and Zinkernagel (Doherty, Blanden & Zinkernagel, 1976) of the role of MHAs in antiviral immunity, was the first real clue to their function: they demonstrated that thymus-derived (T) lymphocytes from mice immunised with a virus could kill virus-infected target cells *in vitro*, but killing was effective only if the lymphocytes could recognise an MHA, as well as a viral antigen, at the target cell surface. In normal circumstances, the MHAs which are recognised are self-MHAs, those present on infected cells during natural viral infections. Thus, immune lymphocytes from a mouse which carries the H-$2D^d$ and K^d MHA alleles, and which has been infected with a particular virus, will kill virus-infected target cells (but not, of course, normal cells or cells infected with another virus) only if the target cells bear the H-2D^d or K^d antigens.

This phenomenon of 'H-2 restriction' has been observed in several other species (humans, rats) and for many antigens, including not only a variety of viral proteins, but other foreign antigens, such as synthetic haptens (e.g. the trinitrophenyl group) (Shearer, Rehn & Schmitt-Verhulst, 1976), and non-MHA cell surface proteins which are antigenically different from those of the immunised host (Bevan, 1977*a*). In fact, the only certain exceptions to the rule of 'restriction' are the immune responses of T lymphocytes against foreign MHC products; these responses are, paradoxically, particularly strong and, as noted in the introduction, can give rise to very rapid skin-graft rejection.

Why cytotoxic T lymphocytes should need to recognise MHAs on target cells is quite unknown. Perhaps the MHAs can pass a cytotoxic signal through the target cell membrane (Zinkernagel, 1978). However, two facts about MHA recognition have been established: first, individual T lymphocytes recognise only one type of MHA; in an H-2 heterozygous mouse, for example, different populations of lymphocytes recognise maternal and paternal MHAs. Secondly, immature T lymphocytes learn to recognise self-MHAs during their development in the thymus. This is most clearly demonstrated by allowing immature H-2-heterozygous T cells to mature in contact with thymus epithelium of maternal (or paternal) origin; they will subsequently show a strong preference for recognising target cells carrying maternal (or paternal) H-2 antigens (Zinkernagel, 1978). It is perhaps important for this learning that thymic epithelial cells carry MHAs, whereas immature thymocytes do not (Rouse, van Ewijk, Jones & Weissman, 1979; Brown, Biberfeld, Christensson & Mason, 1979).

Recognition of both viral antigen and MHA is probably necessary at the earliest stage of the cytotoxic process, namely the binding of the cytotoxic T lymphocyte to the target cell (Kees, Müllbacher & Blanden, 1978; Goulmy, Hamilton & Bradley, 1979; Vadas, Butterworth, Burakoff & Sher, 1979). Two models to explain how recognition takes place have been proposed (see Zinkernagel, 1978).

In the '*neoantigen*' model, virus protein and MHA form a complex which is antigenically different from either single antigen. The neoantigen thus formed is what the cytotoxic T lymphocyte recognises. The principal objection to this model is that different neoantigens must be formed by combination of one foreign antigen with different MHAs; in addition to difficulties of molecular conformation which this proposal seems to raise, the number of neoantigens which must be recognised by T cells becomes exceedingly large (since it approaches the product of the already large numbers of different foreign antigens and of different MHAs).

Whether or not neoantigens are formed, it seems that MHAs can bind specifically to certain viral proteins, notably Semliki Forest virus spike proteins (Helenius *et al.*, 1978). Unfortunately, most of the evidence for viral protein–MHA complexes is derived from immunofluorescent studies of living cells, rather than the purification of such complexes from cell lysates. Complex formation also provides a simple explanation for the dominant genetic control of resistance to virus infection discussed in the next section: lack of suitable virus–MHA interaction will lead to susceptibility to infection, which will be recessive in heterozygotes. The strong reactivity of cytotoxic T lymphocytes towards foreign MHAs can also be explained, by postulating that 'neoantigens' (i.e. viral protein–self MHA complexes) have a tendency to resemble foreign MHAs. There is, in fact, evidence that single cytotoxic lymphocytes can recognise both foreign antigens on syngeneic cells, and unmodified foreign MHAs (Bevan, 1977*b*; Lemonnier, Burakoff, Germain & Benacerraf, 1977).

In the '*dual recognition*' model, foreign antigen and self-MHA are recognised as two separate entities. However, binding of T cells to self-MHAs cannot occur in the absence of the foreign antigen; if this were not so, the self-MHA receptors on all T lymphocytes would be swamped by MHAs on the body's normal tissues. Perhaps formation of a viral–protein–MHA complex (without production of a 'neoantigen') is needed to enable T cell receptors to bind to both entities.

In this model, it is not obvious that MHA polymorphism should be of any selective advantage, but a number of suggestions which take account of MHA polymorphism have been made. In the variant of von Boehmer, Haas & Jerne (1978), the separate T cell receptors for self-MHAs and for foreign antigens are derived from a common pool of receptors which recognise many different polymorphic forms of MHA. 'Self-receptors' are selected during T cell differentiation in the thymus. 'Antigen-receptors' are then derived by obligatory mutation from 'self-receptors'. Thus, the nature of the antigen-receptor depends on the selected self-receptor, and ultimately on the MHA recognised by it. For example, the model accounts for the observation that mice carrying $H-2D^d$ are good responders to radiation leukemia virus by assuming that an anti-$H-2D^d$ receptor could mutate particularly easily to give an anti-leukemia virus receptor (see next section).

Further discussion of these models and their variants (Zinkernagel, 1978) would lead us too far from our main topic, the action of natural selection on MHA polymorphism. Both models suggest that resistance to viral infection depends on MHA structure, and predict that resistance is dominant in appropriate heterozygotes; so both models can account for selection for MHA heterozygosity.

Genetic control of T cell cytotoxicity

The generation of cytotoxic T cells by mice infected with certain viruses, for example, radiation leukemia virus (RLV) (Meruelo *et al.*, 1978; Meruelo, 1979) or vaccinia virus (Zinkernagel *et al.*, 1978), is strongly influenced by their MHA type. In the case of RLV, mice carrying the $H-2D^d$ allele develop cytotoxic lymphocytes, and are resistant to viral infection; this effect is genetically dominant. $H-2D^s$ or D^q homozygotes are susceptible. A striking finding (Meruelo *et al.*, 1978) is that in the thymuses of resistant, but not of susceptible mice, H-2D synthesis is increased severalfold a few hours after intrathymic injection of the virus. H-2K synthesis is increased in both resistant and susceptible animals. The mechanism of this effect is quite unknown. Recently, Heron, Hokland & Berg (1978) showed that addition of interferon to human lymphocytes rapidly increases the amount of HLA antigens detectable on their plasma membranes, so one might suspect that RLV-induced interferon could account for increased H-2 synthesis; if this is so, however, the differential effects of infection on H-2D and K synthesis are difficult to explain. In any case, those infected target cells which have increased H-2D levels are also better at inducing immunity due to cytotoxic T lymphocytes *in vivo* (Meruelo, 1979); so the increased H-2D levels are probably physiologically important in virus resistance. It will be important to establish whether these findings can be generalised to infection with other viruses, both because of their practical value and because they seem easier to explain by formation of a viral protein-MHA complex than by a simple 'dual recognition' model.

Zinkernagel *et al.* (1978) showed that, for certain viruses, some H-2 alleles were better than others at inducing anti-viral immunity. A sort of hierarchy of alleles could be established: thus, in vaccinia-virus-infected H-2D^b mice, good anti-(virus plus H-2D^b) immunity could be induced unless the mice also had the H-2K^k allele (even in the heterozygous form); in this case, cytotoxicity was preferentially directed against H-2K^k (plus the virus). This result could be explained by supposing that viral proteins form more strongly immunogenic complexes with K^k than with D^b proteins; but other explanations are also possible.

In humans, T cell cytotoxicity to influenza-virus-infected cells is also under genetic control (McMichael, 1978). Many HLA-A2-positive individuals do not produce strong cytotoxicity against HLA-A2-carrying target cells. This example is not as clearcut as in the experimental infections of mice, as there do exist high responders to HLA-A2-positive infected cells (McMichael, 1978; Biddison & Shaw, 1979). In addition, although there is no doubt that influenza infection could be a strong selective force, it is not clear how important cytotoxic T cells, as opposed to humoral antibodies, actually are in resistance to influenza-virus infection.

These examples show that different MHAs can be associated with good immune responses to different viruses, and thus provide a rationale for natural selection of MHA heterozygotes and high levels of MHA polymorphism.

MHA polymorphism and disease

That MHA polymorphism is important in the control of some *natural* virus infections seems very likely, although the two main lines of evidence for this assertion, which are derived from natural virus infections in chickens and from human HLA-disease associations, are both incomplete.

The MHC type of chickens has long been known to be correlated with their fertility and longevity. Very probably, these early results were due to the presence, in the tested flocks, of one or both types of common avian tumour virus – Marek's (a DNA virus) or an RNA virus from the avian leukosis/sarcoma virus group – since it has recently been shown that resistance to both Marek's disease (Longenecker *et al.*, 1976; Briles, Stone & Cole, 1977) and the growth of avian sarcomas (Schierman, Watanabe & McBride, 1977; Collins *et al.*, 1977) are strongly associated with particular MHC alleles. Marek's disease in particular (at least until the recent introduction of a successful vaccine) was a major cause of early death in chicken flocks throughout the world, and thus an important agent of natural selection.

In these experiments, however, it is not clear which locus (or loci) in the chicken MHC confers disease resistance; the loci responsible may code for Ia-like antigens rather than MHAs. Since chickens are not highly outbred, birds with a particular MHA tend also to have identical Ia-like antigens and, in the absence of suitable recombinants, disease resistance cannot be assigned to one or the other type of locus. Furthermore, the mechanisms of resistance to the two diseases are not established: the cytotoxic lymphocytes present in infected birds have not been fully characterised (although there is some evidence that they are T cells in Marek's disease) and there is as yet little or no evidence for MHA-restricted cytotoxicity in chickens (Ross, 1977; Sugimoto, Kodama & Mikami, 1978; Bauer & Fleischer, 1980).

Similar difficulties arise when the evidence for disease association with human HLA type is considered. Strong association between HLA type and susceptibility to a number of diseases is well documented (Table 6.5) (Dausset & Svejgaard, 1977; McMichael & McDevitt, 1977); yet usually we cannot be certain that resistance really depends on the MHA locus studied, rather than on other closely-linked loci. However, since humans are more outbred than chickens, particular MHA alleles are found linked to a large variety of alleles of the other MHC loci; therefore, a very clearcut association between disease susceptibility and MHA type would indicate that the MHA gene itself, and not some linked locus, was responsible for the effect.

The strong association of HLA-B27 with ankylosing spondylitis (a type of arthritis leading to fusion of the vertebrae) has been established in

Table 6.5. *Some examples of HLA and disease associations (selected from Bodmer & Bodmer, 1978). Relative risk* $= p(1-c)/c(1-p)$, *where* p *and* c *are the frequencies of the antigen in patients and controls, respectively. Note that in the case of coeliac disease, the main association is not with an MHA, but with an Ia-like product (HLA-Dw3); the weaker association with an MHA (HLA-B8) may simply result from the fact that the Dw3 and B8 genes are often found together on the same chromosome (they are in linkage disequilibrium; see p. 224. The relative risks for psoriasis associated with the listed antigens are about 5–10-fold lower in Caucasians than in Japanese*

Disease	Antigen	Frequency in patients (%)	Frequency in controls (%)	Relative risk
Ankylosing spondylitis	B 27	90	8	103
Reiter's disease	B 27	80	9	40
Coeliac disease	{ D w3	96	27	64
	{ B 8	67	20	8
Psoriasis	{ B 37	35	2	26
(in Japanese)	{ B 13	18	1	22
	{ C w6	53	7	15
Haemochromatosis	A 3	72	21	10

several different populations and is, to date, the clearest example of a MHA-dependent disease association (Table 6.5). The ankylosing spondylitis example is characteristic of HLA-disease association in several ways. First, the disease is rare, afflicts older people, and seems unlikely to have been important in natural selection; secondly, susceptibility is dominant, the opposite of the findings in studies of experimental virus infection; thirdly, the agent causing the disease is unknown. Why do HLA-disease associations, in general, not show the characteristics, dominant resistance to a common disease, which would best qualify them as important for natural selection? A major reason must be the difficulty of finding enough patients to test for recessive susceptibility to any disease, so that there exists a selective bias towards studying diseases to which susceptibility is dominant. At least two possible types of etiology for such diseases exist: the disease may be autoimmune, so that successful resistance to a virus infection involves tissue destruction and is more harmful than the viral infection itself (as in the case of lymphocytic choriomeningitis in mice) (Doherty *et al.*, 1976); or the disease may be caused by an organism which cross-reacts with an MHA, and thus can infect heterozygotes as well as homozygotes (since both would be immunologically tolerant to the infectious agent). In the case of ankylosing spondylitis, it is not clear which of these mechanisms might apply; the disease has some characteristics of autoimmunity, but it has also been proposed that HLA-B27 cross-reacts with a bacterium (*Klebsiella pneumoniae*) which has been found in many (but not all) patients with ankylosing spondylitis (Ebringer, Cawdell, Cowling & Ebringer, 1978; Seager *et al.*, 1979).

Thus the diseases which might be, or have been, important in maintaining HLA polymorphism have not been identified. One strongly suspects that they could have been common viral diseases, such as smallpox or influenza, since there is, to date, no evidence that T cell cytotoxicity, involving MHAs, is important in bacterial or parasitic infections. Cytotoxic T cells may perhaps play a role in preventing growth of non-viral tumours, but resistance to acute viral disease has probably been a much stronger force for natural selection in human populations.

Population genetics of HLA polymorphism

The striking polymorphism of MHAs might itself be taken as *a priori* evidence for the action of natural selection on MHA structure. Unfortunately, as Crow & Kimura (1970) have shown, a high degree of genetic polymorphism can be maintained by mutation alone in the absence of selection. For example, about ten different alleles can be maintained by simple mutation pressure in a stable, randomly breeding population of size 500, subject to mutation at a rate of 5×10^{-4} per locus per generation. The frequency of homozygotes is about 0.5 in this case, so that many of the alleles are rare. Natural selection for heterozygotes (the likely alternative mechanism for maintaining multiple polymorphism) is relatively inefficient in maintaining multiple alleles: for example, if all homozygotes are 5% less fit than heterozygotes in the above population, their frequency is approximately halved, an effect that could be produced simply by doubling the mutation rate.

In practice, it is difficult to apply such calculations directly to real populations, which migrate, inbreed, expand and contract in idiosyncratic ways. However, three lines of evidence suggest that the distributions of multiple HLA alleles in different human populations cannot be entirely explained by a simple neutral mutation model (Bodmer, Cann & Piazza, 1972). As mentioned above, the frequency distribution predicted by this model is such that a few alleles predominate, while many alleles are rare. In all populations studied, the actual frequency distributions of alleles at the HLA-A and B loci depart significantly from this pattern, with many alleles being about equally common, and the total frequency of homozygotes at a given locus is usually around 0.1 (Fig. 6.4).

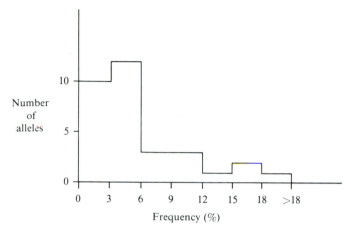

Fig. 6.4. Histogram showing the numbers of HLA-A and HLA-B alleles, whose frequencies in the European population lie between certain values. 15 HLA-A and 19 HLA-B antigens, which account for almost 98% of alleles at these loci, are included. The original allele frequencies are listed in Bodmer & Bodmer (1978).

The neutral mutation model also predicts that populations showing reduced polymorphism at HLA loci, which might arise by inbreeding or genetic drift in small populations, would also be less polymorphic at other loci (e.g. the ABO or MN blood group loci). However, this pattern was not found in population studies of, for example, American Indians, who show significantly less HLA polymorphism than other population groups, but whose blood group polymorphism is not unusual.

Finally, the neutral mutation model cannot account for the existence of only 20 different alleles of (say) the HLA-B locus, if they all differ from one another by perhaps 30 amino-acid substitutions (as the limited sequence data on mouse MHAs seem to suggest). It is possible, however, that the number of different HLA-B products is larger than the 20 which are serologically distinguishable; also, the extent of sequence differences between different HLA alleles is not yet established.

Linked MHC genes. Conservation of the MHC linkage group during the 300×10^6 years of vertebrate evolution might also be considered *a priori* evidence that natural selection has prevented chromosomal rearrangements leading to breakup of the MHC. However, it seems statistically unlikely that any rearrangement leading to non-linkage of MHA genes would have occurred in such a small chromosomal region during this period. The lengths of chromosome much larger than the MHC have been conserved during mammalian evolution (for example, large regions of the X chromosome; or an autosomal region containing various esterase genes, which is about 15 centimorgans long in both mice and rats (Womack & Sharp, 1976)). In fact, the glyoxalase-I gene, which seems unlikely to have any functional requirement for close linkage to the MHC, has remained close to the MHC for at least 50×10^6 years, since it is MHC-linked in both men and mice (Meo, Douglas & Rijnbeek, 1977; Rubinstein, Vienne & Hoecker, 1979).

A second question relating to the action of natural selection on linked MHA loci is more informative: does selection act to maintain the linkage of particular *alleles* of adjacent genes? Since humans are highly outbred, any particular allele of, say, the HLA-A locus can be found linked to many allelic forms of the HLA-B locus. In fact, in a stable population, in the absence of inbreeding and selection, one would expect to find all HLA-B alleles randomly distributed on the chromosomes carrying any given HLA-A allele. However, preferential associations between certain alleles of the two loci do exist: HLA-A1, for example, is found more frequently linked to HLA-B8 than would be expected simply from the relative frequencies of the two individual alleles. This phenomenon (linkage disequilibrium) is found in several different populations, suggesting that natural selection has acted to maintain the HLA-A1/B8 combination at a higher than expected frequency (Bodmer *et al.*, 1972; Bodmer & Bodmer, 1978).

CONCLUSIONS

MHA genes are the most polymorphic known in vertebrates. The polymorphism is expressed as multiple amino-acid sequence differences between different MHAs. The evidence that natural selection acts to maintain this high level of polymorphism forms a patchwork, with many holes, on which the outlines of a coherent picture are visible. The picture has three parts.

First, cytotoxic T lymphocytes need to recognise MHAs, as well as viral antigens, on the surface of virus-infected target cells for lysis of the targets to occur. Why this should be is unknown, but the phenomenon has been clearly established in several mammalian species (mice, rats, humans) and for several different viruses, as well as for non-viral antigens. There are genetic variations in the ability of some MHAs to be recognised in association with some viruses, possibly because MHAs and viral proteins vary in their ability to form molecular complexes at the cell surface. The clearest results are derived from leukemia and other viral infections in mice, but there is also evidence that MHA type affects the cytotoxic response to influenza virus-infected human target cells.

Secondly, the susceptibility of mice, chickens and humans to various diseases is associated with MHA type. Studying inbred mice tells us little about natural selection. In the cases of avian leukosis and Marek's disease, which are important agents of natural selection in chickens, MHA type is strongly correlated with the outcome of viral infection; however, it is not yet known whether T-cell cytotoxicity is involved in disease resistance, or indeed whether the phenomenon of MHA restriction even exists in chickens. The human diseases which show associations with particular MHA alleles are selected for study by a statistically biased procedure, and are, for the most part, rare and of unknown etiology; susceptibility to them is unusual in being dominant. These diseases are very unlikely to be important in the maintenance of human MHA (HLA) polymorphism.

Thirdly, there is nevertheless good evidence, from studies of the population genetics of HLA polymorphism, that natural selection has acted both to alter the frequencies of alleles at individual HLA loci, and to maintain the linkage of particular alleles at adjacent loci. The selective agents, presumably common viral infections, responsible for these effects remain to be identified.

REFERENCES

Bauer, H. & Fleischer, B. (1980). The immunobiology of avian RNA tumor virus induced cell surface antigens. In *Mechanisms of immunity to virus-induced tumors*, ed. J. W. Blasecki (in press). New York: Marcel Dekker.

Bevan, M. J. (1977*a*). Cytotoxic T-cell response to histocompatibility antigens: the role of H-2. *Cold Spring Harbor symposium on quantitative biology*, **41**, 519–27.

Bevan, M. J. (1977*b*). Killer cell reactive to altered-self antigens can also be alloreactive. *Proceedings of the national academy of sciences of the USA*, **74**, 2094–8.

Biddison, W. E. & Shaw, S. (1979). Differences in HLA antigen recognition by human influenza virus-immune cytotoxic T cells. *Journal of immunology*, **122**, 1705–9.

Bodmer, W. F. & Bodmer, J. G. (1978). Evolution and function of the HLA system. *British medical bulletin*, **34**, 309–16.

Bodmer, W. F., Cann, H. & Piazza, A. (1972). Differential genetic variability among polymorphisms as an indicator of natural selection. In *Histocompatibility testing, 1972*, ed. J. Dausset & J. Colombani, pp. 753–67. Copenhagen: Munksgaard.

Briles, W. E., Stone, H. A. & Cole, R. K. (1977). Marek's disease: effect of *B* histocompatibility alleles in resistant and susceptible chicken lines. *Science*, **195**, 193–5.

Brown, G., Biberfeld, P., Christensson, B. & Mason, D. Y. (1979). The distribution of HLA on human lymphoid, bone marrow and peripheral blood cells. *European journal of immunology*, **9**, 272–5.

Cohen, N. & Collins, N. H. (1977). Major and minor histocompatibility systems of ectothermic vertebrates. In *The major histocompatibility system in man and animals*, ed. D. Götze, pp. 313–37. Berlin: Springer-Verlag.

Coligan, J. E., Kindt, T. J., Ewenstein, B. M., Uehara, H., Martinko, J. M. & Nathenson, S. G. (1979). Further structural studies of the murine H-2K^b glycoprotein using radio-chemical methodology. *Molecular immunology*, **16**, 3–8.

Collins, W. M., Briles, W. E., Zsigray, R. M., Dunlop, W. R., Corbett, A. C., Clark, K. K., Marks, J. L. & McGrail, T. P. (1977). The *B* Locus (MHC) in the chicken: association with the fate of RSV-induced tumors. *Immunogenetics*, **5**, 333–43.

Cook, R. G., Siegelman, M. H., Capra, J. D., Uhr, J. W. & Vitetta, E. (1979). Structural studies on the murine Ia alloantigens. *Journal of immunology*, **122**, 2232–7.

Crow, J. F. & Kimura, M. (1970). *An introduction to population genetics theory*, pp. 433–78. New York: Harper & Row.

Dausset, J. & Svejgaard, A. (1977). *HLA and disease*. London: Williams and Wilkins.

Doherty, P. C., Blanden, R. V. & Zinkernagel, R. M. (1976). Specificity of virus-immune effector T cells for H-2K or H-2D compatible interactions: Implications for H-antigen diversity. *Transplantation reviews*, **29**, 89–124.

Ebringer, R. W., Cawdell, D. R., Cowling, P. & Ebringer, A. (1978). Sequential studies in ankylosing spondylitis. Association of *Klebsiella pneumoniae* with active disease. *Annals of the rheumatic diseases*, **37**, 146–51.

Gates, F. T., Coligan, J. E. & Kindt, T. J. (1979). Complete aminoacid sequence of rabbit β_2-microglobulin. *Biochemistry*, **18**, 2267–72.

Götze, D. (ed.) (1977). *The major histocompatibility system in man and animals*. Berlin: Springer-Verlag.

Goulmy, E., Hamilton, J. D. & Bradley, B. A. (1979). Anti-self HLA may be clonally expressed. *Journal of experimental medicine*, **149**, 545–50.

Helenius, A., Morein, B., Fries, E., Simons, K., Robinson, P., Schirrmacher, V., Terhorst, C. & Strominger, J. L. (1978). Human (HLA-A and HLA-B) and murine (H-2K and H-2D) histocompatibility antigens are cell surface receptors for Semliki Forest virus. *Proceedings of the national academy of sciences of the USA*, **75**, 3846–50.

Heron, I., Hokland, M. & Berg, K. (1978). Enhanced expression of β_2-microglobulin and HLA antigens on human lymphoid cells by interferon. *Proceedings of the national academy of sciences of the USA*, **75**, 6215–19.

Jones, P. P., Murphy, D. B., Hewgill, D. & McDevitt, H. O. (1979). Detection of a common polypeptide chain in I-A and I-E sub-region immunoprecipitates. *Molecular immunology*, **16**, 51–60.

Kees, V., Müllbacher, A. & Blanden, R. V. (1978). Specific adsorption of H-2-restricted cytotoxic T cells to macrophage monolayers. *Journal of experimental medicine*, **148**, 1711–15.

Kehry, M., Sibley, C., Fuhrman, J., Schilling, J. & Hood, L. (1979). Aminoacid sequence of a mouse immunoglobulin μ chain. *Proceedings of the national academy of sciences of the USA*, **76**, 2932–6.

Kimball, C. S., Coligan, J. E. & Kindt, T. J. (1979). Structural characterisation of antigens encoded by rabbit RLA-11 histocompatibility genes. *Immunogenetics*, **8**, 201–12.

Klein, J. (1978). H-2 mutations: their genetics and effect on immune functions. *Advances in immunology*, **26**, 56–147.

Klein, J. (1979). The major histocompatibility complex of the mouse. *Science*, **203**, 516–21.

Lachmann, P. J. & Hobart, M. J. (1978). Complement genetics in relation to HLA. *British medical bulletin*, **34**, 247–52.

Lemonnier, F., Burakoff, S. J., Germain, R. N. & Benacerraf, B. (1977). Cytolytic thymus-derived lymphocytes specific for allogeneic stimulator cells cross-react with chemically-modified syngeneic cells. *Proceedings of the national academy of sciences of the USA*, **74**, 1229–33.

Longenecker, B. M., Pazderka, F., Gavora, J. S., Spencer, J. L. & Ruth, R. F. (1976). Lymphoma induction by herpesvirus: resistance associated with a major histocompatibility gene. *Immunogenetics*, **3**, 401–7.

Maizels, R. M., Frelinger, J. A. & Hood, L. (1978). Partial aminoacid sequences of mouse transplantation antigens. *Immunogenetics*, **7**, 425–44.

McMichael, A. (1978). HLA restriction of human cytotoxic T lymphocytes specific for influenza virus. Poor recognition of virus associated with HLA-A2. *Journal of experimental medicine*, **148**, 1458–67.

McMichael, A. & McDevitt, H. O. (1977). The association between the HLA system and disease. *Progress in medical genetics*, **II**, 39–100.

McMillan, M., Cecka, J. M., Murphy, D. B., McDevitt, H. O., and Hood, L. (1978). Partial aminoacid sequences of murine Ia antigens of the I-EC^d sub-region. *Immunogenetics*, **6**, 137–47.

Meo, T., Douglas, T. & Rijnbeek, A. M. (1977). Glyoxalase-I polymorphism in the mouse: a new genetic marker linked to H-2. *Science*, **198**, 311–13.

Meruelo, D. (1979). A role for elevated H-2 antigen expression in resistance to neoplasia caused by radiation-induced leukemia virus. Enhancement of effective tumor surveillance by killer lymphocytes. *Journal of experimental medicine*, **149**, 898–909.

Meruelo, D., Nimelstein, S., Jones, P., Lieberman, M. & McDevitt, H. O. (1978). Increased synthesis and expression of H-2 antigens as a result of radiation leukemia virus infection. A possible mechanism for H-2 linked control of virus-induced neoplasia. *Journal of experimental medicine*, **147**, 470–87.

Milstein, C., Adetugbo, K., Cowan, N. J., Köhler, G., Secher, D. S. & Wilde, C. D. (1977). Somatic cell genetics of antibody-secreting cells: studies of clonal diversification and analysis by cell fusion. *Cold Spring Harbor symposium on quantitative biology*, **4**, 793–803.

Nathenson, S. G., Ewenstein, B. M., Uehara, H., Martinko, J. M., Coligan, J. E. & Kindt, T. J. (1979). Structure of H-2 major histocompatibility products: recent studies on the H-2K^b glycoprotein and on the H-2K^b MHC mutants. In *Current Trends in Histocompatibility*, ed. R. Ferrone & R. A. Reisfeld. New York: Plenum Press (in press).

Orr, H. T., Lancet, D., Robb, R. J., Lopez de Castro, J. A. & Strominger, J. L. (1979*a*). The heavy chain of a human histocompatibility antigen (HLA-B7) contains an immunoglobulin-like region. *Nature*, **282**, 266–70.

Orr, H. T., Lopez de Castro, Parham, P., Ploegh, H. L. & Strominger, J. L. (1979*b*). Comparison of aminoacid sequences of two human histocompatibility antigens, HLA-A2 and HLA-B7. Location of putative alloantigenic sites. *Proceedings of the national academy of sciences of the USA*, **76**, 4395–9.

Pink, J. R. L., Droege, W., Hála, K., Miggiano, V. C. & Ziegler, A. (1977). A three-locus model for the chicken major histocompatibility complex. *Immunogenetics*, **5**, 203–16.

Ploegh, H. L., Cannon, L. E. & Strominger, J. L. (1979). Cell-free translation of the mRNAs for the heavy and light chains of HLA-A and HLA-B antigens. *Proceedings of the national academy of sciences of the USA*, **76**, 2273–7.

Pober, J. S., Guild, B. C. & Strominger, J. L. (1978). Phosphorylation *in vivo* and *in vitro* of human histocompatibility antigens (HLA-A and HLA-B) in the carboxy-terminal intracellular domain. *Proceedings of the national academy of sciences of the USA*, **75**, 6002–6.

Ross, L. J. N. (1977). Antiviral T cell-mediated immunity in Marek's disease. *Nature*, **268**, 644–6.

Rouse, R. V., van Ewijk, W., Jones, P. P. & Weissman, I. L. (1979). Expression of MHC antigens by mouse thymic dendritic cells. *Journal of immunology*, **122**, 2508–16.

Rubinstein, P., Vienne, K. & Hoecker, G. F. (1979). The location of the C3 and GLO (Glyoxalase-I) loci of the IXth linkage group in mice. *Journal of immunology*, **122**, 2584–9.

Schierman, L. W., Watanabe, D. H. & McBride, R. A. (1977). Genetic control of Rous sarcoma virus regression in chickens: Linkage with the major histocompatibility complex. *Immunogenetics*, **5**, 325–32.

Schwartz, B. D., Cigen, R., Berggard, I. & Shevach, E. M. (1978). Guinea pig homologues of the TL and QA-2 antigens. *Journal of immunology*, **121**, 835–9.

Schwartz, B. D. & Cullen, S. E. (1978). Chemical characteristics of Ia antigens. In *Springer seminars in immunopathology*, ed. B. Benacerraf, **1**, pp. 85–110. Berlin: Springer-Verlag.

Seager, K., Bashir, H. V., Geczy, A. F., Edmonds, J. & de Vere-Tyndall, A. (1979). Evidence for a specific B27-associated cell surface marker on lymphocytes of patients with ankylosing spondylitis. *Nature*, **277**, 68–70.

Shearer, G. M., Rehn, T. G. & Schmitt-Verhulst, A.-M. (1976). Role of the murine major histocompatibility complex in the specificity of *in vitro* T-cell-mediated lympholysis against chemically modified autologous lymphocytes. *Transplantation reviews*, **29**, 222–48.

Sugimoto, C., Kodama, H. & Mikami, T. (1978). Anti-tumor immunity against Marek's disease-derived lymphoblastoid cell line (MSB-1). *Japanese journal of veterinary research*, **26**, 57–67.

Trägårdh, L., Wiman, K., Rask, L. & Peterson, P. A. (1978). Aminoacid sequence homology between HLA-A,B,C antigens, β_2-microglobulin and immunoglobulins. *Scandinavian journal of immunology*, **8**, 563–8.

Vadas, M. A., Butterworth, A. E., Burakoff, S. & Sher, A. (1979). Major histocompatibility complex products restrict the adherence of cytolytic T lymphocytes to minor histocompatibility antigens or to tri-nitrophenyl determinants on schistosomula of *Schistosoma Mansoni*. *Proceedings of the national academy of sciences of the USA*, **76**, 1982–5.

Vitetta, E. S. & Capra, J. D. (1978). The protein products of the murine 17th chromosome: Genetics and Structure. *Advances in immunology*, **26**, 147–92.

von Boehmer, H., Haas, W. & Jerne, N. K. (1978). Major histocompatibility complex-linked immune-responsiveness is acquired by lymphocytes of low-responder mice differentiating in thymus of high-responder mice. *Proceedings of the national academy of sciences of the USA*, **75**, 2439–42.

Womack, J. E. & Sharp, M. (1976). Comparative autosomal linkage in mammals: genetics of esterases in *Mus musculus* and *Rattus norvegicus*. *Genetics*, **82**, 665–75.

Ziegler, A. & Milstein, C. (1979). A small polypeptide different from β_2-microglobulin associated with a human cell surface antigen. *Nature*, **279**, 243–4.

Ziegler, A. & Pink, J. R. L. (1976). Chemical properties of two antigens controlled by the major histocompatibility complex of the chicken. *Journal of biological chemistry*, **251**, 5391–6.

Zinkernagel, R. M. (1978). Speculations on the role of major transplantation antigens in cell-mediated immunity against intracellular parasites. *Current topics in microbiology and immunology*, **82**, 113–18.

Zinkernagel, R. M., Althage, A., Cooper, S., Kreeb, G., Klein, P. A., Sefton, B., Flaherty, L., Stimpfling, J., Shreffler, D. & Klein, J. (1978). Ir-genes in H-2 regulate generation of anti-viral cytotoxic T cells. Mapping to *K* or *D* and dominance of unresponsiveness. *Journal of experimental medicine*, **148**, 592–606.

7

Immunoglobulins

J. R. L. PINK

The antibody function of immunoglobulins (Igs) is essential in defence against a large number of bacterial and viral infections. To get some idea of the number of different antibodies which an animal can make, consider the number of Ig-producing lymphocytes in an adult mouse, about 2×10^8. Since each of these cells makes antibody of only one specificity, this is an upper limit to the mouse's antibody heterogeneity (at least at a given moment in its lifespan). An approximate lower limit is derivable from the precursor frequencies of antigen-reactive lymphocytes (Lefkovits, 1974). Typically, about one in 10^5 mouse lymphocytes responds to a given antigen, suggesting that the animal can make at least 10^5 different Ig structures. In addition, antibody responses to single antigens are generally heterogeneous: the number of different antibodies which an individual mouse can make to a single antigen is often of the order of 10 to 100. These figures, and other more indirect considerations, suggest that a mouse can make between 10^6 and 10^7 structurally different Igs (Jerne, 1974; Köhler, 1976; Williamson, 1976; Sigal & Klinman, 1978).

How is this enormous diversity produced? Although many different experimental approaches to the problem have been tried, the most fruitful lines of attack have concentrated on tumours of antibody-producing cells induced in inbred mice (Potter, 1978). These tumours secrete large amounts of homogeneous Igs (myeloma proteins), which can be classified and sequenced in order to establish the nature and extent of Ig variability. In addition, the Ig-encoding DNA from the tumour cells can be isolated, sequenced, and compared with embryonic DNA coding for Igs; such experiments can reveal the extent to which Ig genes are modified somatically during the ontogeny of the immune system.

In many recent books and reviews, Ig variability and evolution are discussed in terms of the now very extensive data on Ig protein structure (over 10 000 amino acids in different Igs have been sequenced) (Nisonoff,

Hopper & Spring, 1975; Putnam, 1977; Fudenberg, Pink, Wang & Douglas, 1978; Dayhoff, 1978). In this chapter I have put more emphasis on current work on the structure of Ig genes; the four sections of the review deal, in order, with their organisation, somatic modification, evolution and polymorphism.

ORGANISATION OF IG GENES

Ig structure

In the symmetrical, four-chain structure of a typical Ig molecule (Fig. 7.1), two *light chains*, which can be of either κ or λ type, are disulphide-bridged to a pair of *heavy chains*; there exist different classes and subclasses of heavy chains (μ, α, $\gamma1$, $\gamma2$ and so on), which confer different biological functions on the corresponding Ig molecules (IgM, IgA, IgG1, IgG2, etc.). The different classes and subclasses of mouse Igs, and their component chains, are listed in Table 7.1. Analysis of human

Table 7.1. *Mouse immunoglobulin chains*

Chain	Approximate molecular weight $\times 10^{-3}$	Number of C domains	Number of different alleles described[a]	Corresponding Ig-class[b]; approximate concentration in serum (mg/ml)[c]
Heavy				
$\gamma1$	50	3	2	G 1:2
$\gamma2a$	50	3	12	G 2a:2
$\gamma2b$	50	3	6	G 2b:2
$\gamma3$	50	3	–	G 3:0.2
α	55	3	5	A:0.5
μ	70	4	3	M:1
δ	60	3^d	2	D:very low
ε	70	4^d	–	E:very low
Light				
κ	25	1	2	–
$\lambda1$	25	1	–	–
$\lambda2$	25	1	–	–

[a] In inbred strains of mice (Lieberman, 1978).
[b] Molecules of a given class have the composition H_2L_2, where H is the corresponding heavy chain, and L a κ or λ light chain. Exceptions are serum IgM: $(\mu_2L_2)_5$. JC, where JC is a 15 000 dalton polypeptide required to form the IgM pentamer; and secretory IgA: $(\alpha_2L_2)_2$. JC . SC, where SC is a 70 000 dalton 'secretory component'.
[c] Concentrations are very dependent on age, strain and environment.
[d] By analogy with human homologues.

and rat myeloma proteins, and normal Igs in other species, suggests that very similar Ig classes exist in all mammals; the numbers and types of subclasses, however, may differ in different species.

All Ig chains consist of an amino-terminal *variable region* (V), about 100 amino acids long, and a carboxy-terminal *constant region* (C) (Fig. 7.1). C regions contain one (in the case of light chains) or more (in the case of heavy chains) *domains*, each domain also being about 100 residues long and containing a single intrachain disulphide bridge (Fig. 7.1); there is an obvious amino-acid sequence homology between different C region domains, and a weak homology between V and C region sequences (Moore & Goodman, 1977). Many heavy chains also contain short 'extra-domain' sequences, about 20–40 residues long, which are located at the carboxy-terminal end of the chain (the *tail* sequences of μ and α chains), or in a region of flexibility at the centre of the chain (the *hinge* sequences present in γ, α, and δ chains) (Putnam, 1977).

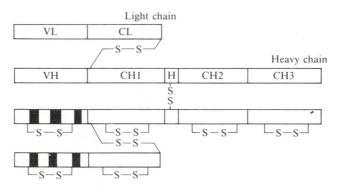

Fig. 7.1.   Schematic illustration of an IgG molecule, containing two light and two heavy chains linked by disulphide bridges (–S–S–). Variable domains (VL and VH) as well as constant domains (CL and CH1, CH2, CH3) each contain an intrachain disulphide bridge (marked for one chain of each type only). The hinge region (H) of the heavy chain, and the hypervariable regions (black regions in VL and VH) are also indicated.

Different antibodies have different V region sequences, and parts of the V regions, its *hypervariable portions* (HV), form the molecule's antigen-binding sites. V region sequences fall into three distinct families, Vλ, Vκ and VH, which are associated with the C regions of λ, κ and heavy chains respectively. Each family of V regions is coded by a gene, or set of genes, which is separate from but closely linked to the corresponding C gene(s). At least in mammals, the DNA sequences coding for λ, κ and heavy chains are very probably located on three different chromosomes (Kunkel & Kindt, 1975; Hengartner, Meo & Müller, 1978). Although there is clear evidence from amino-acid sequence data that the genes coding for these different types of chain have all been derived from a common ancestor, the actual arrangement of Ig genes is best described by considering separately the three types of chain.

Lambda chain genes

About 5% of mouse Ig light chains are of type $\lambda 1$, and about 1% are $\lambda 2$. The V regions of these two chain types are homologous (about 90% sequence identity) while the C regions are less so (70% sequence identity). In contrast to κ chains, and λ chains of other species, mouse λ chains show very limited variability, whether they are isolated from myeloma proteins (Weigert & Riblet, 1977) or normally-induced antibodies (Schilling, Hansburg, Davie & Hood, 1979); indeed, over half of the myeloma λ chains sequenced have identical structures. While this situation facilitates the study of λ genes, one should remember that the mouse λ system is unusual.

DNA rearrangement. Nucleic acid cloning and hybridisation studies show that the number of $\lambda 1$ V and C genes present in haploid embryonic DNA is very small: there is probably only one of each (Tonegawa, 1976; Honjo, Packman, Swan & Leder, 1976). The implications of this result for the mechanism of antibody variability are discussed in a later section. In embryonic cells, the DNA coding for most of the V region of a mouse $\lambda 1$ chain is very distant from the $\lambda 1$ C region gene (Fig. 7.2). In the DNA of a $\lambda 1$-producing plasma cell, however, the V and C regions have been rearranged and are close to each other, but still separated by about 1200 nucleotides (Brack & Tonegawa, 1977). The point between V and C regions where the rearrangement of embryonic $\lambda 1$ DNA occurs is not, in fact, at the classical V–C junction as determined from protein sequence studies, but at a point within the V region which divides it into two parts, a large V piece, and a smaller J piece. The latter codes for 13 amino acids in the V region, but is close to the C region even in embryonic DNA (Fig. 7.2) (Bernard, Hozumi & Tonegawa, 1978; Brack, Hirama, Lenhard-Schuller & Tonegawa, 1978).

A possible mechanism for the rearrangement of $\lambda 1$ DNA during lymphocyte differentiation was suggested by Sakano, Huppi, Heinrich &

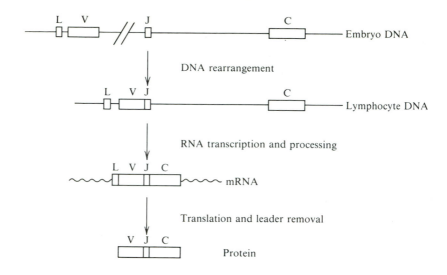

Fig. 7.2. Organisation and expression of mouse λ chain genes. In embryonic DNA, most of the V gene (V), plus a small leader (L) segment, are widely separated from a small portion of the V gene, the J or joining piece, and from the C region gene. V and J are juxtaposed in λ-producing lymphocyte DNA. The sequences intervening between L and V, and J and C genes are removed after transcription, during biosynthesis of mRNA.

Tonegawa (1979) and is illustrated in Fig. 7.3. DNA sequences flanking the V and J regions can form a short base-paired region, giving rise to a looped structure which could be removed enzymatically, with subsequent repair at the V–J junction. These authors point out that prokaryotic insertion elements might be removed from host DNA by a similar mechanism.

RNA processing.   Even after rearrangement, the DNA coding for a $\lambda 1$ chain in a $\lambda 1$-producing cell is still not a continuous structure: it consists of two separate regions coding for the V and C domains of the chain, and a third sequence coding for a short hydrophobic 'leader', about 15 amino acids long, which is present at the amino-terminus of newly synthesised mouse $\lambda 1$ polypeptides. Similar leader sequences are present in all newly synthesised Ig chains (Schechter, Wolf, Zemell & Burstein, 1979); they are removed during or immediately after synthesis of the complete Ig polypeptide, and are probably responsible for ensuring that the chain is synthesised on membrane-bound polysomes.

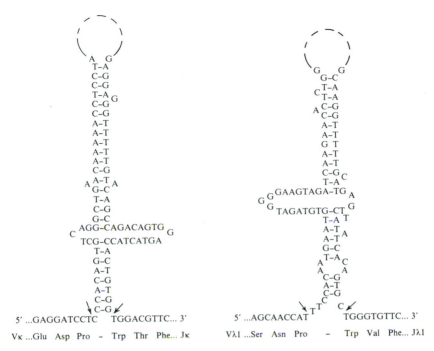

Fig. 7.3. Looped-out, hydrogen-bonded DNA structures which can be formed between flanking DNA sequences of V and J pieces in (left) κ and (right) λ chain genes. Only the coding DNA strand is shown. Dashes connect complementary base pairs. Arrows indicate sites of breakage and repair to create a continuous V–J sequence. In κ chains, the site of V–J joining can vary. Adapted from Sakano *et al.*, 1979a.

The intervening DNA sequences which separate leader from V, and V from C regions, are transcribed with the coding sequences into an mRNA precursor, which is then processed (Gilmore-Herbert & Wall, 1978; Schibler, Marcu & Perry, 1978) to give a light chain mRNA in which leader, V and C regions are truly adjacent. In the DNA, the borders of the intervening sequences are marked by GT and AG sequences (as is the case for intervening sequences in other eukaryotic genes), but how the RNA processing mechanism operates is unknown (Table 7.2).

The organisation of mouse λ2 genes is probably similar to that of λ1 genes; both the λ1 and λ2V genes have been isolated and sequenced (Bernard *et al.*, 1978; Tonegawa *et al.*, 1978). However, the relative locations of λ1 and λ2V and C genes are unknown; it is not even clear that they are all carried on a single chromosome. The evolution of duplicated λ chain genes is considered in a later section (p. 249).

Table 7.2. *Comparison of nucleotide sequences around junctions between coding and intervening sequences in immunoglobulin genes. (Honjo et al., 1979; Tucker et al., 1979a; Sakano et al., 1979b). Solid line indicates identity to the sequence immediately above*

Gene		Coding sequence ⟶	⟶ Intervening sequence ⟶	⟶ Coding sequence	
λ1	J piece	···· GTCCTAG	GT̲GAGTC ···· CCTGCAG̲	GCCAGCC ····	C gene
λ1	Leader	···· AGATCAG	GT̲CAGCA ···· TTTGCAG̲	GGGCCAT ····	V piece
λ2	Leader	··· T —C—	····	—A——G ····	V piece
κ	Leader	···· TTTCAAG	GT̲TAAAA ···· TCCTCAG̲	GGGCTGA ····	V piece
γ1			···· CTTGTAG̲	CCAAAAC ····	CH1
γ2b			···· —C—	—C— ····	CH1
γ1	CH1	···· AAAATTG	GT̲GAGAG ···· TCCACAG̲	TGCCCAG ····	Hinge
γ2b	CH1	···· —C—	···· —TG—	A———— ····	Hinge
γ1	Hinge	···· TGTACAG	GT̲AAGTC ···· TCCTTAG̲	TCCCAGA ····	CH2
γ2b	Hinge	···· —CC—	···· —A—C	CT —TA— ····	CH2
γ1	CH2	···· ACCAAAG	GT̲GAGAG ···· CTCACAG̲	GCAGACC ····	CH3
γ2b	CH2	··· —TT—	—G—C ···· —C—	—GCT —GT ····	CH3

Conserved nucleotide pairs (GT and AG) at ends of intervening sequences are underlined. In a few cases where the sequences published by different authors differed, or the exact point at which the coding sequence ends is not known, sequences were taken to comply with the GT ... AG rule.

Kappa chain genes

The organisation of mouse κ chain genes is more complex than that of λ chain genes (Fig. 7.4). The Cκ region amino-acid sequence is unique, and only one Cκ region gene has been detected by DNA cloning, restriction mapping, and hybridisation experiments (see Seidman *et al.*, 1978*a*, *b*). However, mouse κ chain V regions exhibit great sequence diversity. About 100 Vκ regions have been partially or completely sequenced: the sequences can be classified, according to their degree of relatedness, into groups whose members are at least 90% homologous, with differences largely restricted to HV regions. About 30 different groups have been defined (of which some have only one member), and a statistical estimate of the expected total number of groups is of the order of 50 (Cohn *et al.*, 1974; Potter, 1978).

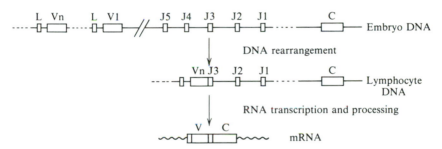

Fig. 7.4. Organisation and expression of mouse κ genes. See Fig. 7.2 for description. C and J pieces are separated by about 3700 bases. The number of V genes is ≥ 50; the spaces between them are probably at least 4000 bases long.

Multiple V genes. Since the sequences of proteins from different V region groups are not very closely homologous, one would expect that different genes would code for members of the different V region groups. Nucleic acid hybridisation and restriction enzyme mapping of embryonic DNA have confirmed this supposition. It is more difficult to decide how many genes code for the closely related members of a single group. Valbuena, Marcu, Weigert & Perry (1978) used saturation hybridisation to show that about 4–6 genes coded for the Vκ 21 group, which constitutes about 8–9% of total κ chains (Weigert *et al.*, 1978); these figures suggest that 50–70 Vκ genes exist in the germline. Independently, Seidman *et al.* (1978*a*, *b*) using restriction enzyme mapping, characterised at least six different embryonic DNA fragments coding for members of the M–41 group, and a similar number coding for the M–149 group. Both Weigert *et al.* (1978) and Seidman *et al.* (1978*b*) assume that the total number of Vκ groups is about 50, and suggest that several hundred different Vκ genes exist in embryonic DNA. At present, the Vκ gene number is still open to uncertainty, and could be as small as 50, or as large as 500.

Like λ chains, κ chains are encoded by separate DNA regions in the embryonic genome: a leader sequence, a V sequence, a J sequence, and the C region (Fig. 7.3). Again, like λ chain DNA, κ DNA undergoes rearrangement, which exactly juxtaposes V and J sequences (Rabbits & Forster, 1978; Seidman & Leder, 1978; Lenhard-Schuller *et al.*, 1978). The rearranging mechanism is presumably similar to that involved in λ V–J joining, since similar matching sequences flank the V and J regions of both chain types (Fig. 7.3).

Multiple J genes. Mouse κ chains differ from λ chains not only in having many more V genes, but also in having more J genes. Max, Seidman & Leder (1979) and Sakano *et al.* (1979*a*) isolated an embryonic DNA fragment containing five different Jκ sequences, interspersed with non-coding sequences 250–300 base pairs long. Amino-acid sequence studies of mouse κ chains show that at least four of these five J pieces have been used in different κ chains. Two further interesting points emerged from the DNA–amino-acid sequence comparisons: first, different Vκ genes can be joined to the same Jκ piece (and vice versa) (Weigert *et al.*, 1978; Rao, Rudikoff, Krütsch & Potter, 1979); and second, the exact point at which the V–J recombination is located lies in the third HV region of the κ gene, but can vary over a stretch of about four nucleotides, so that a single Vκ can be recombined with a single Jκ in several different ways. Table 7.3 illustrates this point by showing how three different sequences, all of which have actually been observed in mouse κ chains, can be generated from one Vκ–Jκ pair (Max *et al.*, 1979; Sakano *et al.*, 1979*a*).

Once the V–J joining event has occurred, transcription and processing of the intervening sequences located between the leader and V segments, and the J and C pieces, probably occur in a similar way to that involved in λ chain production. However, an extra difficulty arises in the processing of κ chain mRNA, which presumably may contain more than one J sequence (Fig. 7.3); it is not clear how the processing system selects for translation the only J sequence which is associated with a V region.

Heavy chain genes

C regions. The heavy chain gene family contains multiple, closely-linked C genes, as well as multiple V and J units. The genes coding for the C regions of mouse $\gamma 1$, $\gamma 2b$ and α chains have been cloned, and complete sequences of the first two of these are available (Honjo *et al.*, 1979; Tucker *et al.*, 1979*a*). These results, besides establishing the separation of heavy chain V and C genes in embryonic DNA, also revealed that the C region genes themselves had an interrupted structure (Fig. 7.5): the Cγ genes consist of four separate coding sequences, which encode, in order, the C$\gamma 1$ domain, the hinge region, and the C$\gamma 2$ and C$\gamma 3$ domains of the chain (Sakano *et al.*, 1979*b*; Tucker *et al.*, 1979*a*; Honjo *et al.*, 1979). The α gene has a similar arrangement (Early *et al.*, 1979), suggesting that the primitive heavy chain ancestral to γ, α and other classes already contained intervening DNA sequences.

A detailed presentation and discussion of these data (over 1800 sequenced nucleotides for each of the γ clones) cannot be presented here, and the following is only a summary of the main conclusions that can be drawn from the $\gamma 1$ and $\gamma 2b$ sequence data (Tucker *et al.*, 1979*a*; Tucker, Marcu, Slightom & Blattner, 1979*b*; Honjo *et al.*, 1979; Rogers, Clarke & Salser, 1979; Sakano *et al.*, 1979*b*).

(1) As expected from the amino-acid sequence data, there is strong (about 60%) homology between corresponding domains of different subclasses and a weaker (about 40%) homology between different domains of the same chain. Surprisingly, one 17-nucleotide region of the $\gamma 2b$ CH1 and CH2 domains can be matched almost perfectly in an out-of-phase reading frame, so that the corresponding amino-acid sequences are quite different; this suggests that a frameshift mutation has occurred, and been tolerated during the evolution of the domains.

(2) In the $\gamma 2b$ gene, there is a 38% homology between the CH1 domain and a region of the longest intervening sequence (that preceding the hinge) plus the hinge itself (Fig. 7.5). This intervening sequence, plus the hinge, therefore are probably derived from what was originally an extra domain, corresponding to that present in μ chains (which, in line with this idea, lack a hinge region). However, this homology was not observed in the $\gamma 1$ gene; instead, Honjo *et al.* (1979) noted a weaker similarity, whose significance is unknown, between the hinge and the start of the C$\gamma 2$ domain.

(3) The sequences at the borders of the intervening DNA show homology to each other (Table 7.2); otherwise, the intervening

Length of coding sequences

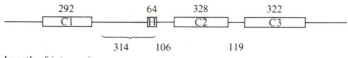

Length of intervening sequences

Fig. 7.5. Organisation of a mouse γ C region gene. The gene contains four sections, coding for the Cγ1, hinge (H), Cγ2 and Cγ3 domains. There is a significant homology between DNA sequences coding for different domains, and also (at least in the γ2b gene) between the domains and the bracketed DNA region, as discussed in the text. Lengths of domains and intervening sequences are given in nucleotide pairs. Tucker *et al.*, 1979*a*.

sequences in γ1 and γ2b chains are less homologous than are the coding sequences. A similar rapid divergence of non-coding DNA sequences has been noted in a comparisons of rabbit and mouse globin genes (van Ooyen, Van den Berg, Mantei & Weissmann, 1979; Konkel, Maizel & Leder, 1979).

(4) Codon usage, as in other eukaryotic genes, is not random. Also, in a comparison of CH3 domains, the rate of substitutions of nucleotides at the third position of a codon (substitutions which do not change the coded amino acid) was higher than the overall rate of nucleotide change; this finding is in accordance with the idea (Jukes & King, 1979) that third position changes are not selected against. However, Tucker *et al.* (1979*b*) disagree with this idea, as they find that the rate of third position substitution is higher only in codons which also contain a first or second position alteration. Resolution of this question requires a more detailed statistical analysis of the nucleotide sequences, which I have not undertaken.

(5) The γ chain amino-acid sequences end with a glycine residue, whereas the DNA sequence codes for an additional lysine at the end of the chains. This discrepancy suggests that the carboxy-terminal lysine is removed in a post-translational modification step.

(6) The structure of the γ chain genes suggests mechanisms for the production of common heavy-chain variants (Adetugbo, Milstein & Secher, 1977; Frangione & Franklin, 1979; Kenter & Birshtein, 1979). These variants include deletions of V or C domains, deletions of the hinge region, and production of hybrid γ2b–γ2a chains; the latter probably result from non-homologous crossing over in a region of the CH2 domain where the nucleotide sequences of the two chains are particularly similar.

The characterisation of such hybrid γ chains in mice (Kenter & Birshtein, 1979) and humans (Kunkel & Kindt, 1975), together with nucleic acid hybridisation studies on mouse DNA (Honjo & Kataoka, 1978), and characterisation of human heavy chain genetic polymorphisms in different populations (Lefranc *et al.*, 1979) suggests that the C genes studied are arranged as follows in the two species (V genes are on the left):

$$\text{Human:} \quad -\alpha?-\gamma4-\gamma2-\gamma3-\gamma1-$$
$$\text{Mouse:} \quad -\mu-\gamma1-\gamma2b-\gamma2a-\alpha-$$

V–C joining in heavy chains. Mouse heavy-chain V and J region diversity, as judged from amino-acid sequence data, is comparable to that of κ chains (Potter, 1978). DNA cloning and hybridisation experiments (Kemp, Cory & Adams, 1979) also suggest that VH and Vκ gene numbers are similar (≥ 50 of each). However, two important complications arise when the nature of V–C joining in heavy chains is considered. The first is that most mature lymphocytes simultaneously express both μ and δ heavy chains. The expression is limited to the μ and δ genes of a single chromosome, and there is good evidence that an identical VH region is shared by the expressed μ and δ C regions (Pernis, 1978). Either a single V gene is copied and located at two distinct sites on the active chromosome, or, more probably, a single RNA transcript including VH, Cμ and Cδ genes can be processed in different ways to give mature μ and δ polypeptides.

A second complication of Ig heavy chain expression is that different heavy chain C regions can be expressed at different times during lymphocyte maturation: for example, a stimulated lymphocyte which secretes IgM may switch to secreting IgG (Wabl, Forni & Loor, 1978). A possible mechanism for such switching (Honjo & Kataoka, 1978) would simply involve deletion of the active μ gene, leaving the appropriate V region next to a γ gene. Analysis of the IgM and IgG produced by a single myeloma patient showed that both V and J regions were identical in the two chain types, even though they were produced by different cells (Wang, Wang & Fudenberg, 1977); thus such a deletion, if it occurs, alters only the C region of the expressed Ig chain.

A change in C region expression also occurs when lymphocytes carrying membrane-bound IgM differentiate to IgM-secreting cells. The secretory μ chain is more hydrophilic than its membrane-bound counterpart (Vassalli *et al.*, 1979); a possible mechanism for this type of switch could involve DNA deletion or RNA processing, such that a region coding for hydrophobic amino acids may be deleted from the μ gene in secretory cells.

Several different mechanisms are involved in the generation of antibody diversity; unfortunately, it is still not possible to estimate the quantitative contribution of each mechanism to the overall heterogeneity of Igs.

Combinatorial VH–VL pairing.   Totally random pairing between the products of 300 different VL and as many VH genes could produce about 10^5 different antibodies. There is no obvious structural bar to this proposal, since the VH and VL contact residues are highly conserved in different V region sequences (Saul, Amzel & Poljak, 1978), but there is no easy way to find out how random VH–VL pairing actually is. In human and mouse myeloma proteins, VL sequences of a particular group are found associated with different VH regions, at frequencies expected from the abundances of the respective groups in the total VH and VL pools; this is quite consistent with the idea of random pairing. On the other hand, light chains from one myeloma protein combine more rapidly with the homologous heavy chains than with randomly selected other heavy chains, suggesting that some selection for good fit occurs during lympho- cyte development (de Préval & Fougereau, 1976). The severity of this presumed selective process is quite unknown.

Somatic mutation.   As noted above, haploid embryonic mouse DNA contains only one $\lambda 1$ V gene. The extent of heterogeneity of normal mouse $\lambda 1$ chains, either from individual mice or from a pool of mice from a single strain, is not yet known, although these figures are in principle derivable from the study of λ-producing hybridomas (λ-producing cells which have been immortalised by hybridisation of spleen cells from a normal mouse with a non-Ig-secreting myeloma cell). However, $\lambda 1$- producing myelomas, which are probably representative of the $\lambda 1$-chains from a pool of normal mice, show very limited diversity; about $2/3$ of $\lambda 1$ chains share a single sequence, that coded by the embryonic $\lambda 1$ gene, while the remaining $1/3$ differ from the basic sequence by one to three point mutations, all in the HV regions of the chains. The total $\lambda 1$ myeloma chain repertoire probably consists of at least 15 different sequences (Weigert & Riblet, 1977). If these figures can be generalised, over 90% of all different VH/VL combinations are products of somatic mutation; however, whether an individual mouse produces most or only a few of these is unknown.

Since the observed mutants in λ chains occur in HV regions, might there be special mechanisms operating to increase somatic mutation rates at these points? A possible basis for enzymic recognition of HV regions is shown in Fig. 7.6, where sequences common to either κ or λ HV regions

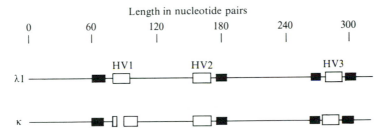

Fig. 7.6. Homologies around and within hypervariable regions (open boxed areas) of mouse κ and λ genes. There are significant homologies between the three HV regions in the embryonic λ1 gene, whose sequences are:

HV1: G T T G T A A C A G C C C C A G

HV2: G T G G T A C C A G C A A C C G

HV3: A T G G T A – C A G C A C C C A

(Tonegawa *et al.*, 1978). A gap has been introduced in the HV3 sequence to maximise homology. Differences between sequences are underlined. The sequences are aligned 5′ to 3′, and are from the non-coding strand for HV1, but the coding strand for HV2 and 3. There are also homologies between the HV regions of κ chain genes (Ben-Sasson, 1979). Regions of significant nucleotide homology (≥ 7 of 8 adjacent bases identical) *between* κ and λ V genes are also indicated (filled boxed areas). A gap has been introduced into the first κ HV region to maximise homology (Wu *et al.*, 1979).

(Ben-Sasson, 1979) are flanked by other sequences which have been highly conserved during the separate evolution of κ and λ chains (Wu, Kabat & Bilofsky, 1979). The V region mutation rate in cultured myeloma cells is less than 10^{-5} per gene per cell division (Adetugbo *et al.*, 1977); but it is likely that a hypermutation mechanism, if one exists, would be active in lymphocyte precursors rather than in their mature progeny. This rate would be sufficient to generate 10^5 different Igs from 200 V genes during the 30-day development of the mouse lymphoid system, if mutant clones could be selected at the expense of unaltered cells (Cohn *et al.*, 1974).

Somatic recombination. As mentioned above, the recombination event which unites Vκ and Jκ sequences is located at a point included in the third HV region of the chain, and can contribute to antibody sequence diversity in two ways: first, different Vκ–Jκ combinations are possible; and secondly, a given Vκ–Jκ pair can give rise to different amino-acid sequences by slight alteration of the exact point in the DNA at which the recombination event is located (Table 7.3; Fig. 7.3).

Recombination could contribute to sequence diversity in another way: unequal pairing and crossover between related V genes could give rise to new sequences, not only during evolution (as discussed below), but during ontogeny, if the crossing-over point was located within the V coding sequence. A very close homology between the intervening sequences of related V genes (as observed for some Vκ, but not VH, genes (Kemp *et al.*, 1979; Seidman *et al.*, 1978*b*)) would facilitate this process, but whether somatic inter-V-region recombination actually occurs is quite unknown; the intervening sequence homologies might simply be due to their very recent origin by gene duplication. If recombination does occur, it must be confined to members of a single group, since there is no

Table 7.3. *Production of different amino-acid sequences by V–J recombination at different sites in κ chain DNA. The amino acid X coded by the sequence resulting from recombination at positions 1 to 4 is Trp, Trp, Arg or Pro respectively. Complementary DNA sequences (see Fig. 7.3) are boxed.*
From Sakano et al., *1979*a

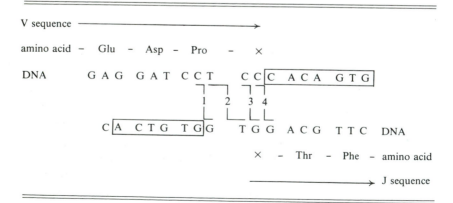

evidence, from amino-acid sequence data, of any generation of new sequences by recombination between genes coding for different sequence groups.

EVOLUTION OF IMMUNOGLOBULIN GENES

Early steps in Ig evolution

Immunoglobulins show a clear amino-acid sequence homology to β_2-microglobulin, major histocompatibility antigens (see previous chapter) and the Thy-1 molecule (Campbell, Williams, Bayley & Reid, 1979) (Table 7.4), which is a basic glycoprotein present at the surface of thymus-derived (T) lymphocytes and certain epithelial cells, and which may be important in forming intercellular contacts during differentiation (Lennon, Unger & Dulbecco, 1979). One might also guess that the antigen receptors present on thymus-derived lymphocytes, which are not well characterised but are not Igs, nevertheless share a common evolutionary origin with the Ig receptors carried by other lymphocytes.

Table 7.4. *Homology (percent amino-acid identity) between rat Thy-1, human IgG1 domains, β2-microglobulin, and major histocompatibility antigen HLA-B7. Data from Campbell et al., 1979. Compare also Table 6.3, Chapter 6.*

Protein or domain	Cκ	Cλ	CH1	CH2	CH3	$\beta_2\mu$	HLA
Homology	20	20	13	22	11	15	11
No. of positions compared (including gaps)	45	45	45	45	45	45	37
No. of gaps introduced	1	2	3	1	1	2	1

Immunoglobulins probably developed their present form and function very early in vertebrate evolution, as no Ig-like molecules have been isolated from invertebrates, while very primitive chordates can synthesise Ig-like antibodies (Litman & Kehoe, 1978; Raison, Hull & Hildemann, 1978). A plausible model for the evolution of Ig genes, which introduces RNA processing and DNA rearrangement systems at early stages in Ig gene diversification, is shown in Fig. 7.7. The gene for a primitive Ig domain duplicates (step 1); the sequences at the appropriate ends of the duplicated genes must be, or must evolve to be, such that V–C joining can

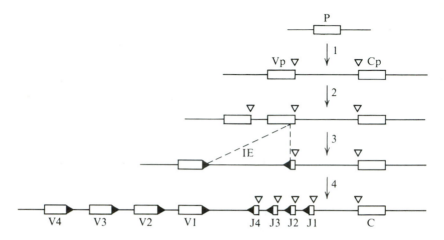

Fig. 7.7.   A hypothetical scheme for the evolution of κ light chain genes (Sakano *et al.*, 1979*a*). P: a primordial gene coding for a single domain. Vp, Cp: precursors of V and C genes. IE: a DNA element thought to be inserted into a Vp sequence, splitting it into V and J segments. Open triangles: sites for RNA processing. Filled triangles: sites for DNA rearrangement.
Step 1: duplication of P and generation of RNA processing sites.
Step 2: only limited duplication of Vp occurs.
Step 3: DNA element is inserted.
Step 4: numerous independent duplications of V and J can occur.

occur during RNA processing (how this requirement for parallel production of processing sites is met is unclear). At this stage, limited further duplication of V genes, with formation of different V–C combinations during RNA processing, could occur (step 2); this arrangement, however, would be limited by the difficulty of translating one V gene in an RNA transcript containing all the V genes in the duplicated family. The crucial further step (step 3 in Fig. 7.7) is the introduction of a DNA insertion element which maintains V and J segments separate in embryonic DNA, but can be removed in one of the Ig gene-bearing chromosomes of a lymphoid cell. The introduction of a DNA rearrangement step allows extensive duplication of V genes (since only one of them needs to be transcribed to make a given chain), and is also compatible with J gene duplication (which may in fact be required to cope with increasing V gene numbers, if some V–J combinations are incompatible). This scheme, due to Sakano *et al.* (1979*a*), is supported by the observation of sequence similarities between IgV and C regions, and between the J piece and the carboxy-terminal end of the C domain.

Evolution of C regions

Clearly the genes coding for heavy, κ and λ chains all had a common origin. Since they are located on different chromosomes, it seems likely that they arose from chromosome doubling or tetraploidisation, very early in the case of the heavy–light chain separation, later for the κ–λ split, which perhaps is coincident with the tetraploidisation known to have occurred with the development of bony fish.

Subsequent C region evolution involves the duplication of domains and the production of extra-domain sequences, as well the duplication of entire C regions.

Domains. The various heavy chain C regions have different numbers of domains (Table 7.1). As discussed above, it seems likely that at least the γ and α chain genes are derived from precursors coding for μ-like chains, by transformation of a domain into a non-coding intervening sequence plus a hinge; this would be consistent with the presence of μ-like, but not γ-like, chains in the serum of primitive vertebrates (Litman & Kehoe, 1978).

As noted by Putnam (1977), the different heavy chain domains have evolved at different rates: for example, human μ and α chains are 50% homologous in their carboxy-terminal domains, but only 20% homologous in the adjacent domain. When human α, γ, μ and ε chains were compared, the CH1 domains showed the most divergence; the rapid divergence of CH1 domains was also observed in a comparison of human, murine and canine μ chains (Kehry *et al.*, 1979), but not in a comparison of mouse γ_1, γ_{2a} and γ_{2b} chains, whose CH1 domains were 75–80% homologous, as against 55–60% homology for the CH3 domains (Tucker *et al.*, 1979*b*). How these different rates are related to the functions of different domains is unclear.

An evolutionary tree of human heavy chain C regions (Fig. 7.8) (Dayhoff, 1978) shows a γ/ε precursor separating from an α/μ precursor slightly before the appearance of the four separate chain types. Evolutionary trees prepared for individual domains are consistent with this scheme, with the exception of the CH1 domain for which an alternative, slightly more parsimonious tree can be constructed. To account for this, Putnam (1977) suggests that heavy chains may have exchanged domains by recombination after their evolutionary divergence; however, it seems just as likely that the anomaly is simply a chance artefact (Dayhoff, 1978).

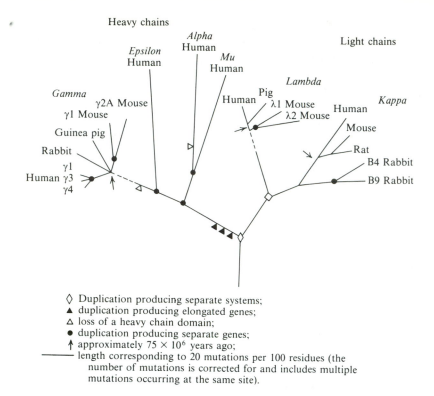

◊ Duplication producing separate systems;
▲ duplication producing elongated genes;
△ loss of a heavy chain domain;
● duplication producing separate genes;
↑ approximately 75 × 10⁶ years ago;
── length corresponding to 20 mutations per 100 residues (the
    number of mutations is corrected for and includes multiple
    mutations occurring at the same site).

Fig. 7.8.   Evolutionary tree of sequenced Ig C regions (Dayhoff, 1978). Branches leading to γ and λ chains are dashed because there exist additional solutions
almost as good as those shown. The early divergence of rabbit from other κ chains
is unexpected, as is the large difference between allelic B4 and B9 rabbit κ
sequences; the two allelic forms may be derived from two different κ genes which
had duplicated at an early stage (see text).

Extra-domain sequences.   The hinge regions of heavy chains evolve
more rapidly than the rest of the C region, perhaps because of a
requirement for different degrees of flexibility in different antibodies
(Fudenberg *et al.*, 1978). This evolution leads to notable changes in the
numbers and positions of interchain disulphide bridges, and can include
loss of the hinge (in human $\alpha 2$ chains) or its duplication (in human $\gamma 3$
chains) (Frangione & Franklin, 1979). The evolutionary origin of the
hinge regions has already been discussed (p. 239).

   The tail sequences of μ and α chains are clearly homologous, but their
origin is obscure.

Duplication and deletion of C region genes. Light-chain C region genes are sometimes unique (e.g. human and mouse κ chains) and sometimes duplicated. For example, human λ chain C genes exist in at least four non-allelic forms (Table 7.5); all of these, as far as is known, share a common V gene pool, although the V–C association is not completely random (Fett & Deutsch, 1975). A similar situation probably holds for guinea-pig λ chains (Brunhouse & Cebra, 1978). On the other hand, mouse $\lambda 1$ and $\lambda 2$ C regions are associated with different V regions, perhaps because in this case a single chromosomal region encompassing both V and C regions was duplicated; if this is the case, however, the resulting V and C region pairs have diverged at quite different rates (see p. 233).

The heavy-chain C region genes form a linked set within which recombination is exceedingly rare (rate $\leq 0.1\%$) (Kunkel & Kindt, 1975). Yet deletion and duplication are sufficiently frequent for duplicated γ chain genes in different species (e.g. the human subclass genes, and the mouse $\gamma 2a$ and $\gamma 2b$ genes) to have arisen in many cases after the divergence of the species (Fudenberg *et al.*, 1978). Evolutionary trees of the C regions (Fig. 7.8) of different species clearly demonstrate this point (Dayhoff, 1978). In addition, chromosomes lacking or carrying extra γ genes are not uncommon in human populations. Homozygotes for $\gamma 1$ deletion have clinical symptoms (Kunkel & Kindt, 1975), whereas $\gamma 3$ deletion homozygotes are healthy (Lefranc *et al.*, 1979), not surprisingly, since $\gamma 1$ and $\gamma 3$ chains constitute, respectively, 70% and 10% of the normal human γ pool.

Evolution of V regions

If the case of mouse κ chains is typical, Ig V regions are coded by large numbers, perhaps hundreds, of linked genes. The final number of V genes obtained by multiple duplications in different species is the result of a series of compromises. For example, the number of inheritable V genes should not be so large that many are totally unused during an individual's lifetime; but neither should it be so small that an individual cannot generate a reasonable repertoire of antibodies against common pathogens (Cohn *et al.*, 1974).

A comparison of V region sequences from different species suggests that the balance thus obtained between excess of unused genes, and lack of useful genes, is not a stable one, but is continually changing as a result of duplication of some genes and deletion of others (Hood, Campbell &

Elgin, 1975). These processes are assumed to result from unequal crossing-over during recombination. Thus an ancestral species which contained the V gene family

$$-VA-VB-VC-VD-VE-$$

might give rise to two descendant species which carried the different V gene sets:

$$-VA1-VA2-VA3-VB1-VC1- \quad \text{and} \quad -VC2-VC3-VC4-VD1-VE1;$$

here the V genes A1, 2, 3 and C1, 2, 3, 4 are derived from VA and VC, respectively, by duplication and point mutation. Note that this result can give rise to the illusion that parallel evolution of V genes has occurred, towards VA-like genes in one species and VC-like genes in the other; such a conclusion would be unjustified, as the only mechanisms required to produce the gene families in question are random duplication, deletion and mutation (Black & Gibson, 1974; Smith, 1974). These features have been incorporated into interesting mathematical models of the evolution of multiple gene families by Perelson & Bell (1977) and Kimura & Ohta (1979). The latter model can be used to predict the average extent of similarity between different V genes, as a function of the mutation and unequal crossing-over rates, and the size of the gene family. The model contains several simplifying assumptions; however, Ohta (1978) used it to conclude that, if V gene numbers were high (500), the differences between HV regions of the same or different species could be accounted for without having to invoke somatic mutation.

IMMUNOGLOBULIN POLYMORPHISM

C region polymorphism

Ig C region genes are very polymorphic (Table 7.1) (Kunkel & Kindt, 1975; Fudenberg *et al.*, 1978). In many cases the polymorphism (*allo- typy*) is easily explained as being the result of one or a small number of point mutations (Table 7.5). In other cases there are multiple amino-acid sequence differences between apparently allelic forms of the same C region: for example, allotypic rat or rabbit κ chains differ by more than 10 or 30%, respectively, of their C region residues (Farnsworth, Goodfliesh, Rodkey & Hood, 1976). Such differences are even larger than those found between different histocompatibility alleles (chapter 6), and seem unlikely to be due to the simple accumulation of point mutations in allelic genes. The origin of these complex allotypes is not known, but they may well arise in the following way: first, a C region gene duplicates, and the duplicated pair diverge in sequence; next, loss of one of the pair occurs, and the chromosome carrying the deletion spreads to some members of the population; finally, a second deletion of the other gene occurs and the resulting chromosome replaces the duplicated genes in the rest of the population. This mechanism, involving duplication and deletion, obviously resembles that postulated above (p. 249) to account for V gene evolution in different species. Alternative mechanisms for the production of complex allotypes are discussed by Gutman, Loh & Hood (1975) and Hood, Campbell & Elgin (1975).

V region polymorphism

V region polymorphism is less easy to investigate, but certainly occurs; for example, in the case of normal mouse light chains (95% κ type), strain-specific differences in peptide mapping pattern, or banding patterns after isoelectric focussing, are inherited as simple Mendelian traits. In addition, V regions associated with a given antibody activity in different mouse strains are often antigenically different, or have different charges (so that the corresponding antibodies have different isoelectric points) (Weigert & Potter, 1977). These antigenic V region markers (*idiotypes*), or isoelectric properties, are again often inherited in a straight-forward Mendelian fashion, as if due to variation in a single V region gene; there is, in these cases, invariably a close linkage between the V region marker and a corresponding C region gene. Almost all cases of inheritable idiotypy involve linkage of the idiotype to the heavy chain genes, probably because heavy chain C region genetics are better studied than those of the light chain; the degree of V–C linkage is difficult to measure with great accuracy, but is reported to vary from $\leq 0.5\%$ to about 5%, depending on the V-gene marker involved. The lower figures are more in line with the crossover frequency (about 0.5%) between rabbit heavy chain V and C regions (Kunkel & Kindt, 1975).

The special case of rabbit heavy chain V region polymorphism deserves further mention. About 70–80% of these V regions carry antigenic determinants, not necessarily the same on each V region, which differ in different rabbits: at least three distinct sets of determinants, a1, a2 and a3, exist (Nisonoff *et al.*, 1975). There are multiple differences in amino acids between normal a1, a2 and a3 chains. A single chromosome expresses a1, a2 or a3 genes, in each case closely linked to the heavy chain C region genes; however, it is not infrequent for rabbits apparently homozygous for one V region type (say a1) to produce very small amounts of molecules of another type (e.g. a2). (In such a case, the a1/a2 ratio would be about 1 : 1000) (Nisonoff *et al.*, 1975; Kunkel & Kindt, 1975).

Several possible explanations for these findings, similar to those proposed for the production of complex C region allotypes (Gutman *et al.*, 1975), have been put forward. One suggestion is that there have occurred separate deletions of different families of genes from precursor chromosomes which carried a complete set of a1, a2 and a3 genes; different deletions will leave behind different families of related genes. A second explanation is that a polymorphic regulatory gene, of unidentified nature, controls the expression of selected V genes present on chromosomes carrying all three V gene families. A comparison of the DNA coding for the relevant V genes in different rabbits should, in principle, distinguish between these possibilities.

Immunoglobulin polymorphism and natural selection

The evidence that any Ig polymorphisms are maintained by natural selection is much weaker than in the case of major histocompatibility antigens (MHAs) (chapter 6), but two of the lines of argument used in that case – the association of polymorphism with immune response in man and experimental animals, and the population genetics of the polymorphisms – support the idea. The third line of argument used for the MHAs, direct association between MHA type and disease, cannot be used, as there is only one brief report of an association between human allotype and disease (typhoid) in the literature (Nevo, 1974).

Several cases of high or low allotype-linked immune responses have been described in mice; the best characterised is that of the anti-dextran response (Weigert & Potter, 1977). However, Biozzi, Stiffel, Mouton & Bouthillier (1975) clearly showed that Ig allotypes are not the only, or even the most important, factor determining the magnitude of the antibody response to several antigens. These authors obtained, by differential selection, lines of mice which gave high or low antibody responses to sheep red blood cells and other antigens; they estimated that polymorphism at many (perhaps 10) loci, including Ig loci, affected the response. In addition, mice which responded well to *Salmonella typhimurium* were actually more susceptible than low responders to infection by the bacterium, probably because resistance, in this case, is more dependent on macrophage activity than on antibody levels.

In humans, weak associations between Gm allotypes (on γ chain C regions) and antibody levels to flagellar or 'H' antigens of *Salmonellae* (Wells, Fudenberg & MacKay, 1971; Nevo, 1974) or to tetanus toxoid (Rivat, Cavelier & Bonneau, 1978; Schanfield, Wells & Fudenberg, 1979) have been reported. In all cases, however, the numbers of sera tested were small and it would be very valuable to have independent confirmation of these results. These findings, if upheld, probably do not imply a direct relationship between γ chain C region structure and antibody response, but rather a linkage disequilibrium (see p. 224) between heavy chain V and C region genes, in conjunction with V region polymorphism affecting antibody activity.

There is some evidence suggesting that such genetic differences in the ability to make certain antibodies can actually result in alteration of Ig allele frequency. The frequencies of some Gm alleles are strikingly different in different racial groups, and also vary even within groups: for example, the frequency of the Gml allele is almost 100% in northwest Europe, but less than 50% in southeast Europe. The variation in Gm frequencies between different groups is higher than that for many other polymorphisms (Piazza *et al.*, 1976). This suggests either that different Gm phenotypes are selectively advantageous in different environments, or that there is stabilising selection against variation in the other polymorphic systems tested (of course, both forms of selection may be occurring together). The hypothesis that alterations in Gm frequencies are due to selection for disease resistance is supported by two studies: Piazza *et al.* (1976) found that there were significant differences in the frequencies of Gm antigens amongst Sardinian villagers living in malarial or non-malarial environments (the same workers also reported similar findings for HLA antigens); more recently, de Vries *et al.* (1979) concluded that 60% mortality in typhoid and yellow-fever epidemics amongst Dutch colonists in Surinam had altered HLA and Gm frequencies more than the frequencies of 20 other polymorphisms.

Finally, there exist interesting cases of parallel polymorphism in the structures of different human and mouse immunoglobulins. For example, human κ chains show genetic polymorphism at residues 153 and 191 (Table 7.5); human λ chains also show variation at almost exactly these positions (although in this case the variants are carried by duplicated rather than allelic genes). As a second example, human $\alpha2$ and mouse α chains both have allelic forms in which the usual disulphide bridge between heavy and light chains is absent; these polymorphic forms very probably have independent origins, since other α chains (e.g. rabbit α or

Table 7.5. *Amino-acid substitutions in* a) *human κ chain allotypes, and*
b) *duplicated human λ chains. Residue numbering is based on κ chain
sequence, with λ chains aligned for maximum homology. λ type numbers
are arbitrary. For κ allotype nomenclature, see Fudenberg* et al., *1978*

a			b					
	Position No.			Position No.				
κ chain allotype	153	191	λ chain type	111	113	152	164	190
Km 1	Val	Leu	λ 1	Ala	Ser	Ser	Thr	Arg
Km 1,2	Ala	Leu	λ 2	Ala	Ser	Ser	Thr	Lys
Km 3	Ala	Val	λ 3	Ala	Ser	Gly	Thr	Arg
			λ 4	Asn	Thr	Gly	Lys	Arg

human $\alpha 1$ chains) are not unusual in this respect. The independent
evolution of similar polymorphisms suggests (although unfortunately, the
argument cannot be made quantitative) that such polymorphisms are
selectively advantageous.

CONCLUSIONS

Immunoglobulins evolved from a domain-sized precursor, which was
probably also ancestral to major histocompatibility antigens and β_2-
microglobulin, and possibly to other proteins such as the Thy-1 antigen as
well. Subsequent steps in Ig evolution can be summarised as follows.

Duplication of the primitive domain produced a split Ig gene, in which
precursor V and C regions were separated by an intervening DNA
sequence; evolution of a splicing mechanism, capable of removing the
corresponding sequence from Ig mRNA, must also have occurred at this
stage. Next, an unusual DNA rearrangement mechanism evolved; a long
DNA stretch was inserted into the V domain, which thus became, in
germline DNA, two separate entities – a large V piece, and a smaller J
(joining) piece close to the C region. V and J pieces can, however, be
reunited in Ig-producing cells. This arrangement is compatible with
independent duplication of V, C and J pieces, and has the advantage that
only one of a large family of duplicated V pieces need be transcribed into
RNA.

Two chromosomal duplications (or rearrangements) then gave rise to
the three separate gene sets coding for heavy, κ and λ chains, the
heavy/light separation preceding the κ/λ divergence. In the subsequent

evolution of Ig constant regions, domain and gene duplications played important roles. Duplication of domains involved concurrent duplication of flanking, non-translated DNA, so that, in heavy chain genes, the sequences coding for individual C region domains, and for the hinge regions, are separated from one another by non-coding regions. Gene duplication and deletion are even more strikingly involved in V region evolution, which may be viewed as a random oscillation of gene numbers between extremes fixed by natural selection. The results are that V gene numbers differ in different chain types and in different species; mouse $\lambda 1$ and $\lambda 2$ V genes are unique, but there exist, probably, about a hundred mouse Vκ genes. Since deletions and duplications have been frequent in V gene evolution, large families of related V genes may be present in one species and absent in another; or even (as may be the case for rabbit VH genes) present or absent in different members of the same species.

In addition to the existence of multiple V genes, there have evolved various somatic mechanisms which increase Ig heterogeneity. These include the use of different heavy–light chain and V–J piece combinations; somatic mutation, perhaps occurring at increased frequency in hypervariable regions; and the generation of some extra diversity through a choice of recombination sites at the V–J joining point. Other mechanisms such as somatic recombination between different, closely-related V genes, have not been demonstrated, but may also contribute to V region diversity.

Igs show extensive genetic polymorphism. In mice and perhaps in humans, correlations between Ig allotype and antibody response have been demonstrated; however, these are generally not as striking or as easy to detect as the correlations between major histocompatibility type and cellular immune or antibody response. Nevertheless, there is some indirect evidence, from studies of the population genetics of human Ig genetic markers, that natural selection has in fact acted to alter the frequencies of some Ig alleles in populations subject to malaria or epidemic typhoid and yellow fever.

REFERENCES

Adetugbo, K., Milstein, C. & Secher, D. S. (1977). Molecular analysis of spontaneous somatic mutants. *Nature*, **265**, 299–304.
Ben-Sasson, S. A. (1979). Immunoglobulin differentiation is dictated by repeated recombination sequences within the V region prototype gene: A hypothesis. *Proceedings of the national academy of sciences of the USA*, **76**, 4598–602.

Bernard, O., Hozumi, N. & Tonegawa, S. (1978). Sequences of mouse immuno-globulin light chain genes before and after somatic changes. *Cell*, **15**, 1133–44.

Biozzi, G., Stiffel, C., Mouton, D. & Bouthillier, Y. (1975). Selection of lines of mice with high and low antibody responses to complex immunogens. In *Immunogenetics and immunodeficiency*, ed. B. Benacerraf, pp. 179–227. Lancaster: MTP.

Black, J. A. & Gibson, D. (1974). Neutral evolution and immunoglobulin diversity. *Nature*, **250**, 327–8.

Brack, C., Hirama, M., Lenhard-Schuller, R. & Tonegawa, S. (1978). A complete immunoglobulin gene is created by somatic recombination. *Cell*, **15**, 1–14.

Brack, C. & Tonegawa, S. (1977). Variable and constant parts of the immuno-globulin light chain gene of a mouse myeloma cell are 1250 nontranslated bases apart. *Proceedings of the national academy of sciences of the USA*, **74**, 5652–6.

Brunhouse, R. F. & Cebra, J. J. (1978). Guinea-pig immunoglobulin light chain isotypes. *European journal of immunology*, **8**, 881–8.

Campbell, D. G., Williams, A. F., Bayley, P. N. & Reid, K. B. M. (1979). Structural similarities between Thy-1 antigen from rat brain and immuno-globulin. *Nature*, **282**, 341–2.

Cohn, M., Blomberg, B., Geckeler, W., Raschke, W., Riblet, R. & Weigert, M. (1974). First order considerations in analyzing the generator of diversity. In *The immune system: genes, receptors, signals*, ed. E. E. Sercarz, A. R. Williamson & C. F. Fox, pp. 89–117. New York: Academic Press.

Dayhoff, M. O. (1978). *Atlas of protein sequence and structure*, Vol. 5, suppl. 3, pp. 197–227. Washington, DC: National Biomedical Research Foundation.

de Préval, C. & Fougereau, M. (1976). Specific interaction between V_H and V_L regions of human monoclonal immunoglobulins. *Journal of molecular biology*, **102**, 657–78.

de Vries, R. R. P., Meera Khan, P., Bernini, L. F., van Loghem, E. & van Rood, J. J. (1979). Genetic control of survival in epidemics. *Journal of immuno-genetics*, **6**, 271–87.

Early, P. W., Davis, M. M., Kaback, D. B., Davidson, N. & Hood, L. E. (1979). Immunoglobulin heavy chain gene organisation in mice: analysis of a myeloma genomic clone containing variable and α constant regions. *Proceedings of the national academy of sciences of the USA*, **76**, 857–61.

Farnsworth,V., Goodfliesh, R., Rodkey, S. & Hood, L. (1976). Immunoglobulin allotypes of rabbit κ chains: polymorphism of a control mechanism regulating closely linked, duplicated genes? *Proceedings of the national academy of sciences of the USA*, **73**, 1293–6.

Fett, J. W. & Deutsch, H. F. (1975). A new λ-chain gene. *Immunochemistry*, **12**, 643–52.

Frangione, B. & Franklin, E. C. (1979). Split immunoglobulin genes and human heavy chain deletion mutants. *Journal of immunology*, **122**, 1177–9.

Fudenberg, H. H., Pink, J. R. L., Wang, A.-C. & Douglas, S. D. (1978). *Basic immunogenetics*, 2nd ed. New York: Oxford University Press.

Gilmore-Hebert, M. & Wall, R. (1978). Immunoglobulin light chain mRNA is processed from large nuclear RNA. *Proceedings of the national academy of sciences of the USA*, **75**, 342–5.

Gutman, G. A., Loh, E. & Hood, L. (1975). Structure and regulation of immunoglobulins: Kappa allotypes in the rat have multiple amino-acid differences in the constant region. *Proceedings of the national academy of sciences of the USA*, **72**, 5046–50.

Hengartner, H., Meo, T. & Müller, E. (1978). Assignment of genes for immunoglobulin κ and heavy chains to chromosomes 6 and 12 in mouse. *Proceedings of the national academy of sciences of the USA*, **75**, 4494–8.

Honjo, T., Obata, M., Yamawaki-Kataoka, Y., Kataoka, T., Kawakami, T., Takahashi, N. & Mano, Y. (1979). Cloning and complete nucleotide sequence of mouse immunoglobulin γ1 chain gene. *Cell*, **18**, 559–68.

Honjo, T., Packman, S., Swan, D. & Leder, P. (1976). Quantitation of constant and variable region genes for mouse immunoglobulin λ chains. *Biochemistry*, **15**, 2780–5.

Honjo, T. & Kataoka, T. (1978). Organisation of immunoglobulin heavy chain genes and allelic deletion model. *Proceedings of the national academy of sciences of the USA*, **75**, 2140–4.

Hood, L., Cambell, J. H. & Elgin, S. C. R. (1975). The organisation, expression and evolution of antibody genes and other multigene families. *Annual review of genetics*, **9**, 305–53.

Jerne, N. K. (1974). Towards a network theory of the immune system. *Annales d'immunologie de l'institut Pasteur*, **125c**, 373–89.

Jukes, T. H. & King, J. L. (1979). Evolutionary nucleotide replacements in DNA. *Nature*, **281**, 605–6.

Kehry, M., Sibley, C., Fuhrman, J., Schilling, J. & Hood, L. E. (1979). Amino acid sequence of a mouse immunoglobulin μ chain. *Proceedings of the national academy of sciences of the USA*, **76**, 2932–6.

Kemp, D. J., Cory, S. & Adams, J. M. (1979). Cloned pairs of variable region genes for immunoglobulin heavy chains isolated from a clone library of the entire mouse genome. *Proceedings of the national academy of sciences of the USA*, **76**, 4627–31.

Kenter, A. L. & Birshtein, B. K. (1979). Genetic mechanism accounting for precise immunoglobulin domain deletion in a variant of MPC 11 myeloma cells. *Science*, **206**, 1307–9.

Kimura, M. & Ohta, T. (1979). Population genetics of multigene family with special reference to decrease of genetic correlation with distance between gene members on a chromosome. *Proceedings of the national academy of sciences of the USA*, **76**, 4001–5.

Köhler, G. (1976). Frequency of precursor cells against the enzyme β-galactosidase. An estimate of the BALB/c strain antibody repertoire. *European journal of immunology*, **6**, 340–6.

Konkel, D. A., Maizel, J. V. & Leder, P. (1979). The evolution and sequence comparison of two recently diverged mouse chromosomal β-globin genes, *Cell*, **18**, 865–73.

Kunkel, H. G. & Kindt, T. J. (1975). Allotypes and idiotypes. In *Immunogenetics and immunodeficiency*, ed. B. Benacerraf, pp. 55–80. Lancaster: MTP.

Lefkovits, I. (1974). Precommitment in the immune system. *Current topics in microbiology and immunology*, **65**, 21–58.

Lefranc, G., Dumitresco, S.-M., Salier, J.-P., Rivat, L., de Lange, G., van Loghem, E. & Loiselet, J. (1979). Familial lack of the IgG3 subclass. Gene elimination or turning off expression and neutral evolution in the immune system. *Journal of immunogenetics*, **6**, 215–21.

Lenhard-Schuller, R., Hohn, B., Brack, C., Hirama, M. & Tonegawa, S. (1978). DNA clones containing mouse immunoglobulin κ chain genes isolated by *in vitro* packaging into phage λ coats. *Proceedings of the notional academy of sciences of the USA*, **75**, 4709–13.

Lieberman, R. (1978). Genetics of the IgCH (allotype) locus in the mouse. *Springer seminars in immunopathology*, **1**, 7–30.

Lennon, V. A., Unger, M. & Dulbecco, R. (1978). Thy-1: A differentiation marker of potential mammary myoepithelial cells in vitro. *Proceedings of the national academy of sciences of the USA*, **75**, 6093–7.

Litman, G. W. & Kehoe, J. M. (1978). The phylogenetic origins of immunoglobulin structure. In *Comprehensive immunology, 5: immunoglobulins*, ed. G. W. Litman & R. A. Good, pp. 205-27. New York: Plenum Medical Book Company.

Max, E. E., Seidman, J. G. & Leder, P. (1979). Sequences of five potential recombination sites encoded close to an immunoglobulin κ constant region gene. *Proceedings of the national academy of sciences of the USA*, **76**, 3450–4.

Moore, G. W. & Goodman, M. (1977). Alignment statistic for identifying related protein sequences. *Journal of molecular evolution*, **9**, 121–30.

Nevo, S. (1974). Association of typhoid fever and response to vaccination with Polymorphic systems in Man. *Protides biologic fluids*, **22**, 649–53.

Nisonoff, A., Hopper, J. E. & Spring, S. B. (1975). *The antibody molecule*. New York: Academic Press.

Ohta, T. (1978). Sequence variability of immunoglobulins considered from the standpoint of population genetics. *Proceedings of the national academy of sciences of the USA*, **75**, 5108–12.

Perelson, A. S. & Bell, G. I. (1977). Mathematical models for the evolution of multigene families by unequal crossing-over. *Nature*, **265**, 304–10.

Pernis, B. (1978). Lymphocyte membrane immunoglobulins: an overview. In *Comprehensive immunology, Vol. 5: Immunoglobulins*, ed. G. W. Litman & R. A. Good, pp. 357–72. New York: Plenum Medical Book Company.

Piazza, A., van Loghem, E., de Lange, G., Curtoni, E. S., Ulizzi, L. & Terrenato, L. (1976). Immunoglobulin allotypes in Sardinia. *American journal of human genetics*, **28**, 77–86.

Potter, M. (1978). Antigen-binding myeloma proteins of mice. *Advances in immunology*, **25**, 141–212.

Putnam, F. W. (1977). Immunoglobulins. In *The plasma proteins, Vol. II*, ed. F. W. Putnam, pp. 1–153. New York: Academic Press.

Rabbits, T. H. & Forster, A. (1978). Evidence for noncontiguous variable and constant region genes in both germ line and myeloma DNA. *Cell*, **13**, 319–27.

Raison, R. L., Hull, C. J. & Hildemann, W. H. (1978). Characterisation of immunoglobulin from the Pacific hagfish, a primitive vertebrate. *Proceedings of the National Academy of Sciences of the USA*, **75**, 5679–82.

Rao, D. N., Rudikoff, S., Krütsch, H. & Potter, M. (1979). Structural evidence for independent joining region gene in immunoglobulin heavy chains from anti-galactan myeloma proteins and its potential role in generating diversity in complementarity-determining regions. *Proceedings of the national academy of sciences of the USA,* **76,** 2890–4.

Rivat, L., Cavelier, B. & Bonneau, J. C. (1978). Correlation between the level of humoral response subsequent to antitetanic immunization. Gm immunoglobulin allotypes and HLA patterns in French blood donors. *Annales d'immunologie,* **129,** 735.

Rogers, J., Clarke, P. & Salser, W. (1979). Sequence analysis of cloned cDNA encoding part of an immunoglobulin heavy chain. *Nucleic acids research,* **6,** 3305–21.

Sakano, H., Huppi, L., Heinrich, G. & Tonegawa, S. (1979*a*). Sequences at the somatic recombination sites of immunoglobulin light chain genes. *Nature,* **280,** 288–94.

Sakano, H., Rogers, J. H., Hüppi, K., Brack, C., Traunecker, A., Maki, R., Wall, R. & Tonegawa, S. (1979*b*). Domains and the hinge region of an immunoglobulin heavy chain are encoded in separate DNA segments. *Nature,* **277,** 627–33.

Saul, F. A., Amzel, L. M. & Poljak, R. J. (1978). Preliminary refinement and structural analysis of the Fab fragment from human immunoglobulin New at 2.0 Å resolution. *Journal of biological chemistry,* **253,** 585–97.

Schanfield, M. S., Wells, J. V. & Fudenberg, H. H. (1979). Immunoglobulin allotypes and response to tetanus toxoid in Papua, New Guinea. *Journal of Immunogenetics,* **6,** 311–5.

Schechter, I., Wolf, O., Zemell, R. & Burstein, Y. (1979). Structure and function of immunoglobulin genes and precursors. *Federation proceedings,* **38,** 1839–45.

Schibler, W., Marcu, K. B. & Perry, R. P. (1978). The synthesis and processing of the messenger RNAs specifying the heavy and light chain immunoglobulins in MPC-11 cells. *Cell,* **15,** 1495–509.

Schilling, J., Hansburg, D., Davie, J. M. & Hood, L. E. (1979). Analysis of the diversity of murine antibodies to dextran B1355: N-terminal aminoacid sequences of heavy chains from serum antibody. *Journal of immunology,* **123,** 384–8.

Seidman, J. G. & Leder, P. (1978). The arrangement and rearrangement of antibody genes. *Nature,* **276,** 790–5.

Seidman, J. G., Leder, A., Edgell, M. H., Polsky, F., Tilghman, S. M., Tiemeier, D. S. & Leder, P. (1978*a*). Multiple related immunoglobulin variable region genes identified by cloning and sequence analysis. *Proceedings of the national academy of sciences of the USA,* **75,** 3881–5.

Seidman, J. G., Leder, A., Nau, M., Norman, B. & Leder, P. (1978*b*). Antibody diversity. *Science,* **202,** 11–7.

Sigal, N. H. & Klinman, N. R. (1978). The B-cell clonotype repertoire. *Advances in immunology,* **26,** 255–337.

Smith, G. P. (1974). Unequal crossover and the evolution of multigene families. *Cold Spring Harbor symposia on quantitative biology,* **38,** 507–14.

Tonegawa, S. (1976). Reiteration frequency of immunoglobulin light chain genes: further evidence for somatic generation of antibody diversity. *Proceedings of the national academy of sciences of the USA*, **73**, 203–7.

Tonegawa, S., Maxam, A. M., Tizard, R., Bernard, O. & Gilbert, W. (1978). Sequence of a mouse germ-line gene for a variable region of an immunoglobulin light chain. *Proceedings of the national academy of sciences of the USA*, **75**, 1485–9.

Tucker, P. W., Marcu, K. B., Newell, N., Richards, J. & Blattner, F. R. (1979*a*). Sequence of the cloned gene for the constant region of murine γ2b immunoglobulin heavy chain. *Science*, **206**, 1303–6.

Tucker, P. W., Marcu, K. B., Slightom, J. L. & Blattner, F. R. (1979*b*). Structure of the constant and 3'-untranslated regions of the murine γ2b heavy chain messenger RNA. *Science*, **206**, 1299–1303.

Valbuena, D., Marcu, K. B., Weigert, M. & Perry, R. P. (1978). Multiplicity of germline genes specifying a group of related mouse κ chains with implications for the generation of immunoglobulin diversity. *Nature*, **276**, 780–4.

van Ooyen, A., Van den Berg, J., Mantei, N. & Weissmann, C. (1979). Comparison of total sequence of a cloned Rabbit β-Globin Gene and its Flanking Regions with a Homologous Mouse Sequence. *Science*, **206**, 337–44.

Vassalli, P., Tedghi, R., Lisowska-Bernstein, B., Tartakoff, A. & Jaton, J.-Cl. (1979). Evidence for hydrophobic region within heavy chains of mouse B lymphocyte membrane-bound IgM. *Proceedings of the national academy of sciences of the USA*, **76**, 5515–9.

Wabl, M. L., Forni, L. & Loor, F. (1978). Switch in immunoglobulin class production observed in single clones of committed lymphocytes. *Science*, **199**, 1078–9.

Wang, A.-C., Wang, I. Y. F. & Fudenberg, H. H. (1977). Immunoglobulin structure and genetics. Identity between variable regions of a μ and a γ2 chain. *Journal of biological chemistry*, **252**, 7192–9.

Weigert, M., Gatmaitan, L., Loh, E., Schilling, J. & Hood, L. E. (1978). Rearrangement of genetic information may produce immunoglobulin diversity. *Nature*, **276**, 785–90.

Weigert, M. & Potter, M. (1977). Antibody variable-region genetics. *Immunogenetics*, **4**, 401–35.

Weigert, M. & Riblet, R. (1977). Genetic control of antibody variable regions. *Cold Spring Harbor symposia on quantitative biology*, **41**, 837–46.

Wells, J. V., Fudenberg, H. H. & MacKay, I. R. (1971). Relation of the human antibody response to flagellin to Gm genotype. *Journal of immunology*, **107**, 1505–10.

Williamson, A. R. (1976). The biological origin of antibody diversity. *Annual reviews of biochemistry*, **45**, 467–500.

Wu, T. T., Kabat, E. A. & Bilofsky, H. (1979). Some sequence similarities among cloned mouse DNA segments that code for λ and κ light chains of immunoglobulins. *Proceedings of the national academy of sciences of the USA*, **76**, 4617–21.

8

Control mechanisms in muscle contraction

ALAN WEEDS & PAUL WAGNER

MRC LABORATORY OF MOLECULAR BIOLOGY, HILLS ROAD,
CAMBRIDGE CB2 2QH, UK

Motility in living systems is almost synonymous with life itself and Nature
has evolved several distinct mechanisms to create movement within cells
and organisms. Here we propose to concentrate solely on those systems
involving actin and myosin. In addition to their well established roles in
muscle contraction, these proteins have been identified in most types of
eukaryotic cells (and possibly in certain prokaryotic ones), where they
may play an essential part in such diverse processes as cytoplasmic
streaming, secretion and phagocytosis, cytokinesis, nerve cell growth and
blood clotting. Another distinctive motile mechanism involves microtubules in cilia and flagella, but although there are similarities between
actomyosin and the tubulin-dynein system, very little is known as yet
about mechanochemical energy transduction in microtubules and its
regulation. In contrast, our understanding of these processes in muscle is
much clearer: indeed the molecular mechanism of muscle contraction has
served as a paradigm for current research into motile phenomena in
non-muscle cells. For this reason we believe that knowledge of the
molecular mechanisms by which calcium ions control actomyosin interaction in different muscle systems will provide insights that will help us to
understand the regulation of cell motility in general. However, cytoplasmic motility requires a more flexible arrangement of the contractile
apparatus and a number of additional actin binding proteins have been
identified in different cells which may be involved in this. Thus additional
regulatory processes must be present to control the location, organisation, distribution and interactions of these various components within
cells. The elucidation of these mechanisms provides a challenge for future
research.

STRUCTURAL FEATURES OF VERTEBRATE STRIATED MUSCLE

Before describing control mechanisms, it is necessary to outline the structural basis of muscle contraction and classification of muscle types. A more detailed account can be found in a review by Offer (1974).

There are two classes of muscles in vertebrate animals: striated and smooth. Striated muscles are made up of parallel arrays of fibres which vary in diameter between 20 and 100 μm and have lengths varying from a few millimetres to as much as half a metre. These elongated multinucleate cells are made by fusion of precursor myoblasts. They are characterised by the presence of transverse striations, visible in the light microscope, which arise from structural periodicities within the fibres themselves. Each fibre contains several hundred myofibrils of 1–2 μm diameter stacked in register. By contrast smooth muscle cells are much smaller, only a few microns in length, and lack this high degree of structural organisation. This has made them more difficult to study and their contractile properties are less well understood. They attach to one another and are found, for example, in the walls of blood vessels and in the intestine, where they are involved in peristaltic movements.

Striated and smooth muscles also vary in their mode of stimulation. In mammals, skeletal muscle fibres are stimulated by a single motoneurone which produces a propagated action potential; i.e. the depolarising current is transmitted throughout the cell membrane and transverse tubules of the sarcoplasmic reticulum to activate even the innermost fibrils of the muscle fibre. Thus these are called 'twitch' fibres, since the whole fibre responds to activation at a single end-plate. Smooth muscles show a much more complex pattern of stimulus. Some show spontaneous activity while others respond to external stimuli. Some smooth muscle cells contract following depolarisation of their cell membranes via the autonomic nervous system, while others show activity in response to hormones like adrenaline as well as to various drugs. Unlike striated muscle cells, they also respond to impulses passing between neighbouring cells. Furthermore neurotransmitters may cause either excitation or inhibition depending on their site of action. Thus the overall activity of the muscle often depends on opposing excitatory and inhibitory effects and their physiological behaviour is therefore both more versatile and more complex. Cardiac muscle, which is striated in appearance, also responds to both acetyl choline and adrenaline and in addition is capable of innate rhythmic contraction: it may therefore be classified as an involuntary striated muscle. Notwithstanding these differences in structural organisation and innervation, all these muscle types are thought to contract by a mechanism involving actomyosin interaction.

The Sliding Filament Hypothesis is based on structural analysis of vertebrate striated muscle. The transverse striations seen in the light microscope arise from the sequential arrangement of A (anisotropic) and I (isotropic) bands along each myofibril. The composition of these bands can be seen in greater detail in the electron microscope (Fig. 8.1). In the centre of each I-band is the sharp Z-line which defines the edge of each sarcomere, the basic contractile unit. Arrays of thin filaments project about 1 μm on either side of the Z-line and overlap with the thick filaments located in the A-bands. These thick filaments are of 1.6 μm length, while transverse sections show them to be arranged in a hexagonal lattice. Cross-bridges, which originate as projections on the thick filaments, attach to the thin filaments, as can readily be seen in longitudinal sections of rigor muscle by electron microscopy at high magnification. These projections are the globular 'heads' of the myosin molecules. The Sliding Filament Mechanism was originally proposed by A. F. Huxley and R. Niedergerke and by H. E. Huxley and the late Jean Hanson. According to this mechanism, shortening occurs when the thick and thin filaments are driven past each other. H. E. Huxley proposed that the driving force was provided by a cyclical movement of the myosin cross-bridges. (Much of the structural evidence for this theory is summarised in H. E. Huxley (1969).) Thus cross-bridges attach to actin monomers and undergo a change in their angle of attachment relative to the actin filament, which produces a translational movement of between 5 and 10 nm. They then detach and move back to their original orientation before reattaching to other actin monomers further along the thin filaments (Fig. 8.2). Thus successive and asynchronous cycles of cross-bridge movement produce increased overlap of the filaments within the sarcomeres and hence overall shortening. This model is further supported by mechanical and X-ray diffraction studies on living muscles. While most of this research has been carried out using frog muscle preparations, our knowledge of the structure and properties of the constituent proteins is based almost exclusively on rabbit muscle (see below).

Structures of actin and myosin and their location

Thin filaments are made up of three proteins: actin, tropomyosin and troponin. Actin, which is the major component, can exist in two different forms. Globular or G-actin, a monomeric protein of 42 000 mol. wt will

A. Weeds and P. Wagner

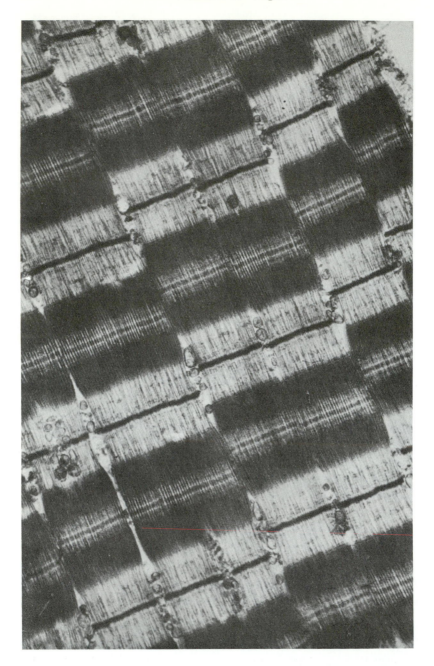

Fig. 8.1. Longitudinal section of rabbit psoas muscle. Myofibrils run from top left to bottom right and the narrow dense lines are the Z-lines which define the ends of the sarcomeres. Each sarcomere contains a dense, broad A-band,

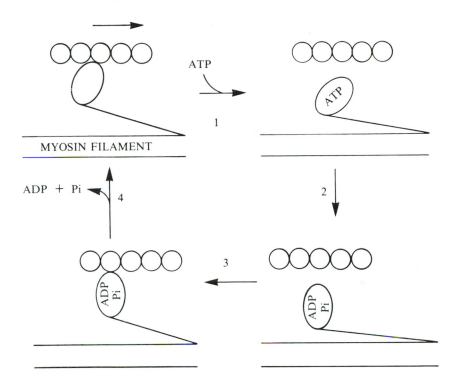

ATP

MYOSIN FILAMENT

ADP + Pi

Fig. 8.2. Simplified scheme of cross-bridge cycle. The four steps in the mechanical cycle are here correlated with the steps in the ATPase cycle (See Fig. 8.6). ATP dissociates the cross-bridge in its angled orientation (1). ATP cleavage occurs on the dissociated cross-bridge which returns to the perpendicular orientation (2). The cross-bridge re-attaches to another actin monomer in the same filament (3) and undergoes the drive stroke by returning to the angled orientation (4).

polymerise into fibrous or F-actin, which is seen as a two-chain helical structure in the electron microscope (Fig. 8.3). Tropomyosin, a two-chain α-helical coiled-coil of 40 nm length, associates end-to-end into a rod-like structure that lies in the grooves on either side of the F-actin helix. Troponin attaches to the actin–tropomyosin complex at intervals of

containing thick filaments. On either side of the A-band is the I-band, containing thin filaments. The additional density in the peripheral regions of the A-bands defines the overlap between thick and thin filaments. (We would like to thank Dr H. E. Huxley for this photomicrograph.)

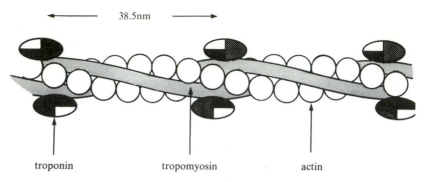

Fig. 8.3. Schematic representation of thin filament structure. F-actin contains a double helical arrangement of G-actin monomers (shown as open circles). The tropomyosin molecules associate end-to-end and lie in the grooves of the actin helix (shown here with exaggerated diameter as shaded ribbons). The troponin complex is represented by an ellipse, with three subunits indicated by the different shadings.

38.5 nm (Fig. 8.3). Tropomyosin and troponin constitute the regulatory complex which controls the interaction between myosin cross-bridges and actin monomers.

Myosin is the main component of the thick filament and comprises over 50% of all myofibrillar protein. It is extracted from muscle mince at high ionic strength, which causes filament disassembly, and which, in the presence of ATP, breaks the complex with actin. Dialysis of this solution to physiological ionic strength produces bipolar filaments similar to natural thick filaments. The precise organisation of myosin molecules in thick filaments is not yet known, but the cross-bridges are located at intervals of 14.3 nm. (Current evidence supports a three-stranded helix of pitch 3×42.9 nm with three cross-bridges at every 14.3 nm.) It is important to realise that the periodicities of thick and thin filaments are different, so that cross-bridges can act asynchronously to develop a fairly steady force as the filaments slide past one another. Just as the thick filaments are bipolar about their centres, so too the thin filaments have opposite polarity about Z-line. In this way cross-bridge cycles in either half of the A-bands produce sliding forces of opposite directionality, i.e. opposing Z-lines are drawn towards the centre of the sarcomere.

Myosin molecules from muscle and nearly all non-muscle sources consist of two globular 'heads' attached to a 150 nm rod-like tail. The molecule may be cleaved under mild conditions by a number of different proteolytic enzymes to generate subfragments which retain various of the properties of the parent molecule (Fig. 8.4). These subfragments have proved particularly useful for studying the substructure of myosin and its enzymic properties. Thus trypsin or chymotrypsin preferentially cleaves monomeric myosin into two pieces, heavy meromyosin (HMM, a water-soluble, two-headed fragment which retains both actin binding and ATPase activity) and light meromyosin (LMM, a rod-like fragment of

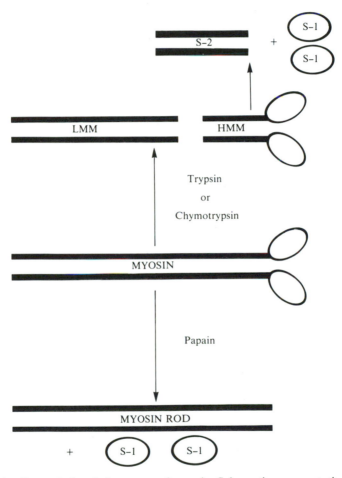

Fig. 8.4. Proteolytic sub-fragments of myosin. Schematic representation of the production of different proteolytic subfragments by various proteolytic enzymes. Heavy meromyosin (HMM) and light meromyosin (LMM) can be most readily produced by trypsin or chymotrypsin while papain gives mainly subfragment-1 (S-1). Further digestion of either HMM or myosin rod will produce subfragment-2 (S-2). This scheme is not intended to be complete but shows the relationship between the major proteolytic fragments and the parent molecule.

about 90 nm length, with the same solubility as myosin itself). LMM comprises the bulk of the myosin tail and is responsible for its filament-forming properties under physiological conditions. When myosin is cleaved with papain, subfragment-1 (S-1) is released. This fragment corresponds to the single globular head of the molecule. Cleavage here also produces the myosin rod, a fragment of the same length as the myosin tail. Further cleavage of myosin with papain, or cleavage of either HMM or myosin rod, will produce subfragment-2 (another rod-like fragment, whose length is about 60 nm but which differs from LMM in being soluble under physiological conditions). All these different rod-like fragments are composed of two polypeptide chains in an α-helical coiled-coil conformation.

The subunit structure of myosin consists of two heavy chains of molecular weight 200 000 which are non-covalently associated with four moles of light chains of about 20 000 mol. wt (Fig. 8.5). These light chains are located in the region of the myosin 'heads'. The different polypeptide chains can be separated by polyacrylamide gel electrophoresis in sodium dodecyl sulphate (Fig. 8.5). Myosins from different types of muscle differ in the electrophoretic mobilities of their light chains under these conditions and this technique has become a standard tool for analysing myosin isoenzymes. (There are also considerable differences in the heavy chains of different myosins, but these cannot be detected by gel electrophoresis.) Here we will consider rabbit skeletal muscle myosin, the bulk of which is of the fast-twitch type. (Muscles required for rapid voluntary movements are termed 'fast' because they shorten more quickly than muscles involved in maintaining posture, which are termed 'slow'. These physiological differences are related to the ATPase activity of the constituent myosins and hence reflect the expression of particular myosin genes.) Although gel electrophoresis shows three light chain bands, two of these (labelled Alkali 1 and Alkali 2 in Fig. 8.5) have closely related amino-acid sequences, but they are distinct from the other class of light chain (termed DTNB in Fig. 8.5). Thus there are two chemical classes of light chains and two moles of each of these per mole of myosin. This generalisation may be applied to all myosins showing the two-headed structure. Our nomenclature for the light chains is an operational one, based on the methods used to dissociate these particular light chains. We use it here because the role of these different light chains is not yet clear and therefore other terminologies are equally unsatisfactory. The Alkali light chains have molecular weights of 20 700 and 16 500 respectively, but the larger one has an anomalous mobility on gel electrophoresis corresponding to an

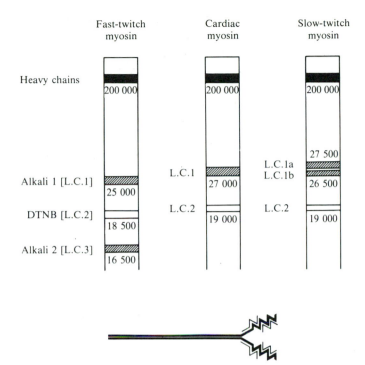

Fig. 8.5. Subunit structure of rabbit muscle myosins. The patterns of light and heavy-chain components obtained by polyacrylamide-gel electrophoresis in the presence of sodium dodecyl sulphate are shown for skeletal and cardiac muscle myosins. Diagonal shading shows chemically related Alkali light chains, while members of the other class of light chains also appear to be related to one another, since these LC2 light chains can all be phosphorylated. The schematic representation of the myosin molecule shows two pairs of light chains associated with the 'heads' of the myosin molecule. The heavy chains are bound together in the 'tail' region in an α-helical coiled-coil (shown here as the parallel black lines).

apparent molecular weight of 25 000. The DTNB light chains (18 500 mol. wt) can be partially dissociated from myosin by reaction with the Ellman reagent (5, 5' dithiobis-(2 nitrobenzoic acid), DTNB). If, following removal of the light chains, the myosin thiol groups are regenerated by reaction with dithiothreitol, full ATPase activity is recovered. Thus the DTNB light chains do not appear to be essential for enzymic activity of myosin. However, there are a number of properties of these light chains which suggest that they are not tightly bound contaminants but may have important physiological functions. They occur in a stoichiometry of 2.0 moles per mole of myosin and are phosphorylated on a unique serine residue by a specific light-chain kinase. Furthermore, the DTNB light chain can substitute for one of the regulatory light chains of scallop myosin in restoring calcium sensitivity to desensitised actomyosin (see later). Similar phosphorylated light chains are also present in cardiac and smooth muscles.

The second class of light chains, here termed Alkali light chains, cannot be dissociated without total loss of enzymic activity. Chemically related Alkali light chains have been identified in myosins from vertebrate cardiac and slow-twitch muscles (Fig. 8.5). All the methods required to dissociate these light chains denature the myosin, making it difficult to analyse their specific function. Recently we have shown that it is possible to exchange light chains between myosins from different muscle types under conditions where the ATPase activity is not impaired (Wagner & Weeds, 1977). The hybrid forms have ATPase activities identical to the parent myosin from which the heavy chain was obtained; thus the light chain does not influence the myosin ATPase in the absence of actin. However, in the presence of actin the ATPase activity of the hybrid form is different from either parent form. These experiments suggest that the Alkali light chains are needed to maintain the myosin in an enzymically active state, and that they may modulate the interaction of actin and myosin in the presence of ATP.

Myosin ATPase and interaction with actin

Because of difficulties arising from the insolubility of myosin at physiological ionic strength, most of the studies on the ATPase mechanism have been carried out using either HMM or S-1. A further complication of working with actomyosin is the phenomenon known as superprecipitation, whereby addition of Mg·ATP causes the protein to precipitate into a tight pellet. It is believed that superprecipitation is due to the disorganised thick and thin filaments sliding together in a manner analogous to contraction in muscle. In this discussion we will consider experiments with S-1 and HMM and make the assumption that these can be extrapolated to myosin itself. This assumption ignores the possibility of co-operative interactions within the thick filament.

The steady-state turnover of Mg·ATP by S-1 alone is very slow (about 0.05 mole of ATP/mole of S-1/s, denoted 0.05 s^{-1}, for rabbit S-1 at 25°C). In the presence of actin this rate is increased maximally about 1000-fold. Thus actin must bypass the rate limiting process of the myosin ATPase mechanism. Our detailed knowledge of the myosin and actomyosin ATPase comes from the pioneering studies of Lymn & Taylor (1971) and also from Trentham (1977). We will summarise these results in terms of the Lymn–Taylor model. Transient kinetic studies have shown that ATP is cleaved at $>150\,\mathrm{s}^{-1}$ by S-1, and the steady-state rate is controlled by a subsequent process involving the release of products, ADP and inorganic phosphate. Actin binds to the S-1 products complex and accelerates dissociation of ADP and inorganic phosphate (Fig. 8.6).

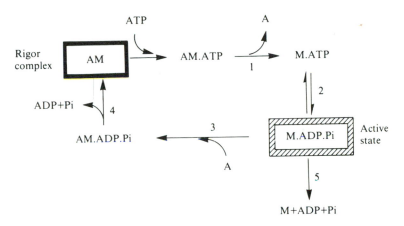

Fig. 8.6. Simplified kinetic scheme for actomyosin ATPase. This kinetic mechanism is based on that of Lymn & Taylor (1971) and Trentham (1977). ATP dissociates the actomyosin complex and is subsequently cleaved by the dissociated myosin 'heads', producing the steady-state intermediate (here denoted the Active state). In the absence of actin, product release is very slow (reaction 5). Actin bypasses the slow process and accelerates the ATPase rate nearly 1000-fold. The two boxes are drawn to emphasise the different states of myosin in the actomyosin complex and the active state.

When actin and S-1 are mixed in the absence of ATP, a tight complex is formed, which can be seen in the electron microscope as a characteristic arrowhead structure. The angled configuration of the S-1 is similar to the orientation of the cross-bridges seen in the electron microscope in thin sections of rigor muscle. A. Weber has termed this complex the 'rigor complex'. Addition of Mg·ATP to this turbid acto·S-1 solution results in an instantaneous decrease in light scattering which is interpreted as dissociation of the complex. The maximum rate of this process at saturating ATP concentrations is at least ten times as fast as the rate of ATP cleavage. Hence it appears that substrate cleavage occurs on the dissociated S-1, which then reassociates with actin to facilitate the release of ADP and inorganic phosphate, thus completing the kinetic cycle. This kinetic cycle may be correlated with the cross-bridge cycle shown in Fig. 8.2.

Further kinetic analysis of the forward and reverse rate constants for various steps in the kinetic mechanism tells us about the reversibility of these processes and the magnitude of standard free-energy changes occurring. It is clear from the work of Bagshaw and Trentham that ATP cleavage is freely reversible, with only a small standard free-energy change (see Trentham, 1977). In contrast, ATP binding to S-1 is essentially irreversible, showing a very large negative standard free-energy change. This free energy is needed to dissociate the actomyosin 'rigor complex'. Thus the primary role of ATP in muscle is to dissociate the cross-bridge from actin. Following this, ATP cleavage produces the steady-state intermediate, whose half-life, estimated from the steady-state rate in the absence of actin, is about 14 s. The steady-state intermediate is predominantly the myosin products complex shown in Fig. 8.6 (though this is in equilibrium with a tightly bound $M \cdot ATP$ complex). Because of its long half-life, this will be the predominant state of myosin cross-bridges in relaxed muscle. However, when muscle is activated, the myosin products complex recombines with actin, which accelerates the release of products and hence the rate of ATP turnover. Thus the myosin·products complex is the 'active state' of the cross-bridge in muscle and is distinct from the state of the cross-bridge in rigor, when no ATP is left. The myosin·products complex can be identified by a number of techniques including protein fluorescence enhancement when ATP binds to S-1, spin labelling studies and the differential chemical reactivity of certain myosin thiol groups in the presence of ATP. The apparent affinity of this complex for actin, estimated from the K_m of the acto·S-1 ATPase (~ 10 μM actin), is much weaker than the dissociation constant of the S-1 'rigor-complex' ($K_d = 0.01$ μM). Regulatory mechanisms of muscle contraction are concerned with the interaction of this 'active state' of the cross-bridge with actin.

GENERAL ASPECTS OF REGULATION

In vertebrate striated muscle, contraction is initiated by an action potential passing down the motoneurone which depolarises the muscle membrane, causing a release of calcium ions from the sarcoplasmic reticulum. The means by which this occurs is not yet understood. As a result the intracellular concentration of free calcium ions rises from <0.1 μM to over 10 μM and these calcium ions act as the intracellular messengers which trigger contraction. They also stimulate glycogen breakdown through the activation of phosphorylase kinase and in this way provide additional ATP as fuel for contraction. When the stimulus is removed, the calcium ions are pumped back into the sarcoplasmic reticulum with concomitant relaxation of the muscle. This description implies that the relaxed state is an inhibited state and that calcium ions relieve this inhibition.

Two general mechanisms may be proposed to maintain the inhibited state: (1) Modification of the myosin head may prevent its interaction with actin. This would be classified as 'myosin-linked' or 'thick filament' regulation. (2) The interaction site on actin may be modified or obstructed by some other protein to prevent the cross-bridge from binding. This would be classified as 'actin-linked' or 'thin filament' regulation. In either case, this inhibition is relieved by calcium ions. We now know that both of these general mechanisms exist and in some animals both are present in the same muscle, that is, dual regulation occurs. Experiments to date have revealed the existence of three distinct regulatory mechanisms, two of which are myosin-linked and the other actin-linked. In addition other mechanisms may be present to modulate actomyosin interaction in conjunction with these basic controls, providing adaptation to the specific requirements of the individual muscle.

All eukaryotic cells maintain their concentration of free calcium ions at below 0.1 μM. In contrast to this, calcium ion concentrations in plasma and body fluids are in excess of 1 mM. Thus all cells contain enzymes which actively exclude calcium ions. Since changes in the electrical properties of cell membranes affect their ion permeability, these changes will often lead to a transient increase in intracellular calcium concentration. These calcium fluxes then serve to transmit information within the cell. For example, in addition to the control processes already mentioned, calcium ions trigger the release of neurotransmitters from vesicles in synapses and are implicated in other forms of exocytosis. They are also involved in regulating the level of cAMP and hence influence its function as a second messenger. In this way cells are able to use calcium ions as a control signal for a large variety of biochemical processes and in addition perhaps to regulate their spatial distribution.

Calcium ions are bound by a number of proteins within cells and readily taken up by mitochondria, which effectively limits the maximum free ion concentration within the cytoplasm. The efficiency of the calcium ion pump in the membrane controls the lower limit of free calcium ions. Thus the concentration range within which calcium ions act as a messenger is between 0.1 μM and somewhat over 10 μM. This means that the target for these ions must have an affinity within this range. Kretsinger (1977) has proposed that the intracellular target will be a calcium binding protein with a dissociation constant between 10^{-7} and 10^{-5} M. Further he has suggested that a family of such proteins exists and that these are evolutionarily related. Some of the evidence for this assertion will be presented in this review.

Role of tropomyosin and troponin in actin-linked regulation

The role of calcium ions in regulating muscle contraction was first demonstrated using vertebrate striated muscle. The ATPase of actomyosin made from purified F-actin and myosin was high whether calcium ions were present or not, but naturally extracted actomyosin showed a low ATPase activity in the absence of calcium (<0.1 μM) and a high ATPase when the calcium concentration was greater than 10 μM. This calcium sensitivity of the ATPase activity depends on the presence of tropomyosin and troponin. As we have already shown, tropomyosin is located in the grooves of the actin helix in a polymerised form with a repeat of 38.5 nm. This repeat distance is seven times the repeat of the actin subunits. Troponin binds to actin·tropomyosin at intervals of 38.5 nm and the molar ratio of these three components is 7:1:1 (actin:tropomyosin:troponin). Of these three proteins, only troponin has specific calcium binding sites, i.e. sites which bind calcium ions in the presence of 1 mM free Mg^{2+} ions. (Although the concentration of free Mg^{2+} ions in the muscle cell is not known precisely, because much of the intracellular magnesium is bound to nucleotides or proteins, it is estimated that this concentration exceeds 1 mM.) It is important to emphasise the distinction between calcium specific sites and non-specific divalent cation sites. While the non-specific sites often bind calcium ions with a higher affinity than magnesium ions, the differences in affinity are small so that under physiological conditions magnesium will always be bound. The affinity of magnesium ions for the calcium specific sites is very low so that only calcium ions will bind to these sites under physiological conditions. The importance of this distinction will become apparent in later discussion of the nature and function of different divalent cation sites.

Troponin is a complex of three non-covalently associated subunits which can be separated by polyacrylamide gel electrophoresis in the presence of sodium dodecyl sulphate. These have been termed troponin-T (Tn-T), the tropomyosin binding component; troponin-I (Tn-I), which inhibits the actomyosin ATPase activity; and troponin-C (Tn-C), the calcium binding subunit. The molecular weights of these proteins from rabbit skeletal muscle, based on amino-acid sequences (with apparent molecular weights from gel electrophoresis in parenthesis) are as follows: Tn-T, 30 500 (37 000); Tn-I, 20 900 (23 000); Tn-C, 17 900 (18 000), showing that the electrophoretic method for determining polypeptide chain molecular weights is sometimes subject to large error! Current evidence suggests that these three subunits are associated in a $1:1:1$ stoichiometry in the native protein complex. While all three components are required together with tropomyosin to confer calcium sensitivity on the actomyosin ATPase activity, information about their interactions and individual roles has come from examination of each subunit and various combinations of them. Tn-I can bind directly to actin and partially inhibit the actomyosin ATPase activity, even in the absence of tropomyosin. Maximum inhibition is achieved at a stoichiometry of one Tn-I per actin monomer. In the presence of tropomyosin the inhibition is much more effective; under these conditions a single Tn-I can inhibit the activity of more than three actin monomers. Addition of Tn-C neutralises this inhibitory effect whether calcium ions are present or not, while the presence of Tn-T is essential to achieve calcium sensitivity of the actomyosin ATPase. Since Tn-I is a basic protein and Tn-C is acidic, this effect of Tn-C might reflect its anionic character. Indeed acidic compounds like heparin have been shown to mimic this neutralizing effect of Tn-C. Perry and his associates have analysed various fragmentation products of both Tn-I and Tn-C to find out whether smaller parts of these subunits retain biological activity. A 22-residue fragment of Tn-I containing nine basic residues was about 50% as active as whole Tn-I in inhibiting the ATPase, and Tn-C again reversed this inhibition. Similarly, a 52-residue fragment of Tn-C, containing the third calcium binding site (see later), substituted for Tn-C in neutralising the inhibitory effect of Tn-I, and like Tn-C itself, this fragment had the capacity of forming a calcium dependent complex with Tn-I. In this way, fragmentation studies have revealed the existence of functional domains within these molecules (Perry, 1979 and references therein).

Calcium ions have been shown to influence the interactions between various troponin subunits. In the absence of calcium, Tn-C binds Tn-I weakly, but this interaction is much stronger when calcium is present. Indeed Perry has shown that this complex is stable even in the presence of 6 M urea. In contrast, the complex between Tn-I and Tn-C binds to actin·tropomyosin strongly in the absence of calcium but much less so in its presence. Furthermore, Tn-T binds Tn-C, but only in the presence of calcium ions. Since Tn-T binds to tropomyosin, its function appears to be that of an anchor, which locates the troponin complex with the correct periodicity and orientation on the thin filament. Thus the effects of calcium ions are to strengthen the binding of Tn-I, through Tn-C to Tn-T and at the same time to lower the affinity of Tn-I for actin (Fig. 8.7). We may now ask how these altered affinities operate the switch to control actomyosin interaction.

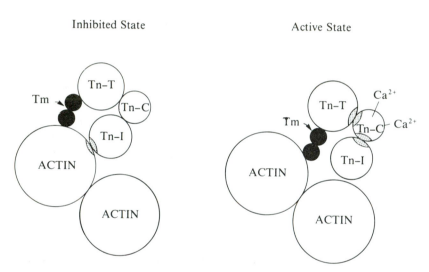

Fig. 8.7. Scheme for the regulation of muscle contraction by troponin. This scheme is based on the steric blocking model of Huxley (see Fig. 8.8) and shows the interactions between actin, tropomyosin and the troponin subunits. The shaded zones denote the presence of specific interactions that are influenced by calcium ions. Thus calcium binding to Tn-C strengthens the interactions between Tn-T and Tn-C and between Tn-I and Tn-C, while the interaction between Tn-I and actin is weakened. (After C. Cohen, *Scientific American*, **233**, 36–45 (1975).)

Based on H. E. Huxley's structural studies, it appears that all the G-actin monomers in F-actin are capable of binding S-1, giving rise to the well-known arrowhead structures. By contrast troponin is located only at 38.5 nm intervals, with a stoichiometry of one mole per seven actin monomers. Since the periodicity of the cross-bridges is not synchronous with the troponin repeat, binding occurs to actin monomers that are located between adjacent troponin units. Thus the inhibitory effects of troponin must be transmitted along the thin filament. The current model for troponin inhibition of actomyosin interaction is based on steric blocking of actin sites by tropomyosin. On the basis of X-ray diffraction studies, Huxley (1972) suggested that in the inhibited state tropomyosin is located in a position where it obstructs the approach of the cross-bridge and that on activation the tropomyosin moves a distance of about 1.5 nm towards the centre of the actin groove, thus exposing the cross-bridge binding site. This model is shown in Fig. 8.8. It is supported by evidence from image reconstructions of electron micrographs (Wakabayashi, Huxley, Amos & Klug, 1975). These reconstructions of actin–tropomyosin paracrystals (the 'ON' state) and paracrystals of actin–tropomyosin–Tn-T–Tn-I (the 'OFF' state) suggested a different location for the tropomyosin in these two states. Recently it has been suggested that the positions of tropomyosin and myosin relative to actin are displaced from those shown in Fig. 8.8, but this possible displacement is such that a steric blocking mechanism still remains a valid hypothesis. Detailed examination of the amino-acid sequence of tropomyosin has shown the existence of groups of acidic residues periodically spaced throughout the molecule. Stewart & McLachlan (1975) have proposed that these 14 repeats of acidic regions may correspond to two sets of seven quasi-equivalent actin binding domains on tropomyosin, one set corresponding to the 'OFF' state and the other to the 'ON' state. Rotation of the tropomyosin through about 90° would produce stable binding to actin by one or other set of these domains. Finally it has been argued that this movement of tropomyosin depends on the interactions between troponin subunits and their modulation by calcium ions as outlined above (Fig. 8.7).

In our earlier description of the actomyosin ATPase mechanism we emphasised the distinction between the 'active state' of the cross-bridge, corresponding to the myosin·products complex, and the rigor forming state which exists in the absence of ATP and has a much higher affinity for actin. Annemarie Weber showed that the presence of rigor complexes abolished calcium control (see Weber & Murray, 1973). Thus using subsaturating concentrations of ATP relative to the S-1 concentration, both S-1·products complexes and rigor complexes will be present simultaneously, and under these conditions the ATPase activity using regulated actin (actin + tropomyosin + troponin) remains high whether calcium ions are present or not. When the ATP concentration is raised to

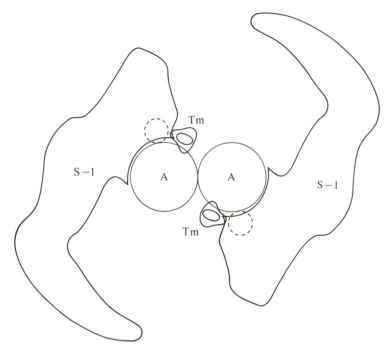

Fig. 8.8. Composite end-on view of actin-tropomyosin-subfragment-1 structure. The positions occupied by tropomyosin in the active state (solid Tm
contours) and relaxed state (dotted contour) are shown relative to that of
subfragment-1. The shape and location of subfragment-1 (S-1) are based on
image reconstruction of decorated actin. (Courtesy of Dr H. E. Huxley and the
Cold Spring Harbor Laboratory, see Huxley (1972).)

saturate all the S-1 present full calcium control is restored, showing that
this effect is not due to irreversible binding of S-1 to actin. This interesting
finding shows that rigor complexes prevent troponin from switching
'OFF' the thin filament when calcium ions are absent, presumably by
obstructing the proper interactions of tropomyosin. Weber further
showed that rigor complexes have a 'potentiating' effect on the ATPase
activity when calcium ions are present. This results from an accelerated
interaction of the myosin·products complexes with regulated actin. Both
of these observations suggest communication between adjacent actin
monomers in the thin filament, i.e. co-operative interactions. Not only do
rigor complexes enhance the ATPase activity of S-1 at neighbouring
actin monomers, but they also influence calcium binding to troponin
itself, showing that this communication operates in both directions to

affect S-1 interaction at neighbouring actin monomers and calcium affinity on neighbouring troponins. None of these effects can be observed in the absence of tropomyosin. Thus these studies complement the structural evidence that tropomyosin is involved in transmitting the signals produced by calcium binding to troponin, to actin monomers which are as much as 20 nm distant from the troponin complex.

This is a brief sketch of the current model of actin-linked regulation in vertebrate striated muscle. Much more evidence is needed, especially at a structural level, before we can test this model further; in particular we require the 3-dimensional structures of both actin and subfragment-1 to provide the detailed geometry of their interactions.

Identification of myosin-linked regulation

Although the discovery of troponin by Ebashi led to the elucidation of calcium regulation in vertebrate striated muscle as we have described, comparative studies on various marine invertebrate muscles soon showed that this is not the universal mechanism for control of muscle contraction. Thin filament preparations obtained from these muscles did not contain troponin. Furthermore, washed myofibrils showed calcium sensitivity of their actomyosin ATPase activity and yet contained only trace amounts of proteins which might correspond to the troponin subunits as identified by gel electrophoresis. Subsequent investigation showed that the calcium ions acted directly on the myosin cross-bridges. A rapid procedure was developed by Lehman & Szent-Györgyi (1975) to identify the presence of myosin-linked or thick filament regulation and they examined muscles from a wide range of different organisms. The test was based on the effect of adding pure rabbit F-actin to myofibrils from the test muscle in the presence and absence of calcium ions. If myosin-linked regulation was present, then in the absence of calcium the ATPase activity would be low whether rabbit actin was present or not. But if the test myofibrils were regulated by troponin, addition of rabbit F-actin would produce a high actomyosin ATPase even in the absence of calcium, due to interaction of myosin in the myofibrils with this added actin. In either case, confirmation required the isolation of native thin filaments and myosin. The occurrence of thick filament regulation in primitive species like molluscs, gastropods and nemertine worms suggested that this form of regulation might have evolved first. Indeed myosin-linked regulation appeared to be much simpler, involving fewer components, which might be considered support for this idea. Furthermore, if it were true that thin filament regulation evolved later, we might expect to find species where both forms of control operated together, while thin filament control occurred only in the more highly evolved species. Lehman & Szent-Györgyi indeed found dual regulation in a variety of insects, crustaceans, annelids and numerous other invertebrates, but actin-linked regulation was not restricted to animals of later stages in biological evolution, since dual regulation occurs in the nematode *Ascaris lumbri-coides*. Thus it seems likely that both forms of calcium regulation arose at very early stages of biological evolution (see later).

Myosin-linked regulation by direct calcium binding

The important discovery that led to the elucidation of the mechanism of myosin-linked regulation was made by Kendrick-Jones & Szent-Györgyi in the early 1970s while investigating regulation in the striated adductor muscle of the bay scallop (*Aequipecten irradians*), (which is better known for its gastronomic qualities than its unusual scientific properties). When myofibrils or myosin from this muscle were washed in the presence of EDTA (ethylene diamine tetraacetic acid, a divalent cation chelating agent), calcium sensitivity was lost, i.e. the actin activated ATPase remained high after calcium ions were removed. Since the myofibrils or myosin could easily be removed by centrifugation, the supernatant fraction was analysed for protein and found to contain a polypeptide chain of 17 000 mol. wt, which was identified as a light chain of the scallop myosin. If the myosin was added back to the light chain fraction in the presence of magnesium ions, the two recombined and calcium sensitivity was restored. Similar restoration of calcium sensitivity was achieved by adding back the 'EDTA' or regulatory light chain to desensitised actomyosin or myofibrils. Thus myosin-linked regulation in molluscan muscles depends on the presence of regulatory light chains which confer calcium sensitivity on the actomyosin ATPase.

Like other myosins, scallop myosin contains two pairs of chemically distinct light chains, but these have the same molecular weight and migrate as a single band on polyacrylamide gel electrophoresis in sodium dodecyl sulphate. However, they can be separated by gel electrophoresis in urea and, as expected, there are two moles of each type of light chain per mole of myosin. The effects of EDTA in dissociating the regulatory light chains suggest that divalent cations are involved in the association between light and heavy chain components. While 50% dissociation of these light chains occurs at only 0.1 μM calcium ions compared with about

10 μM magnesium, under physiological conditions where the free magnesium ion concentration exceeds 1 mM, the sites involved will be occupied by magnesium ions. As will be discussed later, these non-specific, high affinity divalent cation sites are located on the regulatory light chains.

Further details of the regulatory mechanism were worked out by Kendrick-Jones, Szentkiralyi & Szent-Györgyi (1976), who showed that EDTA treatment at 0°C dissociates only one mole of regulatory light chain per mole of myosin. The remaining regulatory light chain is therefore unable to confer calcium sensitivity. They also found that scallop myosin has two calcium specific binding sites ($K_d = 10^{-6}$ M), in addition to the nonspecific sites already mentioned. Following removal of one of the regulatory light chains by EDTA, one of these calcium specific sites is lost, yet the isolated light chain does not bind calcium specifically (i.e. in the presence of 1 mM Mg^{2+} ions). When the regulatory light chain recombines with desensitised myosin to restore calcium sensitivity of the actomyosin ATPase, full calcium binding is also restored. These findings suggest that the calcium-specific site may involve structural elements from both light and heavy chains, though it is equally possible that the calcium specific site is located solely on either the light chain or the heavy chain, but its conformation is perturbed by dissociation of the two.

What is the location of the regulatory light chain and how does it control interaction between the myosin 'head' and actin? Further information has come from studies of proteolytic fragments of scallop myosin. Although heavy meromyosin shows calcium sensitivity, sub-fragment-1 has a high ATPase in the absence of calcium ions and is not regulated, even though it contains the regulatory light chain. Although this shows that the regulatory light chain is located in the head region of myosin, its inability to confer calcium sensitivity on S-1 suggests either that structural elements of myosin outside the head region are required for the light chain to operate properly, or that the S-1 has been damaged during its preparation. If the former explanation is correct, then the light chain may interact with part of the myosin rod or the 'hinge' region between S-1 and the rod. (The presence of a flexible hinge is based on evidence from electron microscopy of single myosin molecules and the high sensitivity of this region to proteolytic enzymes.) Physical charac-terisation of the dissociated light chain by Stafford & Szent-Györgyi (1978) indicates that the molecule can be represented by a general ellipsoidal model of about 10 nm length. Furthermore the structure appears to be stable over a wide range of temperature and pH conditions, which may indicate that the overall shape of the isolated light chain is little different from the associated state. Thus the elongated shape of the regulatory light chain is consistent with the idea that it can span both the myosin head and part of the rod region.

If the regulatory light chain must bind to both head and rod regions of the molecule to operate correctly, the reason for the failure of the remaining regulatory light chain to confer any calcium sensitivity in desensitised myosin is not clear. It may indicate that calcium regulation requires interaction between the two heads, but we must not exclude the possibility that the initial EDTA treatment has affected the interaction of the remaining light chain in a deleterious manner. Indeed Kendrick-Jones *et al.* have reported that the second regulatory light chain, although chemically identical to the dissociated one, is bound with an apparently greater affinity in desensitised myosin. This suggests that the state of the myosin is altered under these conditions.

The molecular mechanism by which the regulatory light chain controls interaction with actin is not known, but current ideas suggest a steric blocking of the actin binding site on the myosin head by the regulatory light chain. However it is also possible that calcium binding may affect interaction between the light chain and rod region of the molecule to generate a more favourable orientation of the cross-bridge. These speculations emphasise the need to know the 3-dimensional structure of subfragment-1 and the precise location of both the light chains and actin binding site. Another aspect of this regulatory system that is at present unclear concerns the presence of tropomyosin in scallop thin filaments. This tropomyosin is not required for calcium regulation *per se* but may have a stabilising effect on the actin monomers in the filaments. Nor can we rule out the possibility that it plays some other role in the control mechanism *in vivo*.

Myosin-linked regulation by phosphorylation

It is now abundantly clear that regulation of metabolic processes by hormones in different tissues depends on the production of intracellular messengers. Best known of these is cyclic AMP, which is synthesised in muscle by adenyl cyclase in response to adrenaline stimulation. Cyclic AMP activates protein kinase which in turn modulates the activity of a number of intracellular enzymes via protein phosphorylation. This is best understood in the regulation of glycogen breakdown and synthesis in muscle, where metabolic flux is controlled by protein phosphorylation.

Thus protein kinase phosphorylates both phosphorylase kinase (which activates glycogen phosphorylase and promotes glycolysis) and glycogen synthase (which is inhibited in its phosphorylated form). Phosphatases are present to reverse these reactions, but these are in turn regulated by specific inhibitor proteins, some of which may themselves be phosphorylated. Thus a complex series of phosphorylation reactions provides fine control of glycogen metabolism. Phosphorylase kinase is also activated by calcium ions within the same concentration range as that required for muscle contraction, and it has recently been shown that this requires a specific calcium binding protein known as calmodulin (see later).

Phosphorylation of myofibrillar proteins was first demonstrated in skeletal muscle by Perry and his colleagues, who showed that both Tn-T and Tn-I could be phosphorylated at specific serine residues (Perry, 1979 and references therein). Although Tn-I is phosphorylated by protein kinase and its level of phosphorylation increases in perfused cardiac muscle in response to adrenaline, England has shown that the correlation between increased contractile force and increased Tn-I phosphorylation is variable. Thus although this phosphorylation may be involved in modifying cardiac contractility, it does not completely explain the inotropic effects of β-adrenergic stimulation. Both skeletal and cardiac muscle Tn-I are phosphorylated when isolated from muscle, but cardiac Tn-I contains an additional phosphorylation site in its extended N-terminal sequence and this site is more rapidly labelled by protein kinase (Perry, 1979). At present the role of phosphorylation of troponin subunits in modulating the effects of calcium ions on actomyosin interaction is unclear, though changes in calcium sensitivity have been reported in the case of cardiac Tn-I (i.e. changes in the calcium ion concentration required to achieve 50% activation of the ATPase). A great deal more research is needed here, though it is reasonable to predict that troponin phosphorylation may provide an additional means of controlling the contractile response.

Perry has also shown that the DTNB light chain of rabbit skeletal muscle myosin can be isolated in a phosphorylated form, and has purified a specific myosin light chain kinase (see Perry 1979, and references therein). This enzyme is active only in the presence of micromolar concentrations of calcium ions. It has been shown to phosphorylate similar light chains from cardiac and slow-twitch muscle myosins and also smooth muscle (stomach) myosin. Perry and his colleagues also purified a light-chain phosphatase from rabbit muscle extracts which removes the bound phosphate group. It is tempting to speculate that an additional control mechanism exists in vertebrate striated muscle resulting from calcium induced phosphorylation of the myosin light chains, but evidence is weak. Bárány, Bárány, Gillis & Kushmerick (1979) have reported an increase in the level of phosphorylation during a single tetanus in frog muscle, while Manning & Stull (1979) have shown that the phosphate content of rat *extensor digitorum longus* muscle continues to rise for several seconds following a one-second tetanus and thereafter declines slowly to the original value. Thus light-chain phosphorylation does not appear to correlate with either tension increase or relaxation, but may be related to the time course of post-tetanic potentiation. These *in vivo* observations coupled with the inability to demonstrate any effect of phosphorylation on the actomyosin ATPase activity *in vitro*, suggest that calcium-induced phosphorylation of skeletal muscle myosin is not essential for regulating the contractile cycle *per se*, but may be involved in modulating force development.

While the function of myosin light-chain phosphorylation in vertebrate striated muscle is not clear, in smooth muscle there is a considerable amount of evidence showing that it plays the essential role in regulation of actomyosin interaction. Studies have been carried out by a number of groups, notably those of Adelstein (1978) and Hartshorne (see Sherry, Gorecka, Aksoy, Dabrowska & Hartshorne, 1978), using a variety of different smooth muscles including chicken gizzard, bovine stomach and guinea-pig vas deferens. These have shown that smooth-muscle actomyosin has a low ATPase activity unless it is phosphorylated (see Fig. 8.9). Light-chain kinases have been isolated which phosphorylate a specific serine residue on the 20 000 mol. wt regulatory light chain (analogous to the DTNB light chain and other light chains of this size from skeletal muscle myosins), and phosphatases are also present which reverse this process. Thus phosphorylation of the myosin light chain promotes actomyosin interaction and ATP hydrolysis, showing that the nature of this control mechanism is to stimulate a dormant state of the myosin cross-bridge. This contrasts with the mechanism in rabbit skeletal muscle or scallop muscle, where activation by calcium ions unblocks an inhibited state.

As in striated muscle, smooth muscle contraction is stimulated by an increase in the level of intracellular calcium ions. Since the light-chain kinase requires calcium ions for activity, a transient rise in calcium concentration produces increased light-chain phosphorylation which stimulates ATPase activity. Following removal of the calcium and

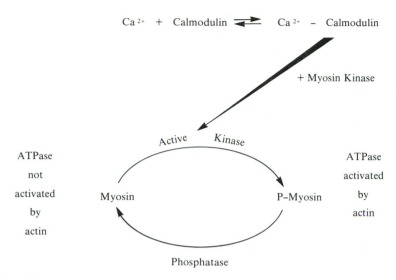

$$Ca^{2+} + Calmodulin \rightleftharpoons Ca^{2+} - Calmodulin$$

+ Myosin Kinase

Active Kinase

ATPase not activated by actin   Myosin   P–Myosin   ATPase activated by actin

Phosphatase

Fig. 8.9. Scheme for the regulation of smooth muscle contraction by myosin phosphorylation. Myosin light chain kinase is activated by binding Ca^{2+}-calmodulin and phosphorylates the regulatory light chain of smooth muscle myosin. The ATPase activity of the phosphorylated myosin is activated by actin and hence contraction is stimulated.

inactivation of the kinase, myosin light-chain phosphatase will release the bound phosphate, producing relaxation. We do not know whether the phosphatase is itself regulated, but its activity in muscle extracts is much less than that of the kinase in its activated state. It is important to note that smooth muscle actomyosin ATPase activity is about an order of magnitude lower than equivalent preparations from fast-striated muscle. Whilst calcium control involving protein phosphorylation is probably slower than a direct calcium binding mechanism, this is not necessarily disadvantageous for the slower contraction rates of smooth muscle.

Myosin light-chain kinases have been isolated from both smooth muscle and non-muscle cells (including platelets and brain). They contain two subunits, a larger one with molecular weight between 80 000 and 125 000 depending on the source, and a 17 000 mol. wt subunit which Hartshorne identified as calmodulin. (This protein, previously mentioned as the calcium regulatory component of phosphorylase kinase, is also known as Modulator Protein or Calcium Dependent Regulatory Protein (CDR). It was first identified in brain as a calcium activator of adenyl cyclase and cyclic AMP phosphodiesterase.) Calmodulin binds four moles of calcium ions with affinities similar to those of Tn-C. Although these proteins are also structurally related (see later), Tn-C will substitute only poorly for calmodulin in activating light chain kinase.

Is myosin light-chain phosphorylation the sole form of myosin-linked regulation in smooth muscle? Several reports have been published that in addition to phosphorylation, direct calcium binding to smooth-muscle myosin may be involved in regulating actomyosin interaction. In the case of gizzard myosin, this now appears to be unlikely since Hartshorne has purified phosphorylated myosin, free from both kinase and phosphatase, and shown it to be equally active in the presence or absence of calcium ions. The same appears to be the case for stomach myosin. However, in the case of guinea-pig vas-deferens myosin, Chacko has reported that the purified phosphorylated myosin is more active in the presence of calcium ions than in their absence, and this effect is particularly marked when tropomyosin is present. This suggests that different smooth muscles may vary in their regulatory mechanisms. But since smooth-muscle contractile proteins are notoriously difficult to purify, with variable enzymic activities, it is premature to accept this claim without more evidence.

We mentioned earlier that the activity of phosphorylase kinase can be controlled either by calcium ions or by phosphorylation via protein kinase. One of the effects of this phosphorylation is to lower the calcium concentration required for activation. Adelstein has recently shown that the large subunit of myosin light-chain kinase from turkey gizzard is itself phosphorylated by cyclic AMP dependent protein kinase. Furthermore, the phosphorylated kinase has a lower activity than the non-phosphorylated form, due probably to a lower affinity for calmodulin. Thus Adelstein has suggested that cyclic AMP may promote relaxation in smooth muscle by reducing the rate of myosin phosphorylation through an inactivation of the kinase. This is yet another example of the versatility of phosphorylation as a regulatory mechanism.

Actin-linked regulation in smooth muscle?

As in the case of scallop muscle, tropomyosin is present in smooth muscle and has been shown to activate the actomyosin ATPase of phosphorylated myosin about 2.5-fold. Yet there is very little evidence for troponin in smooth muscle. Isolation of a Tn-C like protein has been reported, but Perry and his associates have shown that this is probably calmodulin. Evidence for Tn-I like components is very weak indeed. However, a totally different regulatory mechanism has been proposed by Ebashi and his colleagues (Ebashi, Mikawa, Hirata & Nonomura, 1978).

While they acknowledge that a calcium-dependent myosin phosphorylation system exists in smooth muscle, they find that activation of
the actomyosin ATPase does not depend on light-chain phosphorylation
but on the presence of two regulatory proteins, tropomyosin and a new
component called leiotonin. When leiotonin, tropomyosin and actin are
added to myosin, there is an activation of the ATPase activity in the
absence of phosphorylation and this effect is calcium sensitive. Leitonin is
composed of two subunits, whose molecular weights are 80 000 and
18 000. The smaller subunit, termed leiotonin-C, binds calcium ions but·
appears to be distinct from both Tn-C and calmodulin. The action of
leiotonin differs from that of the troponin complex in two important
ways. (1) It acts as an activator since there is no actomyosin ATPase in its
absence; (2) very much less leiotonin is required than troponin (maximum
activation is achieved at about one leiotonin per 100 actin monomers).
The mechanism by which it activates actomyosin ATPase is not known,
but Ebashi does not believe that it acts enzymatically.

On present evidence it is difficult to reconcile these results with the
considerable body of evidence indicating that myosin phosphorylation is
the chief means by which calcium ions activate smooth muscle. While the
use of smooth muscle actin may be relevant, no significant differences
have been observed in the activation of myosin subfragments using
different forms of actin. In addition, the high degree of sequence
homology between different actins makes it unlikely that there are any
large differences in their interactions with myosin. More evidence is
needed concerning the purity of both kinase and phosphatase preparations and of leiotonin, so that the properties of all three can be
investigated more thoroughly. We also need to know more about the
kinetics of these control mechanisms, since the signal for contraction
must be relayed to the actomyosin within about 200 ms and it is not yet
clear whether the activity of the light-chain kinase within the cell is
sufficiently high to modify a substantial fraction of the myosin within this
short time.

STRUCTURAL RELATIONSHIPS BETWEEN
CALCIUM BINDING PROTEINS

In all the regulatory mechanisms discussed here, calcium sensitivity has
depended on the presence of a protein of about 18 000 molecular weight:
Tn-C (and leiotonin-C) in thin filament regulation, myosin regulatory
light-chain in scallop myosin-linked regulation and calmodulin in myosin

light-chain kinase. Since these small proteins are all involved in calcium control processes, it is important to find out whether they are structurally related. Such comparisons are best made on the basis of three-dimensional structures, as has been done, for example, with serine proteases. In the case of the calcium binding proteins, only one crystal structure is available at present: the structure of a calcium-binding protein from fish muscle, known as parvalbumin, has been solved by Kretsinger and his associates and this protein was shown by Collins (1976) to have considerable amino-acid sequence homology with Tn-C. Thus sequence comparisons between various calcium-binding proteins have been related to the 3-dimensional structure of parvalbumin, the essential features of which we outline below.

Carp parvalbumin consists of six α-helical regions (A to F) with five loops between and a non-helical amino-terminal region. There are two calcium binding sites in the loops between helices C and D, and E and F, and these are related to each other by a diad axis of symmetry (Fig. 8.10). Thus each site lies in a pocket between two perpendicular α-helices. Kretsinger has termed this structural unit the 'EF-hand' since it may be broadly compared to a human right hand with the forefinger pointing upwards and the thumb outwards (these the two helices). Examination of the model of carp parvalbumin shows that these domains are stabilised by a number of hydrophobic interactions involving strategically placed apolar residues (Fig. 8.10), which occur with a defined periodicity within the amino acid sequence of 30 residues which constitutes the EF-hand. This periodicity, chiefly within the helices, reflects the requirement that the apolar side chains point inwards towards the hydrophobic 'core' of the molecule. The calcium ions are octahedrally co-ordinated to oxygen atoms of amino acids located in the binding sites. These residues are either aspartic or glutamic acids or their amides, or serine hydroxyl groups, though in the case of the ($-Y$) co-ordinating position, ligation is achieved through the carbonyl oxygen in the polypeptide backbone. Thus from the parvalbumin structure we know that stability of the calcium binding domains depends on a specific distribution of both apolar and oxygen containing amino acids within a sequence of about 30 residues. This specific arrangement of amino acids has been used as a criterion to identify possible calcium binding sites in proteins with homologous amino-acid sequences (see Kretsinger, 1977 and references therein). Comparisons may be made at two levels. First we ask whether the pattern of apolar residues is preserved such that stable helices might exist as in the EF-hand structure. Then we examine the residues in the loops between

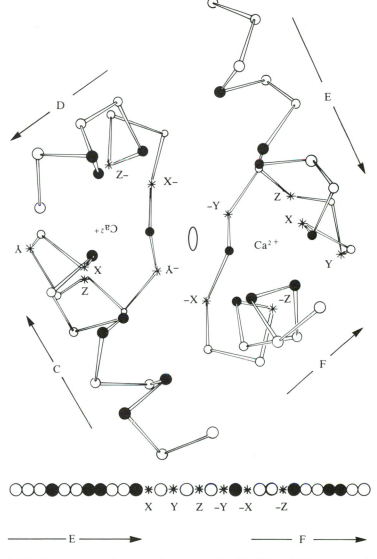

Fig. 8.10. Positions of the α-carbon atoms in the CD and EF regions of carp muscle calcium-binding parvalbumin. Helix C, the CD calcium-binding site and helix D are related by a two-fold axis to the EF-region (symmetry axis denoted by oval area at centre of figure). Helix F is approximately perpendicular to helix E and the calcium site lies between these helices. Solid circles denote α-carbon atoms of apolar residues while the asterisks denote α-carbon atoms of residues ligated to calcium ions. X, Y, Z etc. signify the octahedral co-ordinating positions. The linear sequence of apolar and calcium-ligating residues in the EF-hand is shown below. (Courtesy of Dr R. H. Kretsinger, see Kretsinger (1977).)

the putative helices to see whether they conform to the structural constraints for calcium co-ordination. This information can then be related to experimental studies on calcium binding.

Collins (1976) showed that Tn-C from rabbit muscle contains four EF-hands, which is consistent with known calcium-binding properties. Potter & Gergely (1975) showed that the affinity of the four sites was not identical, since one pair binds calcium ions more tightly than the other. The dissociation constants are 5×10^{-8} M and 5×10^{-6} M respectively in the absence of magnesium ions though, in the presence of these ions, the value for the high-affinity sites is reduced to 5×10^{-7} M. Since these high-affinity sites also appear to bind magnesium ions, they may be classified as non-specific divalent cation sites, while the other pairs of sites are calcium-specific. The role of the non-specific sites is probably structural, maintaining the stability of the molecule.

The presence of four structurally-related domains suggests that this protein, and others like it, may have evolved from a smaller ancestral precursor as a result of gene duplication. Detailed sequence comparisons of the four domains have indicated repeats between pairs of units. Thus the first and third sites are more alike than the first and second, supporting a model in which the structure arose by duplication of a two-site precursor. Can we identify which of the four calcium-binding domains correspond to the calcium specific sites? Sequence analysis has suggested that the calcium-specific sites may be sites I and II, while the non-specific sites correspond to sites III and IV. These predictions were confirmed by recent studies of calcium binding to proteolytic fragments of Tn-C.

Tn-C from cardiac muscle has been shown to bind three calcium ions per mole, with only one calcium-specific site. Analysis of the amino-acid sequence shows deleterious substitutions of amino acids in calcium ligating positions of site I, which account for loss of binding at this site. This difference in calcium binding between skeletal and cardiac muscle Tn-C may be relevant to the physiological differences between these muscles.

Turning to calmodulin, Vanaman, Sharief & Watterson (1977) showed that its amino-acid sequence was very similar to that of Tn-C. Bovine brain calmodulin contains 148 residues compared with 159 in rabbit muscle Tn-C, but there are 79 identities and 35 conservative substitutions, giving homology of nearly 80%. Internal homologies within calmodulin again suggest evolution by gene duplication from a smaller two-site precursor. It binds four moles of calcium with affinities similar to the calcium specific sites of Tn-C. Since these affinities are not affected by the presence of magnesium ions, the sites do not appear to be paired as in Tn-C. The amino-acid sequence of calmodulin from bovine uterus muscle has been determined by Grand and Perry and is virtually identical to the brain protein.

Sequence comparisons have also been made with various myosin light chains. Based on the pattern of apolar residues, scallop regulatory light chain contains four EF-hand domains, but calcium ligating residues are conserved only in site I (Kendrick-Jones & Jakes, 1976). The other potential calcium-binding sites have non-conservative substitutions which might be expected to interfere with cation binding. The one site identified on this basis probably corresponds to the non-specific divalent ion site, which binds magnesium ions under physiological conditions, and is implicated in holding together light and heavy chains. Similarly, there are extensive homologies between the DTNB light chain of rabbit muscle myosin and Tn-C, again showing four EF-hand domains, though only one calcium site (site I) is conserved when potential calcium ligating residues are compared (Collins, 1976). Studies of divalent cation binding to isolated DTNB light chains and to rabbit myosin have indicated the presence of a single non-specific divalent cation site on this light chain. Although the complete amino-acid sequence of the regulatory light chain from chicken-gizzard myosin is not yet known, studies on the amino-terminal end of the molecule have revealed a potential EF-hand domain with the requisite oxygen-ligating residue needed for calcium binding. Like the other myosin light chains, this appears to be a non-specific divalent cation site. These studies suggest a structural relationship between these different light chains, all of which bind divalent cations. This sequence homology also extends to the Alkali light chains of rabbit muscle myosin, which again have been shown to contain four potential EF-hand domains. However, in this case, amino-acid substitutions occur at positions corresponding to those involved in calcium ligation in all four potential sites, so we would predict that this light chain does not bind calcium ions at all. This prediction has been experimentally verified.

From these comparisons we can identify a family of evolutionarily related proteins, which have probably arisen from a smaller ancestral calcium-binding protein. The most primitive calcium-binding ancestor may have contained only a single EF-hand domain, whose stability was improved by dimerisation. This led to the two-site protein, which later duplicated again. Parvalbumin appears to have evolved from this final form by loss of about 40 residues from its amino terminus. This proposed sequence of evolutionary events is supported by the discovery of a calcium-binding protein of 9700 mol. wt from porcine intestine which contains two EF-hand domains and two calcium-binding sites. Barker and Dayhoff have argued, from studies of the recent rate of change in parvalbumin sequences, that the divergence of the myosin light chains, Tn-C and parvalbumin occurred long enough ago to account for the presence of these proteins in all eukaryotic cells.

If evolution occurred in the way suggested, calmodulin appears closest to the ancestral precursor. It has four calcium-specific sites of roughly equal affinity and it interacts with a number of different enzymes in the cell, although with the exception of phosphorylase kinase it does not seem to bind permanently to any of them. Rather it is a peripatetic messenger, found in a wide variety of cell types, whose function is to confer calcium control on the enzyme systems with which it interacts. In contrast, Tn-C has evolved to bind tightly to the other troponin components and perhaps as a consequence has lost two of its calcium-specific binding sites. These are now occupied by magnesium ions, probably involved in stabilising the structure to preserve its all-important subunit interactions. Further loss of divalent cation binding has been shown for the myosin regulatory light chains. Perhaps their interactions with the heavy chains have almost completely replaced the need for divalent cations in stabilising the three-dimensional structure. The Alkali light chains have lost all ability to bind divalent ions and are strongly associated with the myosin heavy chains, requiring denaturing conditions to liberate them.

The structural relationships between these proteins, based on sequence homologies, should not be taken to imply identity of overall shape of the molecules. Current evidence suggests, for example, that myosin light chains are more asymmetric than Tn-C. Thus while evolution of these proteins *may* have preserved their structural domains, the arrangement of these domains may have changed considerably during functional adaptation.

FUNCTIONAL RELATIONSHIPS BETWEEN MYOSIN LIGHT CHAINS

It is now clear that two distinct classes of myosin light chains can be identified, a regulatory class and an 'essential' class (typified by the Alkali light chains, which cannot be dissociated without loss of enzymic activity). Within the regulatory class, calcium regulation of actomyosin interaction in scallop muscle involves direct calcium binding to these light chains, while in smooth muscle, the light chains are phosphorylated by a calcium-dependent kinase. Rabbit DTNB and related 19 000 mol.-wt light chains can also be phosphorylated yet their function remains unclear. Kendrick-Jones has investigated the functional relationship between the light chains in this regulatory class by substituting them for the scallop regulatory light chains in recombination experiments with desensitised scallop myofibrils (Kendrick-Jones *et al.*, 1976). If scallop regulatory light chains are recombined with desensitised myofibrils or myosin, then calcium sensitivity and calcium binding are both restored. If instead, chicken-gizzard regulatory light chains are used, they too will restore both calcium sensitivity and calcium binding. The DTNB light chains of skeletal myosin, but not the Alkali light chains, will also recombine with desensitised scallop myofibrils. However, in this case, restoration of calcium sensitivity occurs without any restoration of calcium binding. But this restoration of calcium sensitivity can be achieved only if myofibrils or actomyosin are used in the experiment. The vertebrate light chains do not restore calcium sensitivity when desensitised scallop myosin is used on its own. Comparative studies with a wider range of this class of light chains have shown that two subclasses exist. Regulatory light chains from muscles which have thick filament regulation will restore both calcium binding and calcium sensitivity in the desensitised scallop system, showing that they substitute fully for one another in this assay. Only partial substitution occurs in the case of myosin light chains from vertebrate striated muscle and other muscles where thick filament regulation cannot be demonstrated. Further it should be remembered that these experiments are carried out under conditions where only one scallop light chain has been removed initially, so we cannot rule out the possibility that the alien light chain acts only passively in these cases, i.e. by occupying the light-chain binding site, it facilitates calcium sensitivity through the remaining scallop regulatory light chain. Szent-Györgyi has recently discovered that it is possible to remove both scallop regulatory light chains reversibly, so it is now feasible to examine hybrids containing two alien light chains.

In spite of the lack of positive evidence, it is still tempting to speculate that the DTNB and related light chains play some regulatory role in actomyosin interaction. Dual regulation, which occurs in a wide variety of insects and other invertebrates, probably improves the precision of the calcium switch and sharpens the transition from relaxed to active state. In considering dual control mechanisms, it is important not to limit our ideas solely to steric obstruction or modification of interaction sites. In the highly organised structural array of filaments that is striated muscle, cross-bridge interaction may be further modulated by controlling the orientation or flexibility of the individual bridges at the hinge region. Thus we should look for additional effects of calcium which may not be all-or none, and for this it is essential not to restrict research to studies on protein preparations *in vitro*. For example, skinned fibres have particular advantages for analysing the fully organised system. Already there is suggestive evidence that the regulatory light chains may interact both in the head and rod regions of myosin. In this way they may control cross-bridge flexibility via the hinge region to facilitate or reduce the rate of interaction. Thus, in vertebrate striated muscle, calcium may modulate cross-bridge movement either by direct binding to myosin light chains or through calcium-stimulated light-chain phosphorylation.

CONTROL OF ACTOMYOSIN INTERACTION IN NON-MUSCLE CELLS

Actin and myosin are present in all eukaryotic cells and appear to be involved in a variety of motile phenomena within them, as outlined at the beginning of this chapter. However, the motile processes occurring within cells are in many ways different from those in muscle. In striated muscle evolution has produced a machine to generate considerable contractile force of a unidirectional nature which operates over relatively small shortening distances. This is achieved by the highly organised lattice of thick and thin filaments and the presence of many individual force generators (cross-bridges). Motile processes in non-muscle cells, e.g. of the type seen in the movement of fibroblasts over surfaces or in cell spreading, are multidirectional, requiring protrusive as well as contractile activity, and operate over relatively much greater distances. These cells do not contain thick filaments of the type seen in muscle, though bundles of microfilaments (thin filaments) have been identified in many cell types both by electron microscopy and by using the light microscope with fluorescent antibodies to actin as a probe. These microfilament bundles may serve to maintain the basic architecture of the cells in addition to their possible roles in motility. It is perhaps significant that the ratio of actin to myosin is much higher in non-muscle cells than in muscle. Thus while in vertebrate striated muscle this molar ratio is about 5, in bovine aorta (smooth muscle) it is about 30, and in platelets or fibroblasts about 100. In addition, the properties of the proteins seem somewhat different. While the presence of myosin filaments is easily demonstrated in striated muscle, they have not been unequivocally identified in non-muscle cells, though Pollard has shown that it is possible to prepare short bipolar filaments from platelet myosin *in vitro* (see Pollard, Fujiwara, Niederman & Maupin-Szamier, 1976). It is also clear from the study of a number of different cell types that much of the actin is present in a non-polymerised form and a number of new proteins have been identified which complex G-actin or depolymerise F-actin. Consideration of actomyosin inter-action and the function of these proteins in cytoplasmic motility must not neglect this new dimension of control involving polymerisation and filament formation *in vivo*. Thus rapid polymerisation of actin, such as has been demonstrated by Tilney in the acrosomal process of *Thyone* sperm during egg fertilisation, may play a role in the protrusive activity in various cells, including membrane ruffling, phagocytosis and the morphological changes that occur in platelets during blood clotting. Detailed discussion of this type of control is beyond the scope of our review: here we will consider control mechanisms involved in the interaction of cytoplasmic actomyosins in comparison to those in muscle (for further information see reviews by Hitchcock (1977) and Korn (1978)).

Superficially, vertebrate cytoplasmic myosins have the same subunit structure as muscle myosins, being composed of two heavy chains of about 200 000 mol. wt and two pairs of light chains of about 20 000 and 16 000 mol. wt respectively. However, although the subunit molecular weights of these myosins may be identical, this does not mean that all cytoplasmic myosins are the same. Just as there are different isoenzymes of myosin in different muscle types, so too different myosins are present for example in brain and platelets, as has been shown by techniques of protein chemistry. The presence of distinctive cytoplasmic myosins in different cell types, probably with a variety of ATPase activities and motile functions, may imply also a number of different regulatory mechanisms.

One type of regulation that is well established in a variety of cell types involves phosphorylation of the 20 000 mol. wt light chain. This has been demonstrated convincingly in platelets, prefusion myoblasts and baby hampster kidney cells with further reports for macrophages, thymocytes and brain. As in smooth-muscle regulation, only the phosphorylated myosin is activated by actin. A calcium-sensitive light-chain kinase involving calmodulin has been identified in a number of cells. Although we can understand how calcium released into the cytoplasm promotes actomyosin interaction in this way, we have no notion how particular calcium-modulated functions within the cell are discriminated and mechanisms controlling calcium release and uptake in non-muscle cells are not understood.

It has been suggested that actin-linked regulation occurs in certain non-muscle cells and there are reports of the presence of regulatory proteins in platelets and brain. However, these proteins have yet to be fully purified and characterised, and to date there is no well-documented example of thin-filament regulation in vertebrate cells. Calmodulin is ubiquitous in vertebrate cells and has been shown to substitute for Tn-C in regulating skeletal muscle actomyosin. Thus it is conceivable that calmodulin and an inhibitor protein analogous to Tn-I might be involved in thin-filament regulation, but as yet there is little evidence to support this. Tropomyosin is present in many cell types though in a form which is somewhat smaller than its muscle counterpart, but both platelet and brain tropomyosin have been shown to substitute for muscle tropomyosin in combination with muscle troponin in conferring calcium sensitivity on muscle actomyosin ATPase. Furthermore, cytoplasmic actins have been used together with muscle myosin, tropomyosin and troponin to demon-strate calcium-sensitive ATPase activity. These experiments with mix-tures of muscle and cytoplasmic proteins suggest that the cytoplasmic components can substitute readily in the muscle contractile system, but they do not prove that such regulation occurs in the cytoplasmic system.

Actin-linked regulation has also been demonstrated in extracts of *Physarum polycephalum*, but the regulatory proteins have not been studied in detail. Thus at present the evidence in support of actin-linked regulation involving troponin-like components is tenuous. Nor is it clear what factors control interaction of tropomyosin and actin in non-muscle cells. In both brain and platelets there is less than 50% of the amount of tropomyosin needed to complex all the actin present and immunofluorescence studies on tissue-culture cells have shown selective localisation of tropomyosin on actin filaments. We do not have sufficient space to discuss the many actin-binding proteins identified in various cell types, but clearly their properties must be studied to elucidate their roles with cytoplasmic actin in cellular motility.

Myosin has also been isolated from several invertebrate non-muscle sources including star-fish eggs and sperm, and squid brain. Actin and myosin are also present in slime moulds and amoebae where their interaction appears to be controlled by intracellular calcium. The most interesting studies have been carried out by Korn on *Acanthamoeba castellani*, which contains at least two and possibly four different myosins (see Korn, 1978). One of these is similar in its subunit structure to muscle myosin, while the others are very unusual in having only one heavy chain and two light chains. This unusual form of myosin does not form bipolar filaments, nor does it appear to require its light chains for ATPase activity. The protein can be phosphorylated but on the heavy chain rather than the light chain. If this protein can really be classified as a myosin (and this seems likely because its ATPase is activated by actin), its unusual properties may reflect novel interactions within the cell and a different regulatory mechanism.

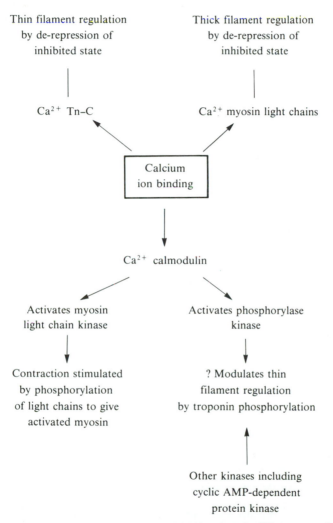

Fig. 8.11. Summary of calcium regulatory mechanisms in different muscle types. Calcium ions regulate actomyosin interaction by binding to specific proteins which are evolutionarily related. Further modulation of contractile activity may be controlled by specific phosphorylation reactions.

While current understanding of the regulation of muscle contraction by calcium ions serves as a paradigm for investigations on cytoplasmic motility, we need to know much more about the factors controlling assembly of both actin and myosin-containing filaments in non-muscle cells. Protein components may be involved in regulating the reversible assembly and disassembly of microfilaments and these must be identified and characterised. We also need to know much more about other actin binding proteins like filamin and α-actinin. If calcium ions are involved in regulating these various protein–protein interactions, the mechanisms must be elucidated. While direct calcium binding may be important in this regard, mechanisms involving protein phosphorylation seem at present more likely.

CONCLUSIONS

Speculation on the evolution of regulatory mechanisms in contractile systems is based on two unifying observations. The first is the universal role of calcium ions as the intracellular messenger, while the second concerns the discovery of a family of structurally related proteins whose affinity for calcium ions is between the intracellular concentration of free calcium in the resting state and the peak transient level in the active state. Of the three distinct regulatory mechanisms discussed here (see Fig. 8.11), the most ubiquitous form seems to be that related to phosphorylation. While it is tempting to speculate that this system evolved first, there is insufficient information about control in the most primitive unicellular organisms. Our challenge for the future is to understand the diversity of actin and myosin interactions in their widest aspects in cytoplasmic motility and to elucidate the control mechanisms involved.

REFERENCES

Adelstein, R. S. (1978). Myosin phosphorylation, cell motility and smooth muscle contraction. *Trends in biochemical sciences*, **3**, 27–30.
Bárány, K., Bárány, M., Gillis, J. M. & Kushmerick, M. J. (1979). Phosphorylation-dephosphorylation of the 18 000-dalton light chain of myosin during the contraction-relaxation cycle of frog muscle. *Journal of biological chemistry*, **254**, 3617–23.
Collins, J. H. (1976). Structure and evolution of troponin C and related proteins. In *Calcium in biological systems, 30th symposium of the society for experimental biology*, ed. C. J. Duncan, pp. 303–34. Cambridge University Press.
Ebashi, S., Mikawa, T., Hirata, K. & Nonomura, Y. (1978). The regulatory role of calcium in muscle. *Annals of New York Academy of Science*, **307**, 451–61.

Hitchcock, S. E. (1977). Regulation of motility in non-muscle cells. *Journal of cell biology*, **74**, 1–15.

Huxley, H. E. (1969). The mechanism of muscle contraction. *Science*, **164**, 1356–66.

Huxley, H. E. (1972). Structural changes in actin- and myosin-containing filaments during contraction. *Cold Spring Harbor symposium on quantitative biology*, **37**, 361–76.

Kendrick-Jones, J. & Jakes, R. (1976). The regulatory function of myosin light chains. *Trends in biochemical sciences*, **1**, 281–4.

Kendrick-Jones, J., Szentkiralyi, E. M. & Szent-Györgyi, A. G. (1976). Regulatory light chains in myosins. *Journal of molecular biology*, **104**, 747–75.

Korn, E. D. (1978). Biochemistry of actomyosin-dependent cell motility. *Proceedings of the national academy of science of the USA*, **75**, 588–99.

Kretsinger, R. H. (1977). Evolution and informational role of calcium in eukaryotes. In *Calcium-binding proteins and calcium function*, ed. R. H. Wasserman *et al.*, pp. 63–72. New York: Elsevier, North-Holland.

Lehman, W. & Szent-Györgyi, A. G. (1975). Regulation of muscular contraction: distribution of actin control and myosin control in the animal kingdom. *Journal of general physiology*, **66**, 1–30.

Lymn, R. W. & Taylor, E. W. (1971). Mechanism of ATP hydrolysis by actomyosin. *Biochemistry*, **10**, 4617–24.

Manning, D. R. & Stull, J. T. (1979). Myosin light chain phosphorylation and phosphorylase *a* activity in rat extensor digitorum longus muscle. *Biochemical and biophysical research communications*, **90**, 164–70.

Offer, G. W. (1974). The molecular basis of muscle contraction. In *Companion to biochemistry*, ed. A. T. Bull *et al.*, pp. 623–71. London: Longman.

Perry, S. V. (1979). The regulation of contractile activity in muscle. *Biochemical society transactions*, **7**, 593–617.

Pollard, T. D., Fujiwara, K., Niederman, R. & Maupin-Szamier, P. (1976). Evidence for the role of cytoplasmic actin and myosin in cellular structure and motility. In *Cell motility*, Cold Spring Harbor conferences on cell proliferation, ed. R. D. Goldman, T. D. Pollard & J. L. Rosenbaum, pp. 689–724. New York: Cold Spring Harbor.

Potter, J. D. & Gergely, J. (1975). The calcium and magnesium binding sites on troponin and their roles in the regulation of myofibrillar adenosine triphosphate. *Journal of biological chemistry*, **250**, 4628–33.

Sherry, J. M. F., Gorecka, A., Aksoy, M. O., Dabrowska, R. & Hartshorne, D. J. (1978). Roles of calcium and phosphorylation in the regulation of the activity of gizzard myosin. *Biochemistry*, **17**, 4411–18.

Stafford, W. F. III, & Szent-Györgyi, A. G. (1978). Physical characterization of myosin light chains. *Biochemistry*, **17**, 607–14.

Stewart, M. & McLachlan, A. D. (1975). Fourteen actin-binding sites on tropomyosin? *Nature*, **257**, 331–2.

Trentham, D. R. (1977). The ATPase reactions of myosin and actomyosin and their relation to energy transduction in muscle. *Biochemical society transactions*, **5**, 5–22.

Vanaman, T. C., Sharief, F. & Watterson, D. M. (1977). Structural homology between brain modulator protein and muscle Tn-C. In *Calcium-binding proteins and calcium function*, ed. R. H. Wasserman *et al.*, pp. 107–16. New York: Elsevier, North-Holland.

Wagner, P. D. & Weeds, A. G. (1977). Studies on the role of myosin alkali light chains. *Journal of molecular biology*, **109**, 455–73.

Wakabayashi, T., Huxley, H. E., Amos, L. & Klug, A. (1975). Three-dimensional image reconstruction of actin-tropomyosin complex and actin–tropomyosin–Tn-T–Tn-I complex. *Journal of molecular biology*, **93**, 477–97.

Weber, A. & Murray, J. M. (1973). Molecular control mechanisms in muscle contraction. *Physiological reviews*, **53**, 612–73.

9

The vertebrate visual pigments

FREDERICK CRESCITELLI

DEPARTMENT OF BIOLOGY, UNIVERSITY OF CALIFORNIA,
LOS ANGELES, CALIFORNIA, USA

The essence of survival for a species lies in its ability to adapt to a changing environment, be that physical, chemical or biological. Such adaptability is one of the central themes of biology, and no system, perhaps, illustrates this better than the visual system. It is a truism to state that vision, in its origin and evolution, has been inextricably linked to light, but it is only when we examine the details of this association that we are struck by the manifold adaptations that have arisen. These adaptations involve the morphological properties of the eye itself; the characteristics of the visual cells within the retina; and the structural and functional relationships of the central pathways. In its entirety the visual system has been shown to be modifiable, within genetic limitations, in accord with the photic environment of the species. Much could be written on this matter but we are here concerned primarily with the visual photopigments of the vertebrate visual cells.

THE VERTEBRATE VISUAL CELL

In its generalised form (Fig. 9.1) the vertebrate photoreceptor is seen to be an elongated cell specialised for the purpose of admitting and absorbing incident light quanta and of responding by generating an appropriate signal that is transmitted to the neuronal layers of the retina where, after due processing, it is passed on to the visual centres of the brain. The biochemical machinery for capturing the quanta, which operates to set off the signal, includes a membrane-bound chromoprotein located in the outer segments. The outer segment is a specialised terminal organelle whose ultrastructure consists of hundreds of double-membraned discs or sacs, some being completely enclosed within the outer segment and others being simple invaginations of the cell membrane, with which they are still connected. In rods closed discs predominate while in cones open sacs are more plentiful.

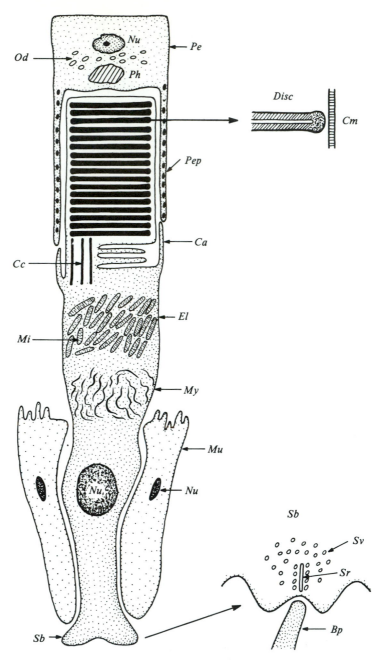

Fig. 9.1. Diagram to illustrate the structure of a typical rod. Cell of pigment epithelium (*Pe*) showing its nucleus (*Nu*), oil droplets (*Od*), and ingested phagosome (*Ph*). Processes of pigment epithelium (*Pep*) extending down along

The outer segment is joined to the inner segment by means of the connecting cilium which serves as a pathway through which substances like the visual pigments, synthesised in the inner segment, are transferred to the outer segment. The inner segment contains the metabolic machinery of the photoreceptor with the mitochondria in the ellipsoid region, the endoplasmic reticulum and Golgi apparatus in the myoid. Proteins, including the visual pigments, are assembled in the ribosomes, passed on to the Golgi apparatus where further additions are made, and then distributed to the specific compartments of the visual cell where they serve their specific functions. The visual pigment moves outwards through the cilium to the base of the outer segment where it is incorporated into the disc membranes that are continuously being formed by infoldings of the plasma membrane.

In the ontogeny of the visual cell the outer segment discs appear before the synaptic apparatus has matured. As a result, light can activate the cell before transmission of information downstream. This has been demonstrated by the recording of a singular negative potential appearing after just a few outer segment discs have been formed and before the synaptic apparatus with its synaptic ribbons and synaptic vesicles has appeared (Fig. 9.1). This sequence of development implies that light is early on afforded a basis for influencing the pattern of retinal development.

Once the outer segments are completed and functioning they do not remain as stable entities. Throughout life the apical tips slough off and are phagocytosed by the cells of the pigment epithelium that surround the outer segments. The engulfed bits of disc material (phagosomes) are then destroyed by the pigment cells. This continuous wear and tear is compensated for by the manufacture of new discs at the basal end of the

---

rod and in intimate contact with it are also shown. Outer segment discs with specialised disc edges and not continuous with outer segment cell membrane (*Cm*). Connecting cilium (*Cc*) joins outer and inner segments, with extensions (calyces) of inner segment (*Ca*) part way up along outer segment. Varying numbers of calyces occur in different species with 27 in *Necturus*; they also occur in cones. The ellipsoid region (*El*) inner segment includes mitochondria (*Mi*) and the myoid region (*My*) contains the endoplasmic reticulum and Golgi bodies. The most inner portion of the rod houses the nucleus (*Nu*) and the synaptic body (*Sb*), in functional contact with bipolar cells and horizontal cells. The region of the synaptic body with one synaptic ribbon (*Sr*) and its synaptic vesicles (*Sv*) is enlarged at right. *Bp* represents a process of a bipolar cell in contact with the membrane of the synaptic body.

outer segment. This process of outer segment renewal is influenced by the light environment and is apparently in phase with the diurnal light cycle. Rods shed their tips early in the light period following darkness whereas cones shed early in the dark period following light.

Rods and cones: the duplicity theory

No better evidence of the adaptability of the vertebrate retina to the photic environment is available than the proposal enunciated by Max Schultze (1866) and commonly named the duplicity (duplexity) theory of vision. According to this theory the vertebrate retina is endowed with two classes of photoreceptors: rods and cones. The rods are part of the scotopic (dim light) and high sensitivity system; the cones, in turn, serve as the receptors for the bright light (photopic) system involved in high visual acuity and the mechanism to discriminate colours. Schultze, of course, employed morphological criteria, along with the nocturnal or diurnal habits of the animals whose retinas he examined, to arrive at this definition of rods and cones. Late in the nineteenth and early in the twentieth centuries, physiological studies, especially of human vision, strengthened the concept of retinal duality which is now a fact of visual science.

The transmutation theory

Schultze was well aware that it was not always a straightforward matter to distinguish rods and cones. Walls (1934) observed this to be a difficulty and proposed the concept of phylogenetic transmutation whereby a common ancestry is ascribed to rods and cones. He supported his idea by the assertion that certain living vertebrates display visual cells having an intermediate morphology, i.e., having both rod-like and cone-like features. Walls proposed that transmutation of cones (photocytes) to rods (scotocytes) has occurred independently in separate vertebrate lines in association with an enforced change in life habit from a photopic to a scotopic one. For Walls the gecko retina was one that best demonstrated the intermediacy of visual cells. In no way does the transmutation theory contradict or negate the duplexity theory. The latter deals with the dual functions of the fully evolved system of a species adapted to the environment; the former relates to the historical events of a species in evolution.

RHODOPSIN

Basic concepts

Rhodopsin is the name given by Willy Kühne to the rose-coloured pigment that he described as Sehpurpur in the rods of many vertebrates. Before Kühne, Franz Boll (1876) had seen this colour, which he named Sehrot, and which he recognised as having significance for vision. The chemical study of rhodopsin was initiated by Kühne (1877) who succeeded in extracting it, using bile salts, and in investigating some of its properties. In 1931 Tansley introduced the plant glycoside, digitonin, a non-ionic detergent, to solubilize rhodopsin. Over the years this has proved to be a mild and effective extractant and most of the fundamental properties of rhodopsin have been worked out with the use of this solubiliser. In recent years a number of other non-ionic and ionic detergents have been introduced to extract visual pigments. A problem in interpreting results with these varied extractants is that the composition and properties of rhodopsin often vary according to the solubiliser.

Rhodopsin is also studied by use of preparations of outer segments in suspension, as sonicated particles, or prepared as vesicles. In this way the behaviour of rhodopsin *in situ* can be compared with the extracted pigment. Often the properties are the same in both, sometimes there are differences and in this chapter the similarities and differences will be pointed out.

The spectral absorbance

The colour of rhodopsin is of functional significance for only the light that is absorbed can do work (Draper's Law). The rhodopsin light reaction serves as a trigger involving, at the absolute threshold of vision, one photon absorbed by one rhodopsin molecule in one rod to excite the rod into action. For vision to occur, however, something like five to eight rods must be excited within a critical period of time and within a critically-sized area of the retina. When Boll discovered rhodopsin it was its colour and the loss of this colour after exposure to light (bleaching) that led him to recognise the significance of Sehrot for vision. With the development of spectrophotometry it was possible to investigate in detail the colour and the changes in colour with light bleaching and other treatments. This has been accomplished with rhodopsin in solution, in rod particles, and within the living retina itself. The ease with which cow eyes are obtained, their relatively high yield of rhodopsin and certain other factors have made bovine rhodopsin the most commonly studied visual pigment and most of the chemical investigations have been carried out with this particular pigment.

The spectral absorbance of bovine rhodopsin in solution reveals the presence of three broad and rather featureless absorption bands (Fig. 9.2) which are typical of rhodopsins in general, using the name rhodopsin to designate the vitamin A_1-based visual pigment of rods characterized by an absorption in the general region of 500 nm. These three bands are the α-band at about 500 nm, the β-band at 340–350 nm and the prominent γ-band at 280 nm. The α- and β-bands are associated with the pigment chromophore, the atomic groups generating the colour. Both bands

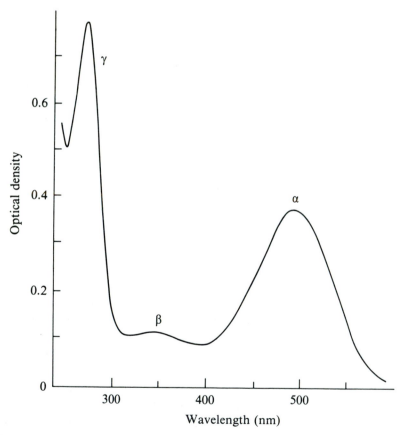

Fig. 9.2. Spectral absorbance of bovine rhodopsin. Adopted from Collins, Love & Morton (1952).

disappear on bleaching. Rhodopsin also possesses optical activity in the visible spectrum in the form of circular dichroism, i.e., the preferential absorption of left over right circularly polarised light. Coupled to the α- and β-absorption bands are α- and β-bands of optical activity but shifted slightly toward shorter wavelengths relative to the absorption bands (Fig. 9.3). They too disappear on bleaching. The γ-band is a protein absorption associated with the aromatic amino acids. Of these three absorption bands the α-band has received the bulk of the attention for many pigment extracts contain impurities absorbing in the blue region of the spectrum and therefore interfering with the recording of the β-absorption.

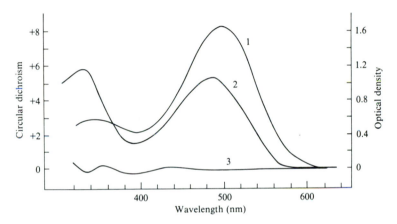

Fig. 9.3. Spectral absorbance (curve 1) and circular dichroism before bleaching (curve 2) and after bleaching (curve 3) of frog rhodopsin. Circular dichroism is given as $(\varepsilon_L - \varepsilon_R)/\varepsilon_{max} \times 10^4$. After Crescitelli, Mommaerts & Shaw (1966).

The examination of rhodopsins from all classes of vertebrates has shown that although these pigments vary slightly regarding the spectral maximum of the α-band (493–505 nm) the shape of this band is the same when absorbance (extinction) is plotted as a function of wave number (frequency) rather than wavelength. This point is illustrated in Fig. 9.4 which gives a frequency plot of rhodopsins from a teleost (500 nm), an amphibian (502 nm), a reptile (502 nm), a bird (502 nm) and a mammal (496 nm). This similarity in form of the α-band when plotted this way permits one to deduce the α-curve of a rhodopsin knowing only the wavelength of maximum absorption (λ_{max}). Dartnall (1953) devised a nomogram by means of which this curve can be constructed directly. While it is true that rhodopsins from all vertebrate classes have this

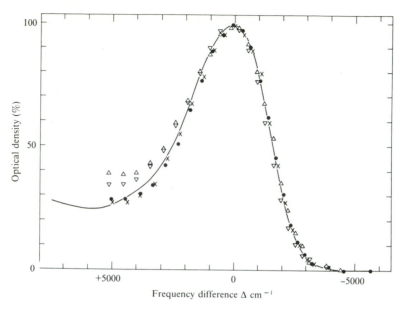

Fig. 9.4. Spectral absorbance of rhodopsins of arctic cod (triangles), frog (filled circles), alligator (crosses), great-horned owl (inverted triangles) and 3-toed sloth (solid line). Normalised absorbances are plotted as functions of wave number (cm^{-1}). The wave numbers are shown as positive and negative deviations from the wave number for maximum absorption which is arbitrarily set at zero for each rhodopsin.

standard form, this does not apply to all visual pigments. The pigment of the green rods of the frog has an α-band that is broader than the standard curve, and iodopsin, absorbing at longer wavelengths than rhodopsin, has a narrower absorption.

The bleaching sequence

When rhodopsin in solution is exposed to light a series of coloured intermediates result (Table 9.1), many of which are so transient that they can only be detected at low temperatures. The sequence that we shall be primarily concerned with in this chapter is the one involving batho-rhodopsin, lumirhodopsin, metarhodopsin I, metarhodopsin II, meta-rhodopsin III, N-retinylidene opsin, retinal and opsin. It will be noted in Table 9.1 that each of these has a characteristic temperature at which it is studied and each has a characteristic absorbance maximum at the temperature indicated. None of these have a pH effect except the

Table 9.1. *Bleaching sequence for bovine rhodopsin*

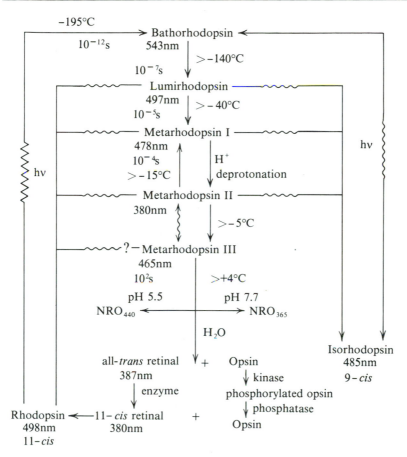

Hypsorhodopsin at 430nm has been described as occurring before bathorhodopsin at the temperature of liquid helium. It is not known whether this intermediate occurs at physiological temperatures and conditions. It is a question whether or not hypsorhodopsin does precede bathorhodopsin.

N-retinylidene opsin (NRO), which absorbs at 440 nm in acid and 365 nm in alkaline solutions, so that the two forms have been called acid and alkaline indicator yellow. It will be noticed especially that neither rhodopsin nor retinal have a pH effect. The bleaching sequence of Table 9.1 has been studied with rhodopsin in solution and with it *in situ* within rod particles. Some of the intermediates have even been detected in the retina, both isolated and intact in the living eye. It can therefore be stated that this bleaching sequence is of interest from a physiological as well as from a biochemical viewpoint.

Structure and properties

The Schiff's base. Wald (1935) initiated the modern phase in the chemistry of rhodopsin with his recognition of the prosthetic group in this protein as a vitamin-A-related substance, a finding that gave meaning to the several reports that related the nutritional deficiency of vitamin A with impairment of the visual threshold. Wald named the prosthetic group retinene and showed that it was related to the carotenoids, which are fat-soluble, unsaturated compounds occurring naturally in plants and animals. The identification of retinene as the aldehyde of vitamin A was accomplished in the laboratory of Morton (Morton & Goodwin, 1944) in Liverpool. The vitamin (now recognized as vitamin A_1) is named *retinol* and its aldehyde is referred to as *retinal* or *retinaldehyde*.

Retinol (vitamin A_1)
$\lambda_{max} = 325nm$

Retinal
$\lambda_{max} = 380nm$

In rhodopsin the retinal is covalently bound as a Schiff's base (aldimine) with the carbonyl group reacting with an ε-NH_2 of a lysine residue. Only one retinal is present in a molecule of rhodopsin. The original evidence for an aldimine structure came from Morton's laboratory where it was shown that amines are able to combine with retinal to form coloured compounds that behave as pH indicators with absorption maxima at 365 nm and 440 nm in alkaline and acid solutions, respectively. This acid bathochromic ('red') shift was explained on the basis of a protonation of the nitrogen of the Schiff's base.

$$C_{19}H_{27}\text{-}\underset{H}{C}\text{=}O + H_2N\ R \longrightarrow C_{19}H_{27}\text{-}\underset{H}{C}\text{=}N\text{-}R + H_2O$$

$$C_{19}H_{27}\text{-}\underset{H}{C}\text{=}N\text{-}R + H^+ \longrightarrow C_{19}H_{27}\text{-}\underset{H\ H}{C}\text{=}\overset{+}{N}\text{-}R$$

$\lambda_{max} = 365nm$ $\qquad\qquad\qquad$ $\lambda_{max} = 440nm$

The relevance of this model chemistry to rhodopsin is obvious when one sees (Fig. 9.5) that the products of rhodopsin bleaching behave as pH indicators. The fact that the aldimine bond is actually present originally and not formed secondarily in the bleaching sequence was proved by blocking free amino groups with formaldehyde first and showing that the usual products of bleaching still appear. The fact that rhodopsin itself is not a pH indicator and is not attacked by reagents that react with Schiff's bases is considered to be due to the fact that the aldimine bond is probably embedded within a hydrophobic pocket of the protein (opsin).

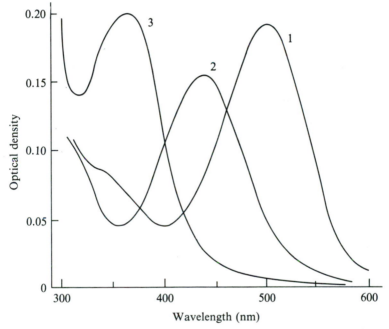

Fig. 9.5. Spectral absorbance (curve 1) of frog rhodopsin and the products of bleaching in acid (curve 2) and alkaline (curve 3) solutions. From the work of Collins & Morton (1950).

The conclusion that it is lysine to which retinal is linked was derived from experiments in which rhodopsin, after bleaching, was treated with sodium borohydride, resulting in a reduction of the N-retinylidene opsin to the stable N-retinyl opsin. Alkaline hydrolysis to break down the

$$C_{19}H_{27}-\underset{H}{C}=N-\text{opsin} \xrightarrow[\text{(NaBH}_4)]{2H} C_{19}H_{27}-\underset{H_2}{C}-\underset{H}{N}-\text{opsin}$$

N–retinylidene-opsin $\qquad\qquad\qquad\qquad$ N–retinyl-opsin protein without cleaving the –C–N– bond of the retinyl opsin leads to the recovery of the group of amino acids at the binding site. It was found that the retinal is attached to lysine and that non-polar amino acids predominate in this group.

The colour of rhodopsin.   Retinal has an absorption maximum at about 380 nm, while that of rhodopsin is at about 500 nm. A persistent question in this field is how to explain this large bathochromic shift in the protein molecule. It is generally accepted that the Schiff's base of rhodopsin is protonated and recent evidence from resonance Raman spectroscopy has confirmed this supposition. In this technique the material (a solution, pellets of outer segments) is illuminated with a pulse of intense light from a laser, using a light frequency in resonance with the vibration frequencies of the chromophoric group. The light scattered from the specimen, usually at 90° to the incident beam, is collected, passed through a spectrometer, detected by a photomultiplier and recorded as a frequency spectrum of the scattered light. The frequency shifts that are produced in scattering are related to the normal mode frequencies of vibrations in bonds such as $C-C$, $C=C$, $C=O$, $C=N$, $CH=NH^+$, etc. The analysis of the spectrum depends on the assignment of specific spectral bands to specific bonds, which is not always an objective matter in the light of the state of the theory of the Raman effect as it applies to large molecules. In the case of rhodopsin the procedure has been to compare the resulting spectrum with those of the various model retinal Schiff's bases, protonated and unprotonated. In order to avoid bleaching the rhodopsin during the measurement, a rapid molecular flow technique has been introduced. The pigment passes through a capillary tube, during the pulse, at such a rate that any bleached molecules move out of the beam's path so that the scattered light is always from unphotolysed rhodopsin. With this technique a band at 1657 cm^{-1} was identified, consonant with a 1655 cm^{-1} band of the protonated Schiff's base of all-*trans* retinal. In addition, in a pellet of rod outer segments subjected to deuterium exchange involving the hydrogen of the Schiff's base, the 1655-band moved to 1630 cm^{-1}. The same result was found with the model protonated Schiff's base. While results such as these have been generally accepted as proof of the protonated Schiff's base structure in rhodopsin, it is only correct to point out that one investigation, using nuclear magnetic resonance spectroscopy, reached the conclusion that the Schiff's base in rhodopsin is unprotonated. In this report the possibility of incorrect assignment of the 1655-band was suggested as well as the fact that the frequency value of this band, as given by several investigators, was somewhat variable, from 1645 cm^{-1} to 1660 cm^{-1}.

If rhodopsin is, in fact, a protonated Schiff's base it is then relevant to ask when in the cycle of bleaching does deprotonation occur. The likely step is the transition meta I to meta II for it is in this step that the spectral maximum shifts from 480 nm to 380 nm. The technique of resonance Raman spectroscopy has been applied to this question with the finding that deprotonation, indeed, occurs at this stage. This of course brings up the further question of meta III which absorbs at 465 nm. Is the molecule reprotonated or is this bathochromic shift produced through some other mechanism? Raman spectroscopy has not yet been applied to this question.

Even if the protonated Schiff's base model is accepted, this alone cannot explain the colour of rhodopsin and many other naturally occurring visual pigments. There must be still another mechanism whereby the spectral absorbance is shifted further toward the red than 440–460 nm. One mechanism that is generally cited as a possible explanation of the adjustment of visual pigments to varying regions of the visible spectrum is that of a secondary, non-covalent electrical interaction between charged or polarisable groups of opsin and the polarised π-electron system of retinal. The protein, acting uniquely in the different visual pigments, stabilises the delocalised π-electrons in various configurations in the different pigments and so leads to different spectral absorptions. Recently, support for this model came from measurements of the changes in absorbance produced by an intense electric field acting on retinal and on its unprotonated and protonated bases with n-butylamine. In this experiment all three molecules became dipolar upon excitation to the first allowed singlet state and the dipole moments of the ground and excited state were approximately parallel to the molecule's long axis. The result was explained on the basis of a displacement of negative charge away from the ring leaving it with a positive charge as shown by the resonance structures.

Accordingly, the charges on the protein that interact with the retinal in the chromophore to stabilize particular resonance structures and set the colour, become the central issue in this problem. Several investigators have pointed out, by molecular orbital calculations, that the location and distance of these charges relative to the retinal can adjust the spectrum over a wide spectral range. Since retinal is not a rigid structure, the ring plane and chain configuration being alterable by twists and bends of specific bonds, it is conceivable that charge interactions of the type proposed could induce such configurational changes. It is now time to test some of these speculations by actual studies of the visual pigments and later in this chapter I will relate what has been attempted with the photopigments of the gecko.

Chemical properties. Rhodopsin is a globular protein, with a single polypeptide chain, with an α-helical content of about 30% and with little β-structure. The molecular weight has been variously estimated as being from 27 000 to 40 600, the higher figure being the one most frequently employed. Because of this uncertainty it has not been possible to estimate directly the molar extinction coefficient (molar absorptivity), but because retinal is present on an equimolar basis the content of this prosthetic group has been determined after bleaching by converting it to the oxime

$$C_{19}H_{27}-\overset{H}{C}{=}O \ + \ NH_2OH \longrightarrow C_{19}H_{27}-\overset{H}{C}{=}NOH + H_2O$$

$$\lambda_{max} = 380nm \qquad\qquad\qquad \lambda_{max} = 367nm$$

with hydroxylamine, and the extinction of the oxime at its maximum (367 nm) is measured. Since the molar extinction of retinal oxime is known, the corresponding extinction of rhodopsin is obtained from the measured ratio of extinctions of oxime to rhodopsin. Alternatively, the retinal may be estimated by the colour reaction with thiobarbituric acid. With such procedures the value of the molar coefficient has been cited as being between 40 000 and 43 000. In all these studies bovine rhodopsin has been routinely employed but Heller (1969) purified rhodopsin from the cow, rat and frog and obtained the same molecular weight, the same molar extinction, the same number of retinals (one) per molecule, the same stability toward sodium borohydride and all three rhodopsins contained carbohydrate moieties. Except for details, the amino-acid composition was generally similar and all three had a relatively high proportion of non-polar amino acids. In addition, Dartnall (1972) reported the same photosensitivity at $2.63-2.85 \times 10^4$ l per cm mole for several rhodopsins and other A_1-based visual pigments, all measurements being made in the presence of hydroxylamine. Employing the extinction coefficient of 4.06×10^4, the quantum yield turned out to be 0.67 which means that for every three photons absorbed, two rhodopsin molecules are bleached. It is also known that the spectral absorbance curve and the photosensitivity spectrum are the same, which indicates that the photosensitivity is invariant throughout the spectrum.

As already indicated, rhodopsin is a glycoprotein, the carbohydrate moiety being an oligosaccharide of N-acetyl glucosamine and mannose. Tryptic digestion of bovine rhodopsin led to the recovery of a peptide of 16 amino acids that contains the carbohydrate in two separate locations, both asparagine residues. The amino-acid sequence of this peptide was found to have an hydrophilic character, suggesting that it might be in contact with an aqueous environment in the disc membrane. One report described the rhodopsin molecule as an elongated (7.5 nm) structure with the carbohydrate at one end, the retinal at the other. This would suggest a molecule with one end hydrophilic, the other hydrophobic. Studies to determine the location of the carbohydrate within the disc membrane gave different results, with some suggesting that it is located at the cytoplasmic (aqueous) face, others placing it at the intradisc face. The function served by this oligosaccharide is not known although the idea was offered that it serves to orientate the visual pigment properly within the membrane when it is being assembled into the discs. Part of the oligosaccharide, possibly the first glucosamine that binds to asparagine, is incorporated into the peptide in the ribosomes of the inner segment while the remaining sugar groups are added in the Golgi apparatus. Whatever the exact sequence, it appears that the glycoprotein is completed before the retinal is joined to the molecule and certainly before the molecule is assembled into the disc. It is possible, then, that the hydrophilic carbohydrate functions to orientate the rhodopsin properly within the membrane. A further suggestion was made that sugars added to rhodopsin within the outer segment serve as recognition sites for the pigment epithelium in the process of phagocytosis. It may be added as an additional, not unimportant, detail that the carbohydrate moiety of rhodopsin is not essential for either the colour or the ability of rhodopsin to regenerate. Oxidation of the carbohydrate with periodate did not affect the 500 nm absorption as long as the –SH groups were protected. Neither did it destroy the ability to regenerate the photopigment with added retinal.

The stereospecific nature of rhodopsin. Retinal is an isoprenic polyene with a β-ionone ring linked in conjugation to a 5 double-bonded chain that terminates in a carbonyl oxygen. Two methyl groups in the chain are at C-9 and C-13. The molecule occurs in a number of *cis* and *trans* configurations of which the 9-*cis*, 11-*cis* and all-*trans* forms are of special interest biologically. The π-electrons constitute a mobile system so that

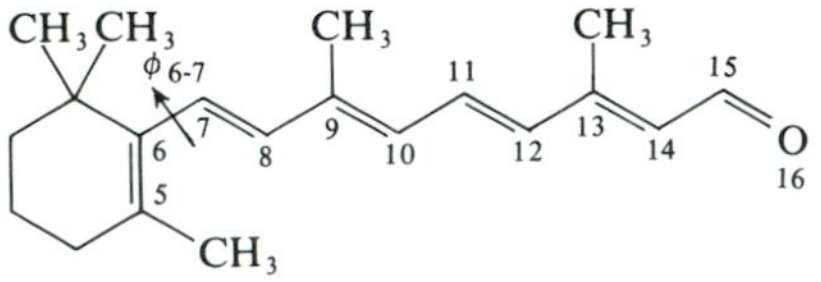

6S-*cis*, all-*trans*

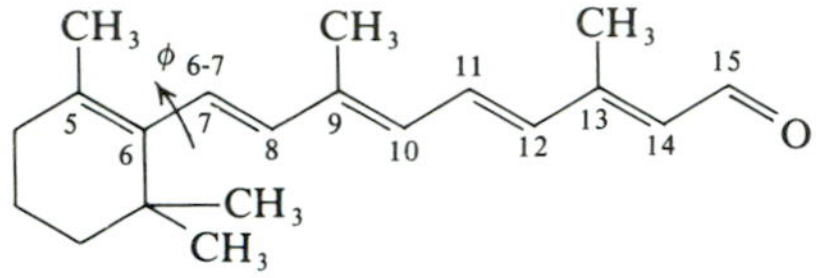

6S-*trans*, all-*trans*

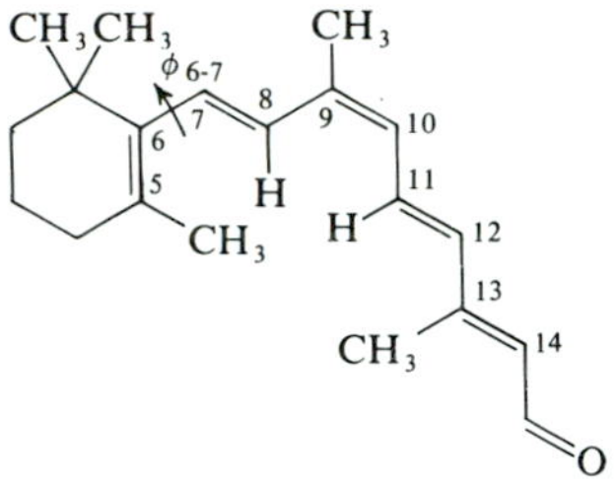

6S-*cis*, 9-*cis*

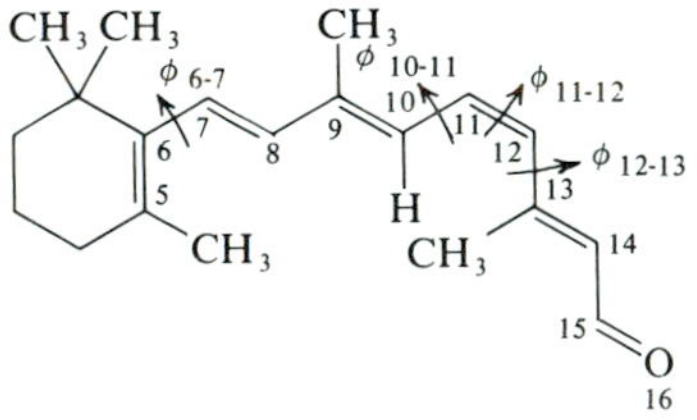

12S-*trans*, 11-*cis*

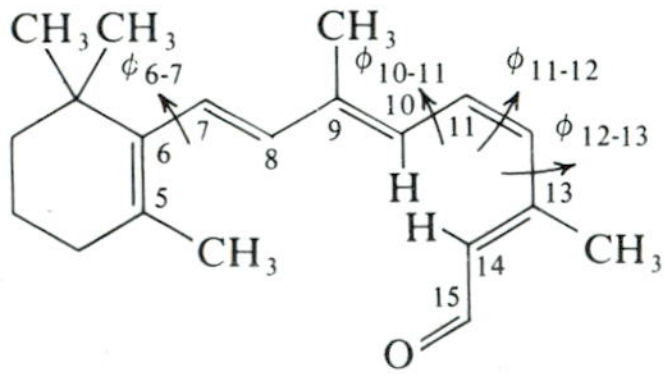

12S-*cis*, 11-*cis*

all bonds have some double bond character. The polyene chain tends toward planarity which is not achieved because of steric hindrance. For example, the 11-*cis* isomer was at one time believed to be so strained, due to hindrance between the hydrogen of C-10 and the methyl of C-13, as to be impossibly unstable. Yet 11-*cis* retinal is the isomer employed by nature in the visual pigments (Fig. 9.6). The strain is apparently relieved

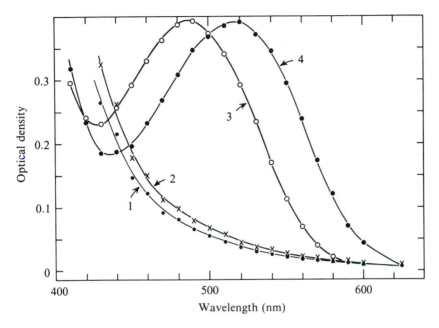

Fig. 9.6. The stereospecificity of the gecko 521-pigment. Lack of regeneration with added all-*trans* retinal (curve 1) and 13-*cis* retinal (curve 2). Regeneration of isopigment at 488 nm with 9-*cis* retinal (curve 3) and of 521-pigment with 11-*cis* retinal (curve 4). (Crescitelli, 1979.)

by twisting about the 12–13 single bond and this twist can be either s-*cis* or s-*trans*. In crystals it is apparently 12s-*cis* but in solution it is now thought to consist of an equilibrium mixture of 12s-*trans* and 12s-*cis*. Steric hindrance also prevents coplanarity between the plane of the ring and the chain and this is relieved by twisting about carbons 6–7, where 6s-*cis* and 6s-*trans* configurations can also occur. Accordingly, the retinal molecule shows up as a dynamically flexible structure whose specific conformation is adjustable to the nature of the environment. This concept is especially intriguing from a biological point of view for it means that the protein environment with all its varied features can induce in its prosthetic group changes not known in either the crystal or the solution. An example of the possibilities is stated by Liebman (1972) under the name of the two-stage theory of Blatz and Liebman. According to this idea the ring orientation relative to the chain can vary according to the degree of twist at the 6–7 position. When the ring and chain are nearly coplanar the maximum conjugation is achieved and visual pigment absorbance is at longer wavelengths. When the twist is such as to rotate the ring 70–80° out of planarity with the chain, reduction of conjugation leads to pigment absorption at shorter wavelength. The degree of twist, assumed to be different in the differently coloured visual pigments, is the result of interaction with the opsin which is variable and unique for the differently coloured pigments.

The *cis–trans* twisting about single bonds also helps in understanding another property of rhodopsin which has already been mentioned. This is the circular dichroism which is a feature of rhodopsin but not of free retinal. There are several possible reasons for this optical activity, two of which are relevant at this point. If 11-*cis* retinal in solution exists as a mixture of s-*cis* and s-*trans* forms, opsin by selecting only one of these forms leads to the synthesis of an optically active rhodopsin. The alternative explanation is that opsin, itself dissymmetric, twists the retinal to conform to its own configuration. Data from resonance Raman spectroscopy suggest that the 11-*cis* retinal in rhodopsin may have a distorted 12s-*trans* form.

Certain other features of the retinal molecule may also play a role in determining the synthesis of a visual pigment. The longitudinal dimension from C-15 to the centre of the ring has been considered to be a critical characteristic, for retinal isomers or analogues having dimensions shorter or longer than 9-*cis* or 11-*cis* retinal (about 1 nm), fail to react with opsin. The methyl group at C-9 appears to be necessary in both the 11-*cis* and 9-*cis* isomers. In both cases the desmethyl analogues in which 9-CH_3 is replaced by hydrogen, led to the synthesis of photopigments that were 'blue' shifted by some 35 nm relative to the locations of rhodopsin and isorhodopsin. This was in contrast to the 13-desmethyl analogues which regenerated photopigments unaltered from their normal positions. Possibly, the 9-CH_3 is involved in a specific non-covalent interaction with a particular site in the opsin. Alterations in the ring structure also modify the behaviour. Failure to regenerate a photopigment resulted when a methyl group at C-1 was replaced with hydrogen or when the 5–6 double bond was replaced by a 4–5 double bond. The relevance of the ionone ring was also indicated by experiments in which it was shown that β-ionone competitively inhibited the regeneration of cattle rhodopsin when 11-*cis* retinal was added to opsin already exposed to β-ionone. A β-ionone binding site was postulated to exist in opsin so that the

β–ionone

regeneration of rhodopsin was pictured as a 2-step reaction, the ring anchoring itself into its hydrophobic pocket, to be followed, if the length is correct, by the formation of the aldimine bond.

Rhodopsin: A membrane-bound system. From the point of view of function the behaviour of the photopigment within the membrane system is of primary interest. There is no general agreement as to the precise location of rhodopsin within the disc membranes and various investigators have reported that it is located at the cytoplasmic face of the disc, the intradisc face, that it is buried within the lipid bilayer or that it spans the entire thickness of the membrane. The shape of the molecule is also subject of differing opinions, some workers believing it to be approximately spherical with a diameter of 4–5 nm, others that it is elongated with a length of 7.5 nm or longer. A rat rod is estimated to house something like 3×10^7 rhodopsin molecules and these constitute 80–90% of the protein in the outer segment. The membrane site of the molecule must be fluid for the rhodopsin in frog rods was observed to undergo rapid rotational diffusion about the axis normal to the membrane. Lateral diffusion was also noted but diffusion along the rod axis, i.e., from disc to disc, did not occur. These results picture the membrane as consisting of a lipid, fluid core in which float rhodopsin molecules that can rotate and translate. The oily nature of the core is to be expected in terms of the high content of polyunsaturated fatty acids that are constituents of the membrane lipids.

Though the rhodopsin molecules can rotate and translate, they cannot tumble. This is indicated by the property of linear dichroism possessed by the rods. Linearly polarized light passing down the long axis of the rod is absorbed equally regardless of the plane of polarization. In contrast, light incident transversely at right angles to the long axis is absorbed unequally depending on the polarisation. Light of the wavelength that is absorbed by rhodopsin suffers greater absorption if the electric vector is perpendicular to the long axis of the rod and less absorption if it is parallel to the axis. This dichroism is due to alignment of the pigment chromophores in the plane of the discs. Clearly. this orientation is biologically efficient, for light passing in the normal direction down the rod is equally absorbed regardless of the polarisation. In this way the quantum absorption is increased over that for a comparable number of rhodopsin molecules randomly oriented as in solution. Accordingly, it appears that the photopigment molecules are held in some constraint whereby they are allowed to rotate and translate but not to tumble. A possible model for this is the embedding of a hydrophobic portion within the lipid layer while the hydrophilic end extends into the aqueous phase.

A question of some importance which claimed the attention of earlier workers seems now to have been laid to rest. It is the comparison of spectral absorbance curves for the pigment in solution and *in situ*. This comparison was first attempted by relating the human scotopic sensitivity curve with the spectral absorbance of extracted rhodopsin. It was believed by some that while the two functions are basically the same, the precise spectral location of the two are not, the action spectrum being displaced bathochromically from the absorbance. Spectral sensitivity and pigment absorbance curves for a number of animals have been compared in the literature with varying degrees of success but in nearly all cases a precise agreement has not been observed. Perhaps this situation is to be expected since the two functions involve different standards of analytical precision. The introduction of microspectrophotometry into visual science now permits the direct examination of pigment absorption curves *in situ* and their comparison with the pigments when solubilised. This has been most successfully accomplished with large rods such as the red rods of the frog and the rods of the gecko. I illustrate the satisfactory agreement of these two absorptions for the 521-pigment of *Gekko gekko* (Fig. 9.7). A second pigment which absorbs at 467 nm has also been found in this lizard by both techniques and the absorbances also agree. The same is true for the rhodopsin of the frog, *Rana pipiens*, which absorbs at 502 nm. It may be assumed, therefore, that results obtained with extracts are applicable to the pigment in the cellular environment and that when disagreement with action spectra is found it is the latter function that requires enquiry.

THE A$_2$ ROD SYSTEM

According to present knowledge the invertebrates have evolved but one system of visual pigments, the A$_1$-system. Vertebrates, on the other hand, possess a second system based on 3-dehydroretinal, the aldehyde of

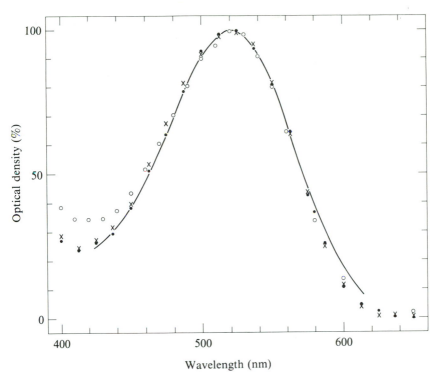

Fig. 9.7. Spectral absorbance of gecko 521-pigment *in situ* (filled circles, crosses) and extracted into digitonin (open circles). Continuous line is a Dartnall nomogram curve for a pigment at 521 nm. (Crescitelli, Dartnall & Loew, 1977.)

vitamin A₂ or 3-dehydroretinol. This vitamin, commonly occurring in the livers of freshwater fishes, has an additional double bond in the ring in

3-dehydroretinol (vitamin A₂)
$\lambda_{max} = 350nm$

3-dehydroretinal
$\lambda_{max} = 401nm$

conjugation with those of the chain so that the spectral absorbance is extended bathochromically by about 20 nm for the alcohol, the aldehyde, and the visual pigments that utilise this prosthetic group.

The discovery of this second system was the work of Wald (1939) who extracted from the retinas of some freshwater fishes a photopigment with maximal absorbance, not at 500 nm, but in the region of 520 nm. This pigment, named porphyropsin, was found to yield a product of bleaching that reacted with the Carr–Price reagent (antimonous chloride, $SbCl_3$) by giving a colour unlike that given by retinal. Eventually, this led to the isolation and identification of 3-dehydroretinal and to its being related to the vitamin A found in the livers of freshwater fishes. In time, studies were made by various investigators of the porphyropsin system that paralleled the researches on rhodopsin that have been discussed.

The A_1- and A_2-pigments are functionally and chemically similar at the fundamental level, differing only in certain details. Although there is some overlap in colour of the two systems, the spectral absorbances of the A_2-pigments as a group are located further toward the red than are the A_1-pigments. The α-band of porphyropsin is broader than that of rhodopsin and a different nomogram has been developed for porphyropsin. The form of the porphyropsin curve is somewhat different as the absorption at shorter wavelengths, relative to the α-peak, is higher than for rhodopsin. Although fewer measurements have been made, the molecular weight and molar absorptivity appear to be lower than in rhodopsin. The same stereospecificity is present and 11-*cis* 3-dehydroretinal appears to be the naturally occurring isomer. The 9-*cis* isomer reacts with opsin to give isoporphyropsin. The products of bleaching are the same as with rhodopsin, and bathoporphyropsin (592 nm), lumiporphyropsin (542 nm), metaporphyropsin I (509 nm) and metaporphyropsin II (408 nm) have been described. Under some conditions carp porphyropsin has been found to lack a meta III intermediate but this is not unique as the gecko P521, an A_1-pigment, also lacks this late intermediate.

The biological distribution of the A_2-system has some interesting features. In a number of species it appears to be coupled with the A_1-system either at the same time in the retina or alternating with it according to a seasonal or metamorphic cycle. In such cases the same opsin is apparently employed, the shift being in the nature of the prosthetic group attached to the opsin, a matter that seems to be directed by the pigment epithelium, which, in turn, may be the subject of a higher authority such as the endocrine system. This point is well illustrated by the dual (A_1–A_2) photopigments of the bullfrog, *Rana catesbeiana*. The bullfrog, like other ranid amphibia, has porphyropsin as a tadpole. During metamorphosis the porphyropsin is progressively replaced by rhodopsin, the thyroid probably being involved. At one time, only rhodopsin was found in the adult bullfrog retina indicating a total metamorphic conversion to the A_1-system. It is very likely that this finding was correct, for the bullfrog, like many freshwater teleosts, may undergo a seasonal change in visual pigment composition. Analyses made in the summertime, for instance, may well show little or no porphyropsin. Recently it was reported that the bullfrog retina displays a segregated visual pigment system, at least in certain seasons, the superior half of the retina with porphyropsin, the inferior half with rhodopsin. The difference in colour of these two retinal regions has been photographed and is very convincing. The interesting feature of this segregation is that it is a segregation in the pigment epithelium. Thus, when the light-adapted (bleached) retina was carefully removed from the eye cup, inverted 180° and placed back on to the pigment epithelium with the superior retinal half lying on the inferior pigment epithelium, and the inferior half on the superior epithelium, the pigments that were regenerated in the dark were reversed from their normal locations in the retina but were normally placed with respect to the pigment epithelium. This can be explained by assuming that the same opsin, within the outer segments, is employed for both rhodopsin and porphyropsin and that the pigment epithelium has the enzyme system for producing 3-dehydroretinal located in the superior half of this tissue so that the opsin in contact with this half receives the 3-dehydroretinal and makes porphyropsin.

THE TRANSMUTED PIGMENTS

Until very recently the transmutation theory of Walls rested principally on morphological and a few tidbits of physiological observations. With the expansion of interest in visual pigments we are now in a position to extend further the ideas of Gordon Walls. The newer knowledge may be enumerated briefly as follows.

(1) Some cells classed as cones, morphologically, have been found, by means of microspectrophotometry, to contain photopigments that resemble rhodopsin and porphyropsin in spectral absorbance. The accessory member of the double cones of the adult frog, *Rana pipiens*, has a 502-pigment, apparently like the rhodopsin of the red rods in the same retina. The tadpole, on the other hand, has within the accessory member a pigment at 527 nm, like the porphyropsin of the red rods of the tadpole. These pigments have not been separately extracted and isolated so nothing can be said as to their properties

relative to frog rhodopsin and porphyropsin both of which have been extracted.

(2) Walls (1942) considered the visual cells in the superior half of the alligator retina to be cones with rod-like outer segments. He wrote of these as being cells undergoing phylogenetic transmutation. Alligator rhodopsin has been extracted and it appears to have one feature unusual for a pigment at 500 nm. After bleaching and adding 11-*cis* retinal to the opsin the rate of regeneration is rapid and like that of the cone pigment, iodopsin, but unlike the rate in frog and cattle rhodopsin. The rhodopsins of the lower and upper halves of the alligator retina have not yet been separately extracted and studied.

(3) The gecko pigments, perhaps, have offered the most coherent argument in support of transmutation at the molecular level and this is most appropriate for it was with these lizards that Walls first conceived his theory. Nocturnal and crepuscular geckos have visual cells with large, cylindrical outer segments whose visual pigments have been analysed in solution and *in situ* (Figs. 9.7, 9.8). A consis-

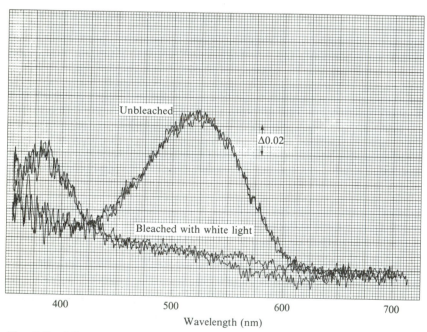

Fig. 9.8. Microspectrophotometer recordings of gecko 521-pigment before and after bleaching with white light. Note the product of bleaching at about 380 nm (Crescitelli, Dartnall & Loew, 1977). Two traces were made to show that pigment was not bleached by recording light.

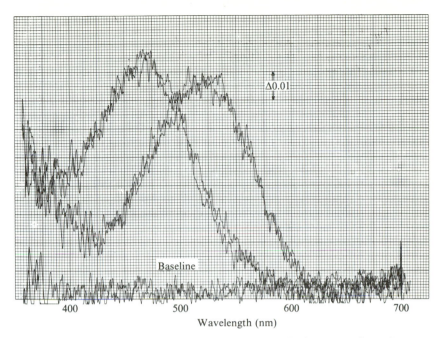

Fig. 9.9. Microspectrophotometric recordings of the two photopigments (P521, P467) in two different cells of *Gekko gekko* (Crescitelli, Dartnall & Loew, 1977). Two traces are shown.

tent finding is of two photopigments, one absorbing in the green region, the second at shorter wavelengths. In the Tokay gecko (*Gekko gekko*) these are at 521 nm and 467 nm (Fig. 9.9). Both are A_1-pigments but it is the $P521_1$, present in extracts as 90–92% of the photopigment density, that has been examined carefully and is, in fact, the pigment of interest. $P521_1$ is thermally more labile than frog rhodopsin; it is attacked by both hydroxylamine and sodium borohydride in the dark, a property most unlike rhodopsin; and its opsin regenerates rapidly with added 11-*cis* and 9-*cis* retinals. Extracted into digitonin (or Triton-X-100) that is deficient in chloride the spectrum is not at 521 nm, but some 15–20 nm 'blue' shifted. The addition of chloride or bromide restores the normal colour (Fig. 9.10). The action of chloride is specific for no other anion except bromide is able to do this. The nature of the cation is irrelevant as long as it is one that does not destroy the pigment. The nitrate ion is also effective but in the reverse sense from chloride. Nitrate is able to shift the spectrum of the chloride-deficient pigment to still shorter

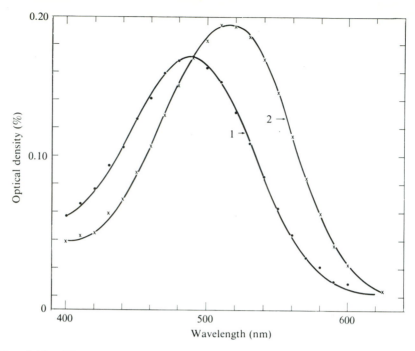

Fig. 9.10.   The chloride ionochromic effect for the 521-pigment of *Gekko gekko*. The chloride-deficient extract (curve 1; circles) and after adding sodium chloride (7.96×10^{-3}M) (curve 2; crosses). (Crescitelli, 1977.)

wavelengths and to antagonise the action of chloride (Fig. 9.11). At a ratio nitrate : chloride of 6 : 1 nitrate completely prevents the chloride effect. None of these properties is shown by the 467-pigment. The relevance of all this for transmutation is that iodopsin, the cone pigment of the chicken retina, has the same properties, although the nitrate effect has not been looked for with iodopsin. It would be surprising if these similarities between an apparent rod pigment ($P521_1$) and a genuine cone pigment, iodopsin ($P562_1$) were purely coincidental, especially as both are so unlike the rhodopsins. It is unlike nature to have evolved two photopigments with such molecular similarities without some phylogenetic relationship between them. Along with the similitudes given here are the morphological attributes pointed out by Walls and certain physiological properties (critical fusion frequency, dark adaptation) of the gekkonid retina that are more like cone than rod behaviour.

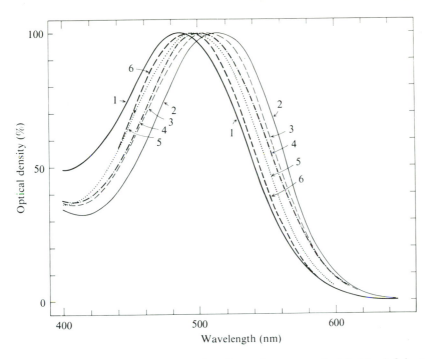

Fig. 9.11. The nitrate ionochromic effect. Spectrum of chloride-deficient extract (curve 1) and after adding sodium chloride (curve 2). Curves 3–6 show effects of successive additions of sodium nitrate. All data corrected for dilution and normalised with maxima at 100%. Crescitelli, unpublished.

Quite apart from the transmutation theory the gecko pigments, such as $P521_1$, are of interest in connection with the physicochemical features of the protein that determine the colour. As explained earlier there are several reasons to suppose that critically placed charges on the opsin are involved in stabilising the retinal molecule to a given configuration. Recently, it has been suggested that perhaps two charges are involved: one near the Schiff's base nitrogen, the second near C-12 and C-14. The chloride and nitrate actions in changing the absorbance of the gecko pigment are best explained in terms of ionic effects that influence these protein sites. Perhaps the opposite effects of these two anions are the result of these two differently sized ions acting each at a separate site in the opsin. The gecko pigment seems to have a more open structure than the rhodopsins, as shown by the accessibility to hydroxylamine and sodium borohydride. Perhaps this is the reason for these ionochromic responses of the gecko pigment.

THE CONE PIGMENTS

Iodopsin

Only iodopsins from chicken and turkey retinas have been successfully extracted in sufficient quantity for study and now spectrophotometry has confirmed what was previously only presumed to be so, that iodopsin is indeed a cone pigment. For this reason it has a special place in our list of vertebrate visual pigments. In certain respects iodopsin, with spectral maximum at 562 nm, is like the rhodopsins, in others it is uniquely different. It is an A_1-pigment with an absorbance curve that is narrower than the standard curve when plotted on a wave number basis. It can be regenerated from its opsin plus 11-*cis* retinal just as rhodopsin can be regenerated, but the rate is significantly faster than that of rhodopsin. This is in accord with the more rapid kinetics of dark-adaptation for cones. With 9-*cis* retinal the opsin of iodopsin forms the isoiodopsin at 510 nm. This is an isoshift of 52 nm (562 to 510 nm) and is to be compared to the rhodopsin isoshift of 12–15 nm, the porphyropsin isoshift of 15 nm, and the gecko P521 isoshift of 33 nm. Accordingly, the cone pigment has the greatest isoshift; the rod pigments, the least; and the gecko pigment is intermediate between them. The same sequence of bleaching intermediates as in rhodopsin has been found with iodopsin: bathoiodopsin (640 nm), lumiiodopsin (518 nm), metaiodopsin I (495 nm) and metaiodopsin II (380 nm). There is no mention in the literature of metaiodopsin III. The molar absorptivity of iodopsin is 40 600 and the molecular weight (36 000) is much the same as that of chicken rhodopsin (39 000). There is one important difference in the rhodopsin and iodopsin systems that is unexplained. When batho-rhodopsin at $-195°C$ is warmed to $-140°C$ it passes to lumirhodopsin (Table 9.1). When this same experiment is made with bathoiodopsin the result is a return, in the dark, to iodopsin. In other words, if one assumes the presence of the all-*trans* configuration in the chromophore of bath-oiodopsin, one has to accept that this reverts back to the 11-*cis* form at $-140°C$, a thermal isomerisation difficult to believe. Either a difference in strain of the all-*trans* isomers at these low temperatures exists between the rhodopsin and iodopsin systems and/or a difference in behaviour of the two opsins accounts for the different result. As indicated above, the protein of iodopsin has significantly different properties compared to the opsin of rhodopsin.

From the biological point of view the closer affinity of iodopsin to the gecko pigments than to rhodopsin is of special interest in respect to the phylogenetic history of visual cells and visual pigments. The similarities between the gecko $P521_1$ and iodopsin have already been pointed out. The thermal behaviour; the sensitivity, in the dark, to hydroxylamine and sodium borohydride and the rapid regeneration rates in the two systems cannot be ignored. It is the ionochromic effect of chloride (and bromide) of the two systems that has particularly impressed me. If, now, a nitrate effect is ever found for iodopsin this will complete the circle of similitudes. Recently, I found another common chemical property for the two systems that is probably related to the more open structure of gecko opsin and chicken cone opsin. In the case of iodopsin, 9-*cis* retinal replaces the 11-*cis* isomer, in the dark, to form the isoiodopsin. I have found, in the case of the gecko $P521_1$, a similar dark exchange with the conversion of about 20% (density-wise) of the 521-pigment to the iso-form. A study of more cone pigments is essential of course but I am more and more struck by the idea first mentioned by Wald, that there are two classes of opsins which he named photopsins and scotopsins. I would only make an addendum that this dual system has had a common phylogenetic origin and has been subjected to all the evolutionary pressures of other biochemical systems. In the case of the visual proteins these pressures could be more immediate in their results because of the immediacy of the photic environment to the photoreceptors.

There are a number of claims for the extraction of cone pigments other than iodopsin but none of these has been confirmed. Early investigators reported such findings for extracts of retinas of diurnal reptiles but the density changes with light were so small and uncertain as to be meaningless. I have extracted the retinas of a number of species of diurnal lizards, known to have a high complement of cones. In none of these have I found evidence of a photopigment. A pigment at 544 nm was reported for the pigeon retina but this has not been confirmed by microspectrophotometry. Instead, this technique has revealed cone pigments at 567 nm, 515 nm and 460 nm. Neither have I been able to extract photosensitive retinal pigments from the cone-dominated retina of the ground squirrel. Recently two new pigments were reported in extracts of the chicken retina, one at 449 nm, the other at 417 nm. They were said to be like iodopsin in the fact of rapid regeneration and in sensitivity to hydroxylamine, but unlike iodopsin in the absence of a chloride effect.

Cone pigments in situ

Pigments of the intact human retina.   The technique of analysing the light reflected back from behind the retina of a human eye (retinal densitometry) has revealed the presence in the fovea of two photosensitive components. These are the red-sensitive erythrolabe and the green-sensitive chlorolabe, as they have been named by Rushton (1972). The protanope was found to lack erythrolabe; the deuteranope, chlorolabe. This explains the protanope's 'red-blindness' and the fact that the deuteranope's spectral sensitivity, though not greatly impaired in the green region, extends further into the red than that of the protanope. These two are not abnormal components of the retina for both erythrolabe and chlorolabe have been found in the normal human trichromat. A third component (cyanolabe) has not been found by this technique and this is understandable since the human fovea is probably deficient in the blue (though not blue-blind) and the macular region contains variable amounts of the blue-absorbing yellow pigment that complicates measurements at shorter wavelengths. Erythrolabe and chlorolabe are probably pigments of colour vision and contained in cones but retinal densitometry cannot prove this localisation. What retinal densitometry has clearly demonstrated is that colour blindness in dichromats can be due to loss of one of the receptors for colour (theory of Koenig) and is not necessarily associated with fusion or mixing of the pigments or channels so that the signal specificity is lost (theory of Fick and Leber). It will be of interest to apply this technique to those rare cases of unilateral colour blindness to establish whether or not the colour deficiency of the abnormal eye is due to unilateral pigment loss.

Pigments of the isolated retina.   By utilising a microspectrophotometer the spectral absorbances of small pieces of human and monkey retinas from the foveal region have been measured, first in the dark-adapted state, and then after bleaching with first red, then yellow light. The difference spectra so obtained revealed two photopigments, one at about 565 nm, the second at 535 nm. The interesting feature of this experiment was the regeneration of both pigments, after bleaching, by the addition of 11-*cis* retinal to the medium, showing the presence of separate opsins linked to 11-*cis* retinal.

Pigments of single cones.   The examination of individual cones by means of modern microspectrophotometers has made it possible to clarify a number of issues that were doubtful or uncertain. First of all, it has been possible to show that erythrolabe and chlorolabe are indeed inhabitants of cones; second, a pigment has been located absorbing in the blue region of the spectrum so that the existence of the hypothetical cyanolabe has been shown and a trichromatic basis for colour vision established. Thomas Young has been proved right. Finally, it has been demonstrated that trichromacy at the receptor level occurs not only in primates but is also present in teleost fishes. Trichromacy appears to be a fundamental invention of nature that evolved early in evolution and has been kept through to man. When deviations from trichromacy occur, as in the case of the protanopic ground squirrels, these represent, not steps in the evolution of colour vision, but genetic deletions that have occurred.

The earliest report of three cone pigments was made by Marks, Dobelle & MacNichol (1964) who showed, with considerable variation, three classes of photopigment cones in human and monkey retinas absorbing at 570, 535 and 445 nm. At about the same time Wald & Brown (1965) were measuring the pigments in the parafoveal cones of the human retina and reported three classes at 440 nm, 525–535 nm and 560 nm. Liebman (1972) listed the three locations for human and monkey cones at 440, 535 and 575 nm.

The cone outer segments of cyprinid fishes, like carp and goldfish, are unusually large (for cones) and have provided us with the most precise recordings for cone photopigments. Three classes of cones have been confirmed for the goldfish retina in the regions of 455 nm, 530–535 nm and 623 nm. These are probably A_2-based pigments, although a direct chemical proof of this is not available. As the goldfish is known to have colour vision, the possession of three separate cone systems suggests a trichromatic system for colour discrimination. The study of the goldfish cone pigments *in situ* has revealed a number of features in common with rhodopsin in rods. One is that there is a segregation of the different photopigments in morphologically different cells. One is reminded here of the fact that there is a very obviously different morphology of the frog red rods, which have rhodopsin, and the green rods which house $P432_1$. The same is true for the goldfish cones which may be longs, shorts, miniatures and doubles. The red-sensitive pigment has been found in the long member of the double and the long single; the green-sensitive pigment in the short member of the double and the long single; and the blue-sensitive pigment in both the short and miniature singles. In addition, the goldfish cones have linear dichroism so that the pigments therein have their chromophores with the electric vector parallel to the plane of the discs. The specific density of the cone pigments within the cells varies somewhat as reported by different investigators. The latest figure of $0.0124/\mu m$ is a bit lower than the figure of 0.0182 for frog rhodopsin but if one considers the smaller molar extinction coefficient of porphyropsin to apply also to the A_2-cone pigments the correction made to the former figure would bring it close to the value of the rhodopsin system. One more similarity between the rod and the cone pigments is with respect to the form of the curve when plotted on a wave number basis. Again one finds that the cone pigment absorbing at longer wavelengths is narrower, confirming that this is a general phenomenon. The study of cone pigments has just begun. With the growing number of microspectrophotometers and improvements in technique much more will be learned about them but the real need at the moment is a few cone pigments that can be brought into solution and studied by the methods that have been employed with cattle rhodopsin. The interesting data obtained with iodopsin should be checked with some other cone system.

SOME BIOLOGICAL SPECULATIONS

As the result of the great spectral diversity of visual pigments in vertebrates a number of questions have arisen regarding the biological meaning of this multiplicity. Unfortunately, these questions can only be approached by pointing out apparent correlations between visual pigments and the environment or phylogeny. Cause and effect relations have been almost impossible to establish.

The ancestral vertebrate visual pigment

Vertebrates have both the A_1- and A_2-pigments and both occur in the oldest of the living vertebrates. The migratory sea lamprey, *Petromyzon marinus*, has rhodopsin in the retina of young adults migrating seaward and porphyropsin in sexually maturing adults going upstream to spawn. Associating the A_2-system with freshwater and assuming that vertebrates originated in freshwater, Wald postulated the idea that porphyropsin represents the ancestral pigment. There are uncertainties about both these assumptions. Firstly, the porphyropsin system, though predominant in freshwater fishes, is not exclusively present there. Some marine fishes, even one of the deep-sea species, have been reported to have A_2-photopigments. Secondly, it cannot be said with confidence that vertebrates originated in freshwater and Denison (1956) made a cogent case for a marine origin.

The significance of the 500-location

Whatever the nature of the ancestral pigment, once rhodopsin at 500 nm arose this became the dominant system. In the list compiled by Lythgoe (1972), incomplete as it is, some 190 species out of 308 are shown to have rhodopsin with maxima between 493 and 505 nm. These 190 species include vertebrates inhabiting the major biospheres, the notable exceptions being the deep sea, freshwaters, sandy shores, muddy bays and estuaries.

Why this preference for the 500 nm location? It is sometimes stated that this is the wavelength of absorption best adapted to the quality of light on earth. When the solar irradiance data of Moon (1940) are converted to quanta, which is the appropriate mode for absorption, it becomes obvious that the spectral maximum for light incident on the earth's surface is not at 500 nm but is shifted well toward the red and even into the infrared if one expresses the flux in terms of quanta per unit frequency interval. Even moonlight, since it is yellower than daylight, does not fit in with the idea of 500 nm being the optimised position. Vertebrates had an origin in an aquatic environment and in shallow waters filled with algae, foliage, etc. the scattering and/or absorption could shift the spectrum toward the blue away from 500 nm or toward the red. At the present time photic measurements have revealed that both types of photic aquatic environments occur. No great case can therefore be made for a 500 nm environment being the directing influence toward which evolution occurred.

There is perhaps another feature of the 500-rhodopsin molecule that should be considered. I have pointed out already the relative thermal and chemical stability of rhodopsin relative to iodopsin and the gecko transmuted pigment. If now the original vertebrates were diurnal, as Walls postulated, and possessed a diurnal, somewhat labile visual pigment, which might serve well under diurnal conditions, the adoption of a nocturnal existence with the need for a more stable system could have led to the evolution of a more stable rhodopsin. There is one fly in this ointment, however, which I have discovered while working with the rhodopsin of the Arctic cod, *Arctogadus borisovi*. This fish is found seasonally in the surface water from the ice flow at Point Barrow, Alaska, the temperature being close to, even below 0°C. Extracted into digitonin this rhodopsin was found to be exceptionally thermolabile, significant and irreversible loss of pigment occurring when the temperature of the extract was raised from 3.3 to 11.6°C. The molecule seems to have become adapted to the lower temperature to which the fish is acclimatised. I have not obtained enough fish to test the stability of the pigment *in situ* but it seems likely that even the rhodopsin molecule is capable of altering its properties in the process of evolutionary adaptation.

The rhodopsin of this cod illustrates another facet of visual pigment behaviour that has been reported by others. This is that the spectral absorbance and photolability can be retained while other properties of the molecule are altered, and this has been demonstrated by several investigators utilising enzymes to break down the rhodopsin to several fragments. I am intrigued by one report (Yutani, Sugino & Matsushiro, 1977) showing that the thermal stability of an enzyme (α-subunit of tryptophan synthetase of *E. coli*) can be modified by the substitution of one amino acid by another. Similar changes at various positions of the rhodopsin polypeptide chain could independently vary different properties of the molecule. This, of course, casts doubt on the idea that the 500 nm location and stability are strictly associated.

Adaptations to the quality of light

The sensitivity hypothesis. The idea of visual pigments with absorbances matched to the spectral quality of the photic environment arose first in connection with the retinal pigments of deep-sea teleosts that inhabit the region of the sea where downwelling light is attenuated and selectively filtered so as to leave a blue-green aquatic world. At greater depths where there is no downwelling light, bioluminescence is present and is known to be predominantly blue-green. It was no great surprise, therefore, when two groups of workers, one in England and the other in California, found visual pigments in deep-sea teleosts that absorbed at significantly lower wavelengths than rhodopsin. For this reason they were named chrysopsins (visual golds). Once this association was revealed for teleosts it was not long before it was extended to other deep-sea or deep diving animals, such as elasmobranchs, chimaerids and mammals. This deep-sea 'blue shift' has also been noted in Crustacea and other invertebrates so that we are dealing here with a phenomenon of general biological applicability.

There is even a finding of a deep-sea teleost with retinal photopigments, not in the region of the chrysopsins, but at longer wavelengths even than 500 nm. This appears to be an exception 'to the rule' but in fact it is an exception that proves the rule. This is the case of *Aristostomias scintillans*, a malacosteid deep-sea teleost, which has two photophores; a suborbital, red-emitting and a postorbital, green-emitting one. Two photopigments were extracted from the retinas of a single fish, an A_1-pigment at 526 nm, and an A_2-pigment at 551 nm. I have seen some unpublished data by the authors of this paper and it is more convincing than the original report. This result suggests a duplex system adapted to the duplex bioluminescence. In a system such as this it is reasonable to suppose that one opsin serves for both pigments, the mechanism for duality being the employment of the two prosthetic groups. One wonders how a deep-sea fish comes by 3-dehydroretinal and why. It is possible, just to speculate, that rod metabolic systems find it more economical to make the dehydrogenase that makes the A_2-retinal from the A_1 than it is to manufacture a different opsin to shift the spectrum as far to the red as 551 nm.

The sensitivity hypothesis is also supported by the occurrence of pigments that are bathochromically shifted from 500 nm in fishes that inhabit waters that tend to be yellowish because of mud, silt and other debris. In such cases marine fishes that are found in harbours, bays, estuaries, etc. have the A_1 system at these longer wavelengths and it is the nature of the opsin that sets the spectrum. In the case of freshwater fishes the A_2 system is employed, either alone or coupled with an A_1-pigment. The two pigments, apparently employing the same opsin, may be present in the retina simultaneously but the relative proportions may be altered seasonally, the A_1-pigment increasing in summer, when the water is clearer, and decreasing in winter as debris enters lakes and streams to make the water yellower. The possession of such a dual system effectively broadens the spectrum of quantum capture and serves the requirements of sensitivity in a labile photic environment. The segregation of porphyropsin and rhodopsin in the bullfrog retina, already pointed out, may be an adaptation to its way of life. This frog sits on a rock or lily pad, often at dusk, with its head half submerged so that the superior half of the retina views the aquatic field, the inferior half the aerial field. The grass frog, *Rana pipiens*, has a different habit of life and does not have the split visual pigment retina. The sensitivity hypothesis has also been employed to account for the visual pigments of fishes during the critical periods of twilight and dawn when the photic environment changes from the day-time spectrum or the moonlight spectrum.

The contrast hypothesis.   It has been pointed out by Lythgoe (1968) that there are ecological situations in which the scotopic visual pigment, rather than being matched to the photic environment, is better offset from it in order to enhance the contrast of the object being viewed. In shallow water a small nearby object somewhat brighter than the water background light has better contrast if the spectral maximum of the visual pigment is not matched to the spectrum of the background light. This is not true if the object is somewhat darker than the background. For distant objects, be they brighter or darker, a pigment system matching the background is best for vision under conditions of poor contrast.

VISUAL PIGMENTS AND VISION

The primary event

The mechanism that couples photochemistry and excitation of the visual cell has been a tantalizing and elusive problem ever since Kühne tried, unsuccessfully of course, to excite a frog nerve by immersing one end in a solution of rhodopsin and illuminating it. When it was discovered that rhodopsin employs 11-*cis* retinal and this is isomerised to the all-*trans* configuration, this event became a central theme in the literature of vision. It was of course realised that this photoisomerisation is followed by a change in conformation of the opsin, and since rhodopsin is a membrane protein any change in the protein with light assumed physiological significance.

Following illumination, rhodopsin goes through the orderly sequence of intermediates pictured in Table 9.1 and since these are known to occur with the pigment *in situ* as well as in solution the question at issue is which step in this sequence is associated with transduction. At physiological temperatures, the transitions up to metarhodopsin I occur before excitation of the retina, as judged by the a-wave of the electroretinogram (ERG), the latter being the sequence of electrical potentials that develop across the retina in response to a flash of light. Accordingly, interest became focused on the meta I to meta II transition. This step is one with a large activation energy and it is assumed that a change in protein conformation occurs at this step.

Attention is still directed to the meta I–meta II transition but before particularizing at this point let us examine what happens at the stages before meta I. There is little to be gained by looking at the steps beyond meta II for these occur late in the sequence and after transduction has occurred. Hypsorhodopsin may be omitted for we know little about it for vertebrate visual pigments and some pigments (iodopsin) appear not to have hypsorhodopsin. The next intermediate, bathorhodopsin, formerly named pre-lumirhodopsin, is presently the centre of attention. Bathorhodopsin has been always treated as if its retinal were in the all-*trans* form, simply because it was assumed that the *cis–trans* photoisomerisation was the initial event. In the last three to four years doubt has been expressed on this point. One objection has been that 6×10^{-12} s (6 ps), the time of formation of bathorhodopsin, is too rapid for isomerisation at the low temperature used in the study of bathorhodopsin. This objection has been dismissed by picosecond measurements of absorbance changes at 561 nm, using the laser pulse technique. The formation time of bathorhodopsin from either rhodopsin or *iso*-rhodopsin is measured to be less than 9 ps and it is argued that since bathorhodopsin is the common product of rhodopsin (11-*cis*) and isorhodopsin (9-*cis*) (Table 9.1) there must be an isomerization of at least one of these. This unexpectedly high speed of isomerization is explained as the result of the interacting effect of the opsin. However one views this matter, there are some results from resonance Raman spectroscopy that are worth considering as alternative views. It has been reported that the Raman spectrum of bathorhodopsin is not identical with that of rhodopsin or of isorhodopsin, and neither is it like that of all-*trans* model compounds. The differences are in the bands at 856, 877 and 920 cm^{-1}. Meta I is the intermediate that does have the Raman spectrum of unstrained all-*trans* compounds.

Several investigators have suggested that conformational changes in the protein can occur in bathorhodopsin and/or lumirhodopsin. Lewis (1978) has expressed the view that a portion of the energy absorbed by the chromophore is transferred to the opsin as conformational energy that can be utilised to initiate the events of excitation without involving a *cis–trans* isomerisation. There is one experiment that suggests a possible structural change in the opsin in going from bathorhodopsin to lumirhodopsin. This involved the production of a photopigment with 7-*cis* retinal in its chromophore after the outer segments were illuminated with light at 530 nm at $-75°C$ and the pigment then allowed to pass through its intermediates until a product absorbing below 470 nm was produced. The outer segments were then extracted and analysis by high pressure liquid chromatography showed the presence of a large complement of 7-*cis* retinal. This isomer has been isolated and has been found to combine with cattle opsin to form a photopigment which absorbs at 450 nm. The production of the 7-*cis* photopigment at $-75°C$ but not at lower temperature (−190°C) suggests that at −75°C the opsin is permitted to have a configuration that allows the fit of the highly hindered 7-*cis* isomer. Accordingly, the protein may undergo a conformational change in going from bathorhodopsin to lumirhodopsin. The circular dichroism has not revealed any protein changes in these early stages for the optical activity in the visible region (Fig. 9.3) persists through meta I and meta II.

Those who have rejected the idea of an 11-*cis* to all-*trans* reaction as the primary event have looked for other mechanisms that might link with the excitatory process and a number of models have been suggested based on the idea of a proton translocation. One model proposes an early deprotonation of the rhodopsin Schiff's base but this is clearly not in accord with the facts. Deprotonation would produce a 'blue' not a 'red' shift and it has already been mentioned that deprotonation occurs in the step from meta I to meta II. One might invoke hypsorhodopsin here as preceding bathorhodopsin. This would involve deprotonation followed by reprotonation but more needs to be learned about hypsorhodopsin in vertebrate pigments before this can be considered seriously. A second model is based on the transfer of a hydrogen to the Schiff's base nitrogen with the formation of a carbonium ion at ring C-5 and a rearrangement of the double bonds. Evidence cited against this model is that there is no change in the $-C=N^+H$ stretching frequency in going from rhodopsin to bathorhodopsin. A third model that proposes a hydrogen transfer from the methyl group at ring C-5 to the opsin is made doubtful by the experiment employing the retinal analogue, 5-desmethyl retinal which lacks the methyl group at C-5. This analogue was synthesised and shown not only to be capable of forming a photopigment with cattle opsin, but also, at low temperature, to form the batho-intermediate from the photopigment. The same result was obtained with the 9- and 13-desmethyl analogues. Other models proposing charge transfer mechanisms to clarify the structure of bathorhodopsin and to substitute an alternative reaction for the *cis–trans* proposal face similar problems. In summary, then, there is no reason not to discard the *cis–trans* idea, if need be, but until a better case is made for alternatives it is best to retain it as a working hypothesis.

The electrical responses

The photic excitation of the retina results in the production of several electrical phenomena which can be recorded by means of electrodes placed either across the retina (mass responses) or by means of intracellular electrodes (single cell responses). The first response recorded subsequent to an appropriate flash of light is the early receptor potential (ERP) which is a diphasic change to a brief, high-intensity flash and is characterised by a latency of but a few microseconds. Accordingly, it appears after bathorhodopsin but before metarhodopsin I. The living retina is not necessary for the generation of an ERP since the same result can be obtained from a retina treated with formaldehyde or glutaraldehyde. The ERP is not related to the movement of ions and its origin is usually attributed to the movement of charges, possibly two in succession, within the oriented rhodopsin molecules. Its role in vision, if any, is not known, and the opinion has even been offered that it is an interesting artefact unrelated to the events of visual excitation.

Following the ERP, but much later, is the complex of potentials known as the ERG, which can be recorded from the intact eye or across the isolated retina. This is a compound potential clearly associated with physiological activity and dependent on a living retina and is altered by changes in the ionic environment and by various chemical reagents. It consists of a series of variations known consecutively as the a-, b-, c-, d- and e-waves. The initial, corneal negative a-wave has a latency in milliseconds (1–30) depending on the light intensity so that in time it is related to the meta I–meta II transition. The a-wave can be recorded with a much weaker light stimulus than that needed for the ERP, a ratio of $10^6 : 1$ giving comparable early and late responses. The a-wave represents activity in the receptors and it can be dissociated from the rest of the ERG by treating an isolated retina with aspartate, the result being an isolated receptor potential or distal PIII. Both the a-wave and the distal PIII require sodium ions and will be abolished, albeit reversibly, by sodium deprivation.

Related to the a-wave and the distal PIII is the interstitial photo-voltage (IPV) recorded by Hagins, Penn & Yoshikami (1970) from slices of living rat retinas, by means of external electrodes with one placed at the outer segment tips, a second at variable distances from the first toward the rod synapses. In the dark a steady current is found flowing inward at the outer segment cell membrane and balanced by an equal outward current from the remainder, more internally located portions of the rods. This current results in an interstitial dark voltage gradient (IDV) along the visual cell, the synaptic region being positive relative to the outer segment. A flash of light applied locally to the outer segments evokes a transient reduction of the IDV which, in form, is reminiscent of the a-wave and the distal PIII. This photovoltage (IPV) is apparently generated locally at the point of absorption of light by the rhodopsin. Passive current change from this local IPV source can apparently flow down into the synaptic region and is intense enough to produce excitation at the synapse.

The most directly recorded sign of excitation has been obtained by means of microelectrodes inserted into the visual cell, some into the outer segment itself. By means of such intracellular recordings a resting potential of some 10–30 mV, inside negative, has been detected across the cell membrane. This potential is associated with a sodium-ion dependent 'dark' current due to sodium ions entering the outer segment as the result of an out–in sodium ion concentration gradient across a membrane that is permeable to these ions in the dark. Light has been shown to reduce this sodium ion influx and to hyperpolarise the membrane. The outer segment appears to be the specialised region where hyperpolarisation to light occurs, leading to the passive flow change measured by the Hagins group (Hagins, 1972). This hyperpolarising transient is related to the a-wave and to the distal PIII as well as to the IPV. It is also related to the increase in resistance of the membrane outer segment to light that has been observed with intracellular electrodes. Accordingly, light is assumed to act by reducing the sodium-ion dependent 'dark' current by blocking sodium-ion channels of the cell membrane. The remarkable feature of this mechanism is that one photon absorbed by one rhodopsin molecule anywhere in one disc is able to reduce the interstitial current by some 3% and thereby excite the cell. It has been calculated that for a dark flux of $2 \times 10^9 \, Na^+ \, rod^{-1} \, s^{-1}$ the excitation of one rhodopsin molecule prevents some 10^7 sodium ions from entering the cell. The maintenance of the sodium-ion gradient necessary for continued activity is assumed to be due to a sodium pump. Ouabain, for example, has been found to suppress the ERG, to abolish the distal PIII and to reduce the IPV. Ouabain is generally regarded as an inhibitor of the sodium–potassium pump but in these cases the action of ouabain is believed to be the result of a reduction in the sodium ion concentration gradient rather than to the immediate inhibition of the pump.

The transmitter hypothesis

The link between absorption of a photon and the hyperpolarisation of the outer segment membrane has been difficult to forge. In the case of cones which are assumed, at least in some portions of the outer segment, to have open sacs that are continuous with the cell membrane the mechanism is a local one, the cone pigment absorbing the light and subsequent events taking place at the site of absorption. Rods, however, have closed discs with the disc edges (Fig. 9.1) not continuous with the cell membrane. It has been necessary, therefore, to assume that a transmitter, released after light is absorbed by a rhodopsin molecule, diffuses or is transported to the cell membrane adjacent to the disc edge where it then blocks the sodium channels. The most important candidate for the role of a transmitter is the calcium ion, which is altogether fitting considering the comparable functions that this ion has been called to fill in neuromuscular transmission, in excitation–contraction coupling in muscle, and in other biological systems. I can now summarise some of the experiments made to implicate calcium ions in photoreception.

The disc calcium. It appears that rod discs do indeed concentrate calcium ions above the level in the cytoplasm and that this gradient is maintained by means of a calcium pump. The calcium level within the discs, as reported by several investigators, varies considerably, from 0.1 to 10 calcium ions per rhodopsin molecule. Whatever figure one employs it does appear that there is more than enough calcium to qualify for the calcium hypothesis. No one knows how this calcium is held within the disc: whether bound to rhodopsin, bound to some other membrane protein or constituent, or in the intradisc space, whatever the nature of this space. Many proteins are being discovered that bind this ion, some being specific for calcium ions and many being physiologically important. In 1976 Kretsinger listed 70 such proteins.

Release of calcium ion by light. Both positive and negative results have followed experiments to test whether calcium ion is released from rod discs by illumination. The report has been made that light is even able to cause release, and then to stimulate uptake of calcium ions by vesicles prepared from rod discs.

Effect of altering the external calcium ion concentration. Experimental decrease of the calcium content external to the visual cells has produced results that apparently meet the expectations of the calcium hypothesis. Both the a-wave and the IPV, both of the rat, increase as the result of lowering the calcium ion concentration in the medium. Using intracellular recordings, it has been shown that toad outer segments depolarise and give larger photic receptor hyperpolarising swings with a calcium decrease. Conversely, increasing the external calcium ion concentration produced the opposite results: depression of IDV and IPV, an hyper-polarisation of toad rods and a reduction of the intracellular toad rod response to light. In addition, the ionophores X537A and A23187, which are known to increase the permeability of the membrane to calcium, hasten and increase the effects of lowered calcium ions, and make more effective the responses to elevated concentrations. These results are assumed to be due to alterations of internal calcium ion concentration as the external concentration is changed. A more direct experiment has been reported in which the internal concentration of this ion is changed by injecting calcium ion or the calcium-sequestering reagent EGTA into the outer segments of a toad rod. The first led to an hyperpolarisation; the second to a depolarisation.

These results support, but by no means do they prove, the calcium hypothesis. When one considers the multitude of intracellular reactions, metabolic and physiological, that this ion participates in, one hesitates before employing the results as evidence of a function in excitation of the photoreceptor. In recent years, for example, the interrelations of calcium ions with the cyclic nucleotides in the regulation of diverse cellular activities has attracted greater and greater attention. Even the photoreceptor mechanisms have received their share of this attention. Though the story is still vague and incomplete it deserves a few words.

The cyclic nucleotides. All the familiar components of the cyclic nucleotide family have been found in the outer segments. This includes both the adenosine 3′, 5′-monophosphate (cAMP) and guanosine 3′5′-monophosphate (cGMP) although the latter is considered to play a possible role in the light reactions of the retina. It has also been demonstrated that the enzymes that synthesise and hydrolyse cGMP, i.e., guanylate cyclase and cGMP-phosphodiesterase, respectively, are present as well as guanosine triphosphate (GTP) and GTPase. It appears that a complete system for the regulation of cGMP is present in the outer segment. It has even been asserted that some rhodopsin molecules (1 in 1000) bind the cGMP-phosphodiesterase (cGMP-PDE) in the disc membrane.

In addition to the cyclic nucleotide system listed above phosphorylation reactions involving rhodopsin and other proteins of the outer segments have been described. Phosphorylation of rhodopsin is accomplished by means of a kinase that binds to rhodopsin after it is bleached, an opsin kinase. The kinase apparently is membrane-bound and acts independently of cyclic nucleotides. To the events of Table 9.1 may be added the phosphorylation of rhodopsin with light and then the dephosphorylation in the dark involving a phosphatase.

The experimental finding that triggered interest in cyclic nucleotides on the part of visual scientists was the finding that cGMP decreases with light and recovers with return to darkness. This was followed by the demonstration that light activates the cGMP-PDE, an activation that requires GTP. Accordingly, light absorbed by rhodopsin lowers the cGMP level by activating the enzyme that hydrolyses this cyclic nucleotide.

The possibility that calcium ion may be involved in these reactions is suggested by the report of a considerable increase in cGMP in the retinas of dark-adapted mice when the calcium level in the medium was lowered by omitting calcium and adding EGTA. This rise in cGMP was reversed by adding this ion back in normal concentrations. This effect of calcium ion deficiency was not seen in light-adapted retinas, and dark-adapted retinas having high levels of cGMP, due to low calcium ion concentration, lost their elevated cGMP when bleached photically. The reverse experiment – raising the level of calcium ions above normal – had no measurable effect. It appears that the response to low calcium occurs in the visual cells, for dystrophic mice lacking rods showed little change.

The critical experiments, of course, are physiological ones that test the effects of cGMP and agents that inhibit or accelerate the relevant enzyme systems. Already, such experiments have been made and they have not dispelled the idea that cGMP may be involved in photoreceptor excitation. One study employed the toad rod and intracellular recordings of the membrane potentials in the dark and as the result of illumination. The effects were assayed of Bu_2cGMP (the more permeable dibutyryl derivative), of IBMX (isobutylmethylxanthine, a PDE inhibitor) and the prostaglandins PGE_1 and PGF_2, which are known to alter the intracellular cyclic nucleotide level and calcium. The addition to the medium of Bu_2cGMP, IBMX or PGF_2 caused a membrane depolarisation and an increase of the response to light, simulating the effects of lowered calcium ion concentration. In contrast, PGE_1 led to hyperpolarisation and a decreased response to light, effects like those of increased internal calcium ion concentration. A correlation was revealed between the effects of these reagents on cGMP and the responses to lowering and raising calcium concentrations. These experiments do not tell us whether excitation is the result of a decrease in cGMP brought on by raising intracellular calcium concentration, or the result of an increase in intracellular calcium due to the decrease in cGMP or some more complex interaction of the two mechanisms. In any case, the mechanism of exciting the cell is far from solved, especially the role played by the visual pigment after the photon is absorbed. Current experiments with calcium-sensitive microelectrodes may serve to clarify the role of calcium in photic excitation.

Regeneration of rhodopsin: the cellular events

When the retina is exposed to light and the all-*trans* retinal is detached from the opsin this is the start of a sequence of reactions that eventually end up in the regeneration of rhodopsin in the outer segments and the recovery of the sensitivity (dark-adaptation) of the retina with its full complement of visual pigment. In this sequence of reactions the liver stores of vitamin A (the vitamin-A blood transport system), the retinal pigment epithelium and the visual cell are all involved in an interrelated system of exchange that is not yet clarified.

Retinol is formed principally through the activity of cells in the intestinal epithelium from provitamins, chiefly carotenoids, in the diet. It is then absorbed as retinyl esters and transported to the liver where it is stored, and from where it can be called upon for transport to other cells as needed. Free retinol is somewhat unstable but in the blood the all-*trans* vitamin is transported in a stabilised form bound to a specific low molecular weight (21 000–22 000) protein named retinol-binding protein (RBP). In man, at least, RBP is non-covalently complexed to a thyroxine-binding prealbumin (mol. wt 64 000). The binding of retinol to RBP is a non-specific binding for RBP also reacts with 9-*cis*, 11-*cis* and 13-*cis* retinols and retinals. The RBP carrier serves to transport the water-insoluble vitamin in a stable form.

In the choriocapillaries the retinol, still attached to the RBP, leaves the blood stream and attaches to specific sites of the basal membrane of the pigment epithelium. Retinol is then released from its RBP and enters the cells of the pigment epithelium where it probably attaches to a specific intracellular protein. Storage of retinol in esterified form, mainly in the oil droplets of the pigment epithelium, when these are present, is an important function of these cells. Both all-*trans* and 11-*cis* retinyl esters have been reported to occur in the epithelium. Another source of vitamin A for the pigment epithelium is that from the outer segments. This is produced from the oxidation of the all-*trans* retinal released after bleaching rhodopsin. This retinol from the outer segments simply joins the pool of retinyl esters which can be utilised, when needed, by the visual cells. It is not definitely established where the bulk of the isomerisation of all-*trans* back to 11-*cis* occurs and at one time it was believed that an enzyme, retinene isomerase, occurs in the pigment epithelium. It is also suggested that an enzyme system is present in the visual cells so that all-*trans* retinol (or retinal) can be isomerised within the visual cell itself and utilised there in the regeneration of rhodopsin. There is also some evidence of a specific retinal-binding protein within the visual cell that serves to stabilise it until utilised. To a certain extent, therefore, the visual cells appear to be self-contained, independent entities that can recycle the rhodopsin prosthetic group in the light–dark reactions of the cell. This is supplemented, when necessary, by retinal, either bound to a carrier system or as retinyl ester, entering the outer segment from the pigment epithelium. This is but a beginning in the study of the transport mechanisms and of the interrelations between the pigment epithelium and visual cells. Whatever is learned with this system will be of general interest to the mechanism of transport of lipid-soluble compounds across cell membranes.

CONCLUSIONS

This chapter is an attempt to bring together knowledge in a subject that is rapidly changing as new techniques and new investigators enter the field. Much has been learned in recent years but I have tried to point out, sometimes implicitly, that many of the published ideas are tentative, uncertain and subject to change as studies continue. The technique of resonance Raman spectroscopy has added a fresh note to the visual pigment field but care must be employed in accepting ideas uncritically because they are fresh. Conclusions that are based on approaches from several directions are more satisfying, at least to me, than those arrived at by a technique that is yet uncertain in its theoretical foundation. In addition, the use of paper chemistry models, when applied to the interpretation of protein behaviour must be viewed with caution until means are found to test these models realistically.

It may be concluded that any attempt to make out a phylogenetic or evolutionary progression in visual pigment biology is fruitless at this time. Neither in the colour, the wave number spectral absorbance curves, the molecular weight, the photosensitivity, the products of bleaching, nor any other fundamental property of the visual pigments can one discern any evolutionary or taxonomic tendencies. Lamprey rhodopsin is comparable to human rhodopsin in relation to its function in vision and no major change seems to have occurred in this molecule since lampreys first emerged 400 million years ago from their ancestral Anaspidian stock. Nature hit upon the correct molecule early in vertebrate evolution and it has been retained through to the primates.

And yet upon this general theme there have been evolutionary diversions in apparent adaptation to particular environmental changes and specific problems. The adoption of a second vitamin A system (either the A_1 or A_2) very early in vertebrate phylogeny is one such diversion, apparently serving the purpose of extending the spectral response capability while utilising the same protein system. The distribution of the A_2-pigments is especially puzzling. Present in cyclostomes (lampreys), they have not been found in elasmobranchs, but are common in teleosts in which they occur mainly, though not exclusively, in freshwater fishes. In amphibia they tend to occur in aquatic forms but here too there are erratic exceptions. The tadpoles of frogs and treefrogs have porphyropsin which is replaced by rhodopsin in the adults except for the bullfrog which has retained some porphyropsin in the superior half of its retina. Toad (bufonid) tadpoles, on the other hand, even when taken from the same pond as treefrog (hylid) tadpoles, have only rhodopsin. Reptiles have not been examined sufficiently to warrant any general statement except that the freshwater turtle, *Pseudemys*, may have porphyropsin. Then suddenly, in birds and mammals the A_2-system disappears. The tendency seems to be for the 3-dehydroretinal pigments to occur in aquatic vertebrates especially those inhabiting freshwater but the distribution is so erratic that any attempt to isolate a single factor as the causative agent seems hopeless. In the case of invertebrates, A_1-pigments have been extracted from species inhabiting both freshwater and the sea.

A definite and consistent difference has been uncovered in the properties of iodopsin, now known to be a cone pigment, and the rod pigment rhodopsin. In this dichotomy the transmuted pigments of geckos appear to align themselves more with iodopsin than with rhodopsin. This statement is based on some six or seven physical and chemical properties that have been investigated. Compared to rhodopsin, both iodopsin and the gecko pigments have structures that are accessible to reversible attack by physical and chemical agents. These changes are of considerable interest in connection with several chemical models that have been proposed in attempts to understand the colour and other properties of visual pigments. The great need now is the examination, by extraction, of at least one other genuine cone pigment to certify the existence of fundamental molecular differences between rod and cone photochemical systems.

This work was aided by a grant from the National Eye Institute, N.I.H., U.S. Public Health Service.

REFERENCES

This list is divided into two sections, the first being those specifically referred to in the text, the second being reviews that have the original references to most of the uncited text statements. This method has been chosen in the interest of saving space and making more interesting reading.

Text references

Boll, F. (1876). Zur Anatomie und Physiologie der Retina. *Akademie der Wissenschafften, Berlin Monatsberichte der Koenglich Preussischen,* **12**, 783–8.

Collins, F. D., Love, R. M. & Morton, R. A. (1952). Studies on rhodopsin. 4: Preparation of rhodopsin. *Biochemical Journal,* **51**, 292–8.

Crescitelli, F. (1977). Ionochromic behavior of gecko visual pigments. *Science,* **195**, 187–8.

Crescitelli, F. (1979). The gecko visual pigments. The behavior of opsin. *Journal of general physiology,* **73**, 517–40.

Crescitelli, F., Dartnall, H. J. A. & Loew, E. R. (1977). The gecko visual pigments: a microspectrophotometric study. *Journal of physiology,* **268**, 559–73.

Crescitelli, F., Mommaerts, W. F. H. M. & Shaw, T. L. (1966). Circular dichroism of visual pigments in the visible and ultra-violet spectral regions. *Proceedings of the national academy of sciences of the USA,* **56**, 1729–34.

Dartnall, H. J. A. (1953). The interpretation of spectral sensitivity curves. *British Medical Bulletin,* **9**, 24–30.

Dartnall, H. J. A. (1972). Photosensitivity. In *Handbook of sensory physiology,* ed. H. J. A. Dartnall, *VII/1,* 122–45. Heidelberg: Springer-Verlag.

Denison, R. H. (1956). A review of the habitat of the earliest vertebrates. *Fieldiana, geology,* **11**, 359–457.

Hagins, W. A., Penn, R. D. & Yoshikami, S. (1970). Dark current and photocurrent in retinal rods. *Biophysical journal,* **10**, 380–412.

Hagins, W. A. (1972). The visual process: excitatory mechanisms in the primary receptor cells. *Annual review of biophysics and bioengineering,* **1**, 131–58.

Heller, J. (1969). Comparative study of a membrane protein. Characterization of bovine, rat and frog visual pigments 500. *Biochemistry,* **8**, 675–8.

Kretsinger, R. H. (1976). Calcium-binding proteins. *Annual review of biochemistry,* **45**, 239–66.

Kühne, W. (1877). Über den Sehpurpur. *Untersuchungen aus dem Physiologischen Institut der Universitat Heidelberg,* **1**, 15–103.

Lewis, A. (1978). The molecular mechanism of excitation in visual transduction and bacteriorhodopsin. *Proceedings of the national academy of sciences of the USA,* **75**, 549–53.

Liebman, P. A. (1972). Microspectrophotometry of photoreceptors. In *Photochemistry of vision. Handbook of sensory physiology*, ed. H. J. A. Dartnall, *VII/1*, 481–528. Heidelberg; Springer-Verlag.

Lythgoe, J. N. (1968). Visual pigments and visual range underwater. *Vision research*, **8**, 997–1012.

Lythgoe, J. N. (1972). List of vertebrate visual pigments. In *Photochemistry of vision. Handbook of sensory physiology*, ed. H. J. A. Dartnall, VII/*1*, 604–24. Heidelberg: Springer-Verlag.

Marks, W. B., Dobelle, W. H. & MacNichol, E. F. (1964). Visual pigments of single primate cones. *Science*, **143**, 1181–3.

Moon, P. (1940). Proposed standard solar-radiation curves for engineering use. *J. Franklin Institute*, **230**, 583–617.

Morton, R. A. & Goodwin, T. W. (1944). Preparation of retinene in vitro. *Nature*, **153**, 405–6.

Nilsson, S. E. G. & Crescitelli, F. (1969). Changes in ultrastructure and electroretinogram of bullfrog retina during development. *Journal of ultrastructure research*, **27**, 45–62.

Rushton, W. A. H. (1972). Pigments and signals in colour vision. *Journal of physiology*, **220**, 1–31P.

Schultze, M. (1866). Zur Anatomie und Physiologie der Retina. *Archiv fur mikroskopische Anatomie*, **2**, 175–86.

Tansley, K. (1931). The regeneration of visual purple: its relation to dark adaptation and night blindness. *Journal of physiology*, **71**, 442–58.

Wald, G. (1935). Carotenoids and the visual cycle. *Journal of general physiology*, **19**, 351–71.

Wald, G. (1939). The porphyropsin visual system. *Journal of general physiology*, **22**, 775–94.

Wald, G. & Brown, P. K. (1965). Human color vision and color blindness. *Cold Spring Harbor symposia on quantitative biology*, **30**, 345–61.

Walls, G. L. (1934). The reptilian retina. *American journal of ophthalmology*, **17**, 892–915.

Walls, G. L. (1942). The vertebrate eye and its adaptive radiation. *Bulletin No. 19, Cranbrook Institute of Science*, Bloomfield Hills, Michigan, USA.

Yutani, E., Sugino, Y., & Matsushiro, A. (1977). Effect of a simple amino acid substitution on stability of conformation of a protein. *Nature*, **267**, 274–5.

General references

Barlow, H. B. & Fatt, P. (eds) (1977). *Vertebrate photoreceptors*. London; Academic Press.

Bonting, S. L. (ed) (1976). *Visual process*. Oxford; Pergamon Press.

Callender, R. & Honig, B. (1977). Resonance Raman studies of visual pigments. *Annual review of biophysics and bioengineering*, **6**, 33–55.

Crescitelli, F. (ed) (1977). The visual system in vertebrates. In *Handbook of sensory physiology*, *VII/5*, Heidelberg; Springer-Verlag.

Ebrey, T. G. (1975). Molecular aspects of photoreceptor function. *Quarterly reviews of biophysics*, **8**, 120–84.

Greengard, P. (1978). *Cyclic nucleotides, phosphorylated proteins, and neuronal function.* New York: Raven Press.

Honig, B. & Ebrey, T. G. (1974). The structure and spectra of the chromophore of the visual pigments. *Annual review of biophysics and bioengineering*, **3**, 151–77.

Hubbell, W. L. & Bownds, M. D. (1979). Visual transduction in vertebrate photoreceptors. *Annual reviews of neuroscience*, **2**, 17–34.

Rosenkranz, J. (1977). New aspects of the ultrastructure of frog rod outer segments. *International review of cytology*, **50**, 25–157.

Tomita, T. (1970). Electrical activity of vertebrate photoreceptors. *Quarterly review of biophysics*, **3**, 179–222.

Wald, G. (1960). The significance of vertebrate metamorohosis. *Circulation*, **21**, 916–38.

Warshel, A. (1977). Interpretation of resonance Raman spectra of biological molecules. *Annual review of biophysics and bioengineering*, **6**, 273–300.

Worthington, C. R. (1974). Structure of photoreceptor membranes. *Annual review of biophysics and bioengineering*, **3**, 53–80.

Index

Acanthamoeba, myosins of, 296
Acinetobacter spp., strict aerobes, 121, 123, 124, 126
actin, found in most types of eukaryote cell, 261, especially in thin filaments (I-bands) of striated muscle, 263
  globular (G-) form of, polymerising into two-chain helical fibrous (F-) form, 263, 265
  interaction of, with myosin, *see* myosin-actin interaction
actin-linked control of myosin-actin interaction, troponin in, 276–9
adenine, formed in prebiotic conditions, 21–2
aerobes, strict (bacteria), 120, 121
Agrobacterium tumefaciens: virulence of, in producing plant tumours, related to presence of plasmid, 128
alanine: specified by one of earliest tRNAs, and present in prebiotic conditions, and in meteorites, 67, 68
aldehydes, formed in prebiotic conditions, 20
aldolase (for glutarate), common origin of hydrolase and? 142
algae, brown, green and red, and diatoms, 171
  cyanobacteria in origin of, 155
  ferredoxins of, 189, 190, 194
  photosynthesis in, 156
  thylakoids in different classes of, 174
alligator, rhodopsin in retina of, 324
amides: development of a family of enzymes for metabolising previously unusable, in *Pseudomonas*, 139–41
amino acids
  changes of, in a protein, detected by electrophoresis, 6, 7
  clustering of genes for synthesis of, into operons, 133, 134

  in ferredoxins, same as those formed in prebiotic conditions, 189
  formed in prebiotic conditions, 20, 21, 22
  four specified by earliest tRNAa, also most abundant in prebiotic conditions, and in meteorites, 67–8
  frequencies of different types of substitution of, in proteins, 94
  suggested original genetic code for hydrophilic and hydrophobic, 64
  synthetic pathways for, similar in all living organisms, 133
amoeboids, containing mitochondria: as ancestors of eukaryotes, 153
cAMP: in phosphorylations, 282, 286; in visual cell, 242
Anabena, cyanobacterium, 168, 183
anaerobes, facultative (bacteria): fermenters, 119–20, and fermenters and oxidisers, 120
anaerobes, strict (bacteria), 117
  as earliest organisms, 117, 151
  fermenters (Clostridia), and methane-producers, 119
  fermenters closely related to green sulphur bacteria, 169
  ferredoxins of, 189, 191
annelids, both actin- and myosin-linked control of myosin–actin interaction in, 280
antibiotics
  plasmids carrying resistance to, 125–7
  types of, inhibiting synthesis on cytoplasmic ribosomes, but not on chloroplast ribosomes, 181
antibodies, immunoglobulins as, 230
  generation of diversity in: by random pairing of variant light and heavy chains of Ig, 242; by somatic mutation, 242–4, 255; by somatic

Rhodospirillum, 161
Rhodospirillum capsulata, 166
Rhodospirillum rubrum, 166
chromatophores of, 161
ferredoxin of, 188
Rhodospirillum spheroides, 168
ribitol dehydrogenase: mutants of *K.
aerogenes* producing high levels of,
and others possessing higher affinity
for xylitol, 136, 138
ribose, formed in prebiotic conditions,
23–4
in experiments on nucleotide synthesis
in prebiotic conditions, 27; deoxy-,
less effective, and likely to be a later
product of evolution, 33
ribosomes, possible survivors of prebiotic
era, appearing to be universal in
biosphere, 17
in chloroplasts, 181; protein synthesis
on, 181–2
of prokaryotes and chloroplasts, similar
in size and antibiotic sensitivity, 181–
2, 183
of prokaryotes and eukaryotes, 153
rRNAs in structure of, 61
ribulose bisphosphate carboxylase, in
stroma of chloroplasts, 171
fixation of carbon dioxide by, 179
large subunits of, coded for by
chloroplast DNA, smaller subunits by
nuclear DNA, 182
large subunits of, synthesised by system
from *E. coli* supplied with
chloroplast RNA, 183
RNA, 25
accuracy and error rate in synthesis of
(prebiotic, phage, *E. coli*), 50–1
experiments on replication of, by
bacteriophage Qβ, 52–3; maximum
chain lengths for, model and
observed in Qβ, 54, 61
processing of, for light chains of Ig:
kappa, 238, 246; lambda, 234–5
single-stranded, tertiary structure of, 61
mRNA, 69
of chloroplasts, translated by *E. coli*
system, but not by wheat cytoplasmic
ribosomes, 183
rRNAs
of chloroplast, form hybrids only with
chloroplast DNA, 183
considered to be very old, 61
evolutionary tree based on ferredoxin,
cytochrome *c*, and 5S, 196, 197
sequences of 16S, in Rhodospirillaceae,
197

tRNAs, possible survivors from prebiotic
era, universal in biosphere, 17
accuracy and error rate for enzyme-free
replication of, 50
as catalysts in transmission of amino
acids, 61
of chloroplasts, 181, 183
discriminator base in (14th from 3′
end), 62
earliest four of, for glycine, alanine,
aspartic acid, valine? 67, 68
primary sequences in, from twenty
organisms, suggest divergent
evolution from common ancestor,
rather than phylogenetic tree, 34, 35
quest for origin of synthetases for, 70
reconstruction of primordial sequence
of, 36–8, 67
two possible configurations for
anticodon loop of, 65
RNA polymerase
of bacteriophage Qβ, 52, 53, 54
in first replication-translation
machinery, 77
RNA replicase of bacteriophage, in
invasion of bacterial cell, 59

Saccharomyces cerevisiae, tRNA
sequences in, 36, 37, 67
Salmonella spp., 121
plasmids in: catabolic, 127; drug-
resistance, 125
Salmonella typhimurium, plasmids in, 126
scallop (*Aequipecten irradians*), light
chains of myosin in
control of muscle contraction by
direct binding of calcium to, 280–2
interchange of, with other species, 293
sea water: solids of, as catalyst for
nucleoside formation in prebiotic
solid-state conditions, 27
self-replication of molecules in prebiotic
conditions, 29–38
basic features of, 40–4
in compartments, 18, 73–5, 76
co-operation in, 55–60
dynamics of, 38–40
errors in, 17, 44–5; as source of
variability, 76
experiments on, 51–5
suppression of competition in, 17–18,
55
template-induced, necessary
prerequisite for Darwinian evolution,
76
serine proteases
perhaps preceded by a single form able